Alkaloids: Chemical and Biological Perspectives

Related Titles of Interest

Books

GAWLEY
Asymmetric Synthesis*

RILEY *ET AL.*
Pharmaceutical & Biomedical Applications of Liquid Chromatography

WILLIAMS
Synthesis of Optically Active Alpha-Amino Acids 2nd Editon*

WONG
Enzymes in Synthetic Organic Chemistry

WOOD
Practical Biochemistry for Colleges

* In preparation

Journals

Bioorganic & Medicinal Chemistry

Bioorganic & Medicinal Chemistry Letters

Journal of Pharmaceutical & Biomedical Analysis

Tetrahedron

Tetrahedron Asymmetry

Tetrahedron Letters

ALKALOIDS: CHEMICAL AND BIOLOGICAL PERSPECTIVES

Volume Nine

Edited by

S. WILLIAM PELLETIER

Institute for Natural Products Research

and

The Department of Chemistry
University of Georgia, Athens

PERGAMON

U.K.	Elsevier Science Ltd, The Boulevard, Langford Lane, Kidlington, Oxford, OX5 1GB, U.K.
U.S.A.	Elsevier Science, Inc., 660 White Plains Road, Tarrytown, New York 10591-5153, U.S.A.
JAPAN	Elsevier Science Japan, Tsunashima Building Annex, 3-20-12 Yushima, Bunkyo-ku, Tokyo 113, Japan

First Edition 1995

Library of Congress Cataloging in Publication Data
A catalogue record for this serial is available from the Library of Congress

British Library Cataloging in Publication Data
A catalogue record for this book is available from the British Library

ISBN 0 08 042089 3

Printed in Great Britain by Redwood Books, Trowbridge

Dedicated to
the memory of

Richard Helmuth Frederick Manske
(1901—1977)

who while still a graduate student determined the structures of harmine and harmaline and synthesized both alkaloids. Subsequently he carried out a systematic investigation of plants of the Fumariacae and isolated new alkaloids of the aporphine, benzophenanthridine, phthalideisoquinoline, protopine and protoberberine ring systems. He also isolated many spirobenzylisoquinolines, the complicated cancentrine alkaloids and alkaloids of the Canadian Lycopodiaceae. His studies on the isoquinoline alkaloids are classic. *The Alkaloids*, which he began in 1950 and continued editing until his death, is not only a permanent monument to his energy, learning and scholarship, but also an enormously useful resource for alkaloid chemists.

Contributors

N. Aimi, Faculty of Pharmaceutical Sciences, Chiba University, 1-33 Yayoi–cho, Inage-ku, Chiba 260, JAPAN

Amrik Singh Chawla, Department of Pharmaceutical Sciences, Panjab University, Chandigarh–160014, INDIA

James M. Cook, Department of Chemistry, University of Wisconsin Milwaukee, P. O. Box 413, Milwaukee, Wisconsin 53201-0413, U.S.A.

Linda K. Hamaker, Department of Chemistry, University of Wisconsin Milwaukee, P. O. Box 413, Milwaukee, Wisconsin 53201-0413, U.S.A.

Thomas Hartmann, Institut für Pharmazeutische Biologie der Technischen Universität Braunschweig, Mendelssohnstrasse 1, D-38106 Braunschweig, GERMANY

Vijay K. Kapoor, Department of Pharmaceutical Sciences, Panjab University, Chandigarh–160014, INDIA

L. A. Mendonza, I.N.T.A., Centro de Investigaciones de Recursos Naturales, Castelar, ARGENTIA

P. Obitz, Johannes Gutenberg– Universität Mainz, Institut für Pharmazeutische Biologie, Staudinger Weg 5, D-55099 Mainz, GERMANY

Ronald D. Plattner, National Center for Agricultural Utilization Research, Agricultural Research Service, U. S. Department of Agriculture, Peoria, Illinois 61604, U.S.A.

Richard G. Powell, National Center for Agricultural Utilization Research, Agricultural Research Service, U. S. Department of Agriculture, Peoria, Illinois 61604, U.S.A.

S. -i. Sakai, Faculty of Pharmaceutical Sciences, Chiba University, 1-33 Yayoi–cho, Inage-ku, Chiba 260, JAPAN

J. Stöckigt, Johannes Gutenberg – Universität Mainz, Institut für Pharmazeutische Biologie, Staudinger Weg 5, D-55099 Mainz, GERMANY

M. E. Wall, Research Triangle Institute, P. O. Box 12194, Research Triangle Park, North Carolina 27709, U.S.A.

M. C. Wani, Research Triangle Institute, P. O. Box 12194, Research Triangle Park, North Carolina 27709, U.S.A.

Ludger Witte, Institut für Pharmazeutische Biologie der Technischen Universität Braunschweig, Mendelssohnstrasse 1, D-38106 Braunschweig, GERMANY

Preface

Volume 9 of *Alkaloids: Chemical and Biological Perspectives* presents timely reviews on several important alkaloid topics.

Chapter 1 by Monroe Wall and M. C. Wani, the discoveries of taxol, present a concise review of this medicinally important alkaloid. Current clinical uses of taxol include treatment of both ovarian and breast cancer. The account of the discovery of taxol about twenty-five years ago and the envolment of this complex naturally-occurring alkaloid to a useful chemotherapeutic agent is an exciting story. This chapter treats the isolation and structure elucidation of taxol, supply problems, mechanism of action and clinical aspects of taxol.

Chapter 2 surveys the synthesis of the macroline-related sarpagine alkaloids. Interest in these compounds from *Alstonia* species originated as a result of folk tales describing the medicinal properties of these plants. Over the past few years over seventy indole alkaloids have been isolated from *Alstonia* species. Many of these alkaloids are related to the sarpagine/ajmaline alkaloids and feature a common macroline unit. The chapter summarizes synthetic work on macroline/sarpagine alkaloids and of ajmaline and related alkaloids. The biological properties of these alkaloids are also summarized.

Chapter 3 treats the *Erythrina* alkaloids. The *Erythrina* genus consists of 108 species of orange or red-flowered trees, shrubs and herbaceous plants that occur throughout the tropical and semi-tropical regions of the world. The chapter discusses the isolation, structure elucidation, biosynthesis, synthesis and pharmacology of this very important class of alkaloids.

Chapter 4 summarizes work on the biosynthesis, transport, storage and metabolism of the pyrrolizidine alkaloids in plants, the role of the *N*-oxides in plants and the involvement of pyrrolizidine alkaloids in plant-insect relationships. The chapter demonstrates the pyrrolizidine alkaloid to be a complex, highly specific and functionally important system of plant secondary constituents.

Chapter 5 presents an interesting discussion of the production of alkaloids from cell cultures of *Aspidosperma quebracho-blanco*. Many of the Apocynaceae plants are very important in traditional medicine, being collected from wild forests rather than grown in plantations, thus making the supply of medicinal agents uncertain. As a consequence, many attempts have been made to establish cell culture systems of these rare plants to provide a reliable supply of plant material. This chapter provides an account of the establishment of tissue and cell cultures of *Aspidosperma quebracho-blanco* plants and the isolation and structure of alkaloids isolated from these cultures.

Chapter 6 summarizes work on the fumonisins, toxic alkaloids produced by *Fusarium moniliforme* and related species. The cancer promoting activity of these alkaloids has caused widespread interest in this new class of mycotoxins, particularly since the fungi that produce them are found throughout the world and *F. moniliforme* is a prevalent fungi associated with human and animal dietary staples such as corn. This chapter treats the chemistry, biosynthesis and biological activity of the fumonisins.

Each chapter in this volume has been reviewed by at least one expert in the field. Indexes for both subjects and organisms are provided.

The editor invites prospective contributors to write him about topics for review in future volumes in this series.

S. William Pelletier
Athens, Georgia
May 1994

Contents of Previous Volumes

Volume 3

Volume 4

Volume 5

Volume 6

Contents

Chapter One

Taxol

M. E. Wall and M. C. Wani
Research Triangle Institute,
P.O. Box 12194,
Research Triangle Park,
North Carolina 27709-2194
U.S.A.

CONTENTS

1. INTRODUCTION

Natural Products chemists and phytochemists have always been impressed by the fact that compounds found in nature display an almost unbelievable range of diversity in terms of their structures and physical and biological properties. Most of these compounds are secondary metabolites whose functions in plants, fungi, and marine organisms are still not widely understood. Currently it is believed that many of these compounds act to defend the host from the harmful effects of toxins, carcinogens, or mutagens found in the plant or plant's environment [1,2] or from attack by external predators [3].

The novel diterpenoid, taxol [4] (**1**, Figure 1), is a secondary metabolite found in several Taxus (Taxaceae) species. Utilizing bioactivity-guided fractionation, it was isolated from extracts of the wood and bark of *Taxus brevifolia* [5] more than twenty-five years ago in Wall's laboratory at the Research Triangle Institute (RTI). The structure was elucidated a few years later [4,5]. Because of the low yield, scarcity of supply, modest activity in the L1210 mouse leukemia assay, then considered by the National Cancer Institute to be the most important animal *in vivo* test, and possibly because of limited funding, further work on taxol was not conducted for almost ten years. Currently taxol is considered by many oncologists to be one of the best cancer chemotherapeutic agents discovered in many years [6-9]. This chapter will review the discovery, structure activity relationships, biological and clinical activity of taxol, and present certain aspects of synthetic approaches to taxol.

Figure 1

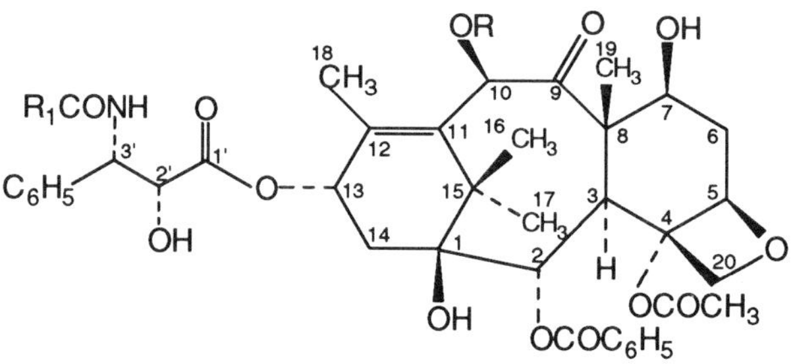

Taxol (**1**), R = Ac, $R_1 = C_6H_5$
Taxotere (**30**), R = H, R_1 = t-BuO
Cephalomannine (**33**), R = Ac, R_1 =

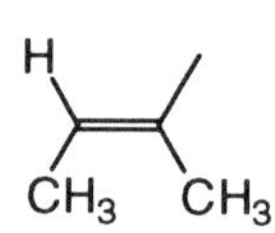

2. DISCOVERY OF TAXOL

2.1 Initial Procurement

A screening program for antitumor agents in the plant kingdom was initiated in 1960 under Dr. Jonathan L. Hartwell. In this program plant samples collected at random were supplied by the U.S. Department of Agriculture (USDA) under an interagency agreement with NCI. In August 1962, USDA botanist, Arthur S. Barclay, and three college student field assistants collected 650 plant samples in California, Washington, and Oregon, including bark, twigs, leaves and fruit of *Taxus brevifolia* in Washington State.

T. brevifolia is a slow growing tree which is found only in localized coastal areas of the above-mentioned West Coast states. It had never received any chemical investigation until it was assigned to our laboratory by Dr. Hartwell. The assignment of the plant to our group was not entirely serendipitous since some of these samples had been shown to have 9KB cytotoxicity. (KB cells are derived from a human epidermoid carcinoma of the nasopharynx. Activity against KB cells is defined as the amount of test material in µg/mL required to reduce the cell growth by 50%.) At that time there were only three groups working under contract to NCI in the laboratories of Dr. Jack Cole, University of Arizona, the late Dr. S. Morris Kupchan, then at the University of Wisconsin, and Wall's laboratory at RTI. The other groups at this time were not particularly interested in plants with 9KB activity. We had noted an excellent correlation in our camptothecin studies between L1210 *in vivo* activity and cytotoxicity against 9KB cells [10] (cf. Table 2 in Ref. 10). Accordingly, we had requested Dr. Hartwell to assign to our laboratory as many 9KB actives as possible. From this arose the assignment to our laboratory of *T. brevifolia* as well as a number of other plants also highly active in 9KB and several highly active novel compounds were found in these cases. These include colubrinol, a maytansine analog [11], carminomycin, related to daunomycin [12], and holacanthone, an active quassinoid [13].

Initial samples of *T. brevifolia* arrived at our laboratory by 1964. In April 1966, prior to our isolating the pure material, Wall requested in a letter to Jonathan Hartwell [14] that the extracts we had sent "receive a special priority with the biological screeners as I regard it as one of the most important samples we have had in a long time." A month later, in a Progress Report to NCI [15], Wall stated, "At present, a major effort by our group is being placed on this plant (*Taxus brevifolia*)."

2.2 Isolation Of Taxol

By November 1966, we were able to report the isolation of a purified fraction, and we presented some physical constants. The actual isolation was completed by June 1966. The method finally adopted after several unsuccessful trials is shown in Charts 1 and 2. It involved our standard ethanol extraction, partition of the ethanolic residue between water and chloroform, followed by a large number of Craig countercurrent distribution treatments, the last of which involved a 400-tube Craig countercurrent distribution (cf. Charts 1 and 2). In this manner, approximately 0.5 gram of taxol was isolated starting with 12 kg of air dried stem and bark. The yield was about 0.004%. All the various steps were monitored by an *in vivo* bioassay which, at that time, involved the inhibition of the growth of the Walker WM solid tumor. As is shown in Chart 2, increased purification is accompanied by lower T/C and dose values. The isolation hence was carried out laboriously, but in a manner in which losses by the treatment or changes in the chemical constitution of the eventual product were minimized because of the mild countercurrent distribution methodology. Much simpler procedures have been subsequently developed both at RTI and elsewhere. Later, we increased the yield to 0.02% [4]. In 1966 we were able to compare crude chloroform extracts from various samples of *T. brevifolia* collected in Alaska, California, Washington, Idaho, Oregon, and Montana and present the data on their cytotoxicity and WM inhibition. The isolation and some of the physical and biological properties of taxol were publicly presented for the first time in 1967 [16]. It should be mentioned that little interest in the compound was shown either by the public, or the National Cancer Institute for more than a decade until the novel mechanism of action was discovered (cf. Section 5).

Chart 1

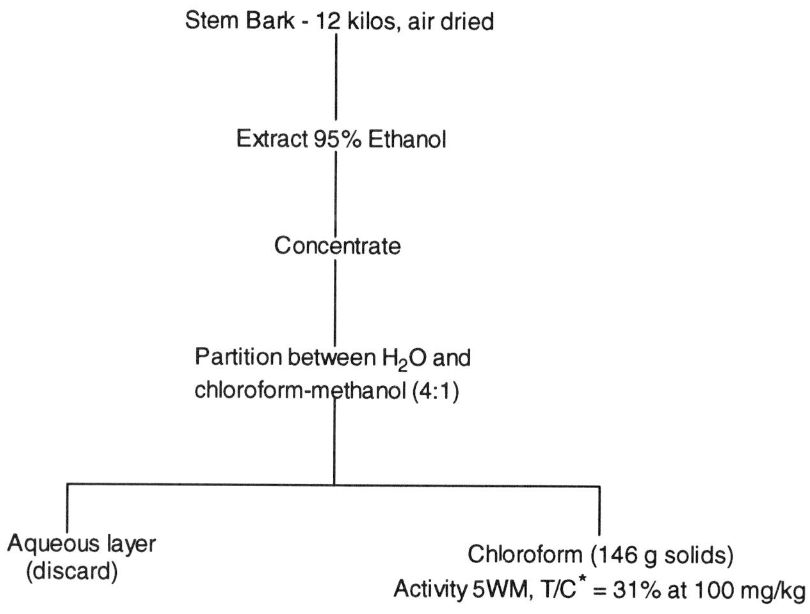

*T/C = Mean Tumor Weight of Treated Animals (T) ÷ Mean Tumor Weight of Control Animals X 100. A plant is considered active if it reduces the mean tumor weight of treated animals to less than 42% of the mean tumor weight of the control group.

3. STRUCTURE DETERMINATION OF TAXOL

As soon as we had isolated taxol in pure form, a great deal of work on the structure of the compound was carried out by available spectroscopic methods. Although methods for ultraviolet, infrared, and mass spectrometry were at a reasonably advanced stage in the late 1960s, NMR was relatively primitive compared to the sophisticated instrumentation and methods now available. Nevertheless, the data obtained was useful. The physical properties of taxol are: mp 213-216° dec; $[\alpha]^{20}{}_{D}$ - 49° (c 0.013, MeOH); UV λ_{max} (MeOH) 227 nm (ε 29,800); IR νmax 3300-3500 (OH, NH) 1730 (ester), 1710 (ketone), 1650 (amide) cm^{-1}; M^+ at m/z 853, calcd. for $C_{47}H_{51}NO_{14}$.

From biogenetic evidence and some of the spectroscopic evidence, it seemed probable that the compound, which we named taxol, was comprised of a taxane nucleus to which several esters were attached. A number of taxane compounds were reported in the literature [17-19].

Chart 2

Chloroform (146 g solids)

11 Tube Craig CCD
System: hexane, acetone, *t*-butanol, water (5:4:4:2)

Tubes 6, 7, 8
41 g solids, activity 5WM
T/C 30% at 45 mg/kg

30 Tube Craig CCD
System: methanol, water, CCl_4, $CHCl_3$ (8:2:7:3)

Tubes 8-18
14 g solids
T/C 30% at 23 mg/kg

400 Tube Craig CCD
System: methanol, water, CCl_4, $CHCl_3$ (8:2:7:3)

Tubes 170-189
2.4 g solids, T/C 16% at 15 mg/kg

Trituration with benzene
0.5 g Taxol, T/C 16% at 10 mg/kg
24% at 5 mg/kg
Yield circa 0.004%

3.1 NMR Studies

The ^1H-NMR data reported for a compound called baccatin V (**2**, Figure 2) isolated in Halsall's laboratory [19] was compared with the ^1H-NMR spectrum of taxol and showed very close agreement with the spectrum of **1** (cf. Table 1). Baccatin V (**2**) is one of the few naturally occurring taxanes with the four-membered oxetane ring. As shown in Table 1, the chemical shifts for the four methyl groups in **1** and **2** are very similar, as are the chemical shifts for a number of protons.

3.2 X-Ray Studies

At this stage of our structural characterization, it seemed that **1** consisted of a nucleus similar to **2**, but was considerably more complex. Mass spectrometry indicated that **1** contained at least 16 more carbon atoms than **2**, plus an additional nitrogen moiety, MW 853 ($C_{47}H_{51}NO_{14}$) versus **2**, MW 586 ($C_{31}H_{38}O_{11}$). Because of the extremely limited quantity of taxol available and its evident structural complexity, attempts were made to prepare derivatives for X-ray analysis. Although a number of crystalline, halogenated compounds were obtained, none had properties suitable for X-ray analysis.

Taxol was therefore subjected to a mild base-catalyzed methanolysis at 0°[20] yielding a nitrogen containing α-hydroxy ester (**3**) $C_{17}H_{17}NO_4$, a tetraol (**4**) $C_{29}H_{36}O_{10}$, and methyl acetate as shown in the equation below:

$$\underset{\mathbf{1}}{C_{47}H_{51}NO_{14}} + 2CH_3OH \rightarrow \underset{\mathbf{3}}{C_{17}H_{17}NO_4} + \underset{\mathbf{4}}{C_{29}H_{36}O_{10}} + CH_3CO_2CH_3$$

Compound **3** was converted to a p-bromobenzoate **5** and **4** to a 7,10-bisiodoacetate **6** and the full structures of the halogenated derivatives of **3** and **4** were determined by X-ray analysis. For full details of the X-ray analysis and physical constants of compounds **3** - **6**, cf. reference 4.

The structures of the methyl ester **3** and the tetraol **4** (10-deacetylbaccatin III) (Figure 2) were derived from the X-ray structures of **5** and **6**, respectively [4]. Compound **3** is the methyl ester of *N*-benzoyl-β-phenylisoserine. It is of interest to note that baccatin V (**2**), isolated in Halsall's laboratory [19] has a structure identical to **4** except being epimeric at C-7. Subsequently, the structure of baccatin III (**7**), also isolated from *T. baccata*, was shown to be the 10-acetyl analog of **4** [19]. Taxol (**1**) and compounds **2** and **7** are among the rare taxanes found in nature with the four-membered 4,5-oxetane ring.

3.3 The Structure Of Taxol

The final structure of **1** requires the placement of the two hydrolyzed ester functions of **1** on the tetraol **4**. The reasonable assumption that no rearrangement of **1** occurs under the mild methanolysis conditions is supported by the fact that the methanolysis of a related taxane derivative, taxinine, gives dideacetyltaxinine which is converted back to taxinine upon reacetylation [21].

Taxol could not be oxidized by neutral, activated manganese dioxide prepared by the Goldman procedure[22] indicating that the two esters were located at the allylic positions 10 and 13. The chemical shifts of the protons at C-10 and C-13 were also in accord with this observation. Moreover, under the same conditions, the 7,10-diacetate (**8**, Figure 2) which was obtained from compound **4** was smoothly oxidized to the corresponding conjugated ketone **9**. Oxidation of **1** with activated manganese dioxide under mild basic conditions (pH of the aqueous suspension 8.0)

Figure 2

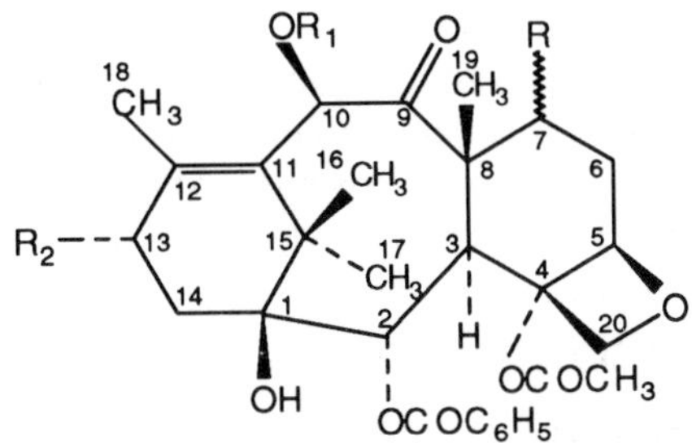

Baccatin V (**2**), R = α-OH, R_1 = Ac, R_2 = OH
10-Deacetyl-baccatin III (**4**), R = β-OH, R_1 = H, R_2 = OH
7,10-Bisiodoacetate (**6**), R = β-$OCOCH_2I$, R_1 = $COCH_2I$, R_2 = OH
Baccatin III (**7**), R = β-OH, R_1 = Ac, R_2 = OH
7,10-Diacetate (**8**), R = β-OAc, R_1 = Ac, R_2 = OH
7β-Acetyl Conjugated Ketone (**9**), R = β-OAc, R_1 = Ac, R_2 = O
7β-Hydroxy Conjugated Ketone (**10**), R = β-OH, R_1 = Ac, R_2 = O

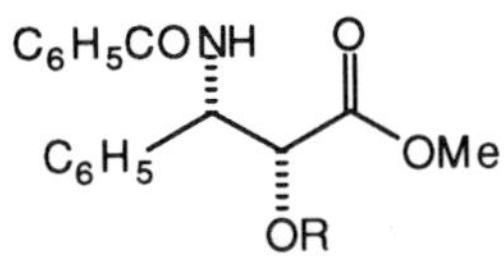

Methyl Ester (**3**), R = H
p-Bromobenzoate (**5**), R = COC_6H_4Br

Table 1. NMR Data (δ Units, J Values in Hertz)[a]

	Compound		
Position	1	Baccatin V[b] (2)	10
C-17 CH_3	1.14 (s)	1.04 (s)	1.20 (s)
C-16 CH_3	1.22 (s)	1.10 (s)	1.26 (s)
C-19 CH_3	1.67 (s)	1.62 (s)	1.66 (s)
C-10 OAc	2.20 (s)	1.99 (s)	2.18 (s)
C-18c	1.80 (s)		2.08 (s)
C-4 OAc	2.36 (s)		2.28 (s)
C-3 H	3.80 (d, J=6)	4.02 (d, J=6)	3.96 (d, J=6)
C-20 2 H	4.20 (d, J=8), 4.30 (d, J=8)	4.38 (s)	4.16 (d, J=6), 4.36 (d, J=6)
C-5 H	4.92 (d, J=10)	4.99 (m)	4.96 (broad d, J=10)
C-2 H	5.68 (d, J=6)	5.74 (d, J=6)	5.71 (d, J=6)
C-13 H	6.20 (broad t, J =8)	6.18 (broad t, J=8)[c]	
C-10 H	6.28 (s)	6.83 (s)	6.46 (s)

[a]s = singlet, d = doublet, t = triplet, m = multiplet, q = quartet.
[b]Della de Casa Marcano and Halsall [19].
[c]In baccatin-V 13α-acetate.

in acetone yielded the 7β-hydroxy conjugated ketone **10** (Figure 2). The molecular composition by high resolution mass spectrometry was in accord with the formula $C_{31}H_{36}O_{11}$, suggesting that it was formed by the loss of the nitrogen-containing α-hydroxy ester function and oxidation of the liberated allylic α-hydroxyl group. Several independent lines indicate that the hydrolyzed ester function was at C-13. The ultraviolet (λ_{max} MeOH, 272 nm, ε 4800) and infrared spectra (ν_{max} $CHCl_3$ 1680 cm^{-1}) are in complete accord with this structure and rule out the alternative Δ^{11}-9,10-dioxo formulation [18]. In addition, the ^{1}H-NMR spectrum of **10** clearly shows the presence of a singlet due to the C-10 proton at δ 6.46 as required by formulation **10**.

4. APPROACHES TO THE SUPPLY OF TAXOL

The efficacy of taxol in treating refractory ovarian cancer is well established. Currently, the bark of the Pacific Yew Tree (*Taxus brevifolia*) is the only approved source of taxol used in the treatment of ovarian cancer. Unfortunately, the procedure for the extraction of taxol from the bark is difficult, low yielding, expensive, and results in destruction of trees. Recently, taxol has also shown promise in treating breast and lung cancers. Therefore, the demand for taxol is increasing very rapidly, and there is an urgent need to develop alternative methods of bulk production of taxol. To save the tree from being extinct, other renewable sources such as needles, twigs, and leaves are being explored. Efforts are also being made by large corporations to cultivate yew on a large scale. Researchers world-wide are also working to produce taxol by other methods such as tissue culture, fungal culture, semisynthesis, and total synthesis.

4.1. Taxol From Natural Sources

Needles As A Renewable Source. The isolation of taxol from renewable sources such as needles and leaves is a very attractive solution to the supply problem because millions of ornamental yews and cultivars are available in nurseries all over the U.S. However, a considerable amount of developmental work remains to be done before the use of needles for bulk production of taxol. It has been found that taxol and related taxanes in needles, unlike the bark, undergo rapid degradation. Therefore, the best method of harvesting and storage of needles prior to extraction needs to be investigated. It will also be desirable to have the extraction facility close to the cultivated plants to prevent deterioration.

Plant Tissue Culture. Plant tissue culture is another potential source of taxol and related taxanes. On May 23, 1991, USDA was issued a patent for the production of taxol and taxol-like compounds by tissue culture, and USDA licensed the technology to Phyton Catalytic, Ithaca, NY. In this process, cells isolated from tissue of a *Taxus* species are grown in large vessels containing artificial growth medium. Cell cultures can be established from superior varieties of plants and growth medium can be varied to increase the production of taxol and related compounds. In theory, this process appears very attractive but in practice it is difficult to establish stable cell lines that produce the desired compounds in high yields. Furthermore, there are problems with large-scale culture. For example, slow growing callus is susceptible to infection by fungus and bacteria. Complex media consisting of highly purified carbon sources are expensive. In spite of these problems, Phyton Catalytic is optimistic about the commercialization of this process for taxol production.

Another West Coast biotechnology company, ESCAgenetics Corporation, San Carlos, CA is also pursuing cell-culture technique for bulk production of taxol. It is claimed that by its proprietary "phytoproduction system" for secondary metabolites, ESCAgenetics has been able to out-do production levels specified in the USDA patent without infringing the patent itself. Even if taxol production by cell culture becomes a commercially viable process, FDA approval for the process will be required. Additionally, impurities in taxol derived from the cell culture technique will differ from those present in bark-derived taxol and therefore the new product will also require FDA approval.

Fungal Culture. Recently, the production of taxol (**1**) by *Taxomyces andreanae*, an endophytic fungus associated with *Taxus brevifolia* has been reported [23]. Of the 200 microbes screened to date, only *T. andreanae* has the ability to produce taxol. The fungus was isolated from the inner bark of one tree in an old-growth cedar forest in northern Montana. The fungus currently produces taxol in nanogram quantities and would require a considerable scale-up to be of commercial use. However, it may be possible to increase the production of taxol by giving the fungus more oxygen or by genetically engineering a better strain. Thus the taxol-generating fungus may offer a promising alternative to the supply of the drug. Recently, Cytoclonal Pharmaceutics Inc., Dallas, TX, has acquired an exclusive license for worldwide rights to the fungal system from Montana State University Scientists.

4.2. Taxol By Synthesis

Total Synthesis. As mentioned above, total synthesis and semisynthesis are also being actively investigated in many laboratories as viable alternatives to the large scale supply of taxol. Taxol is a very complex molecule with 11 asymmetric centers and therefore 2048 diastereomers. The total synthesis of taxol poses a formidable challenge even to the most brilliant organic chemists. However, there is no doubt that a total synthesis of taxol will be accomplished in the not-too-distant future. In fact, it is reported that a team led by Stanford University Professor, Paul Wender, Ph.D., is very close to accomplishing this objective [24]. However, it remains to be seen whether the total synthesis by the Stanford group or others will be a viable commercial process.

Semisynthesis. Unlike total synthesis, semisynthesis from readily available precursors has great potential for the cost-effective production of taxol on a large scale. The synthesis consists of coupling an appropriately protected derivative of an acid to the C-13 hydroxyl group of baccatin III (**7**). The starting material, 10-deacetylbaccatin III (**4**) can be obtained in high yield (1 g/Kg) from the fresh leaves (a renewable source) of *Taxus baccata*, a European yew. A detailed discussion of different approaches to the synthesis of the C-13 ester side chain and its coupling to baccatin III (**7**) to produce taxol is beyond the scope of this review. These approaches have been recently reviewed by Kingston [25]. It should, however, be pointed out that because of the great promise of taxol in cancer chemotherapy, numerous research groups are actively involved in developing new and efficient approaches to the synthesis of the C-13 side chain and semisynthesis of taxol. Therefore, it is quite likely that many more syntheses of the side chain and semisynthesis of taxol will appear before the publication of this article. In this review, only the most promising current approaches to both these compounds are covered.

A French group [26] accomplished the semisynthesis of taxol (**1**) (Chart 3) from 10-deacetylbaccatin III (**4**). Triethylsilylation of **4** under carefully optimized conditions gave 7-triethylsilyl-10-deacetylbaccatin III (**11a**) in 84-86% yield which, upon acetylation, yielded

Chart 3

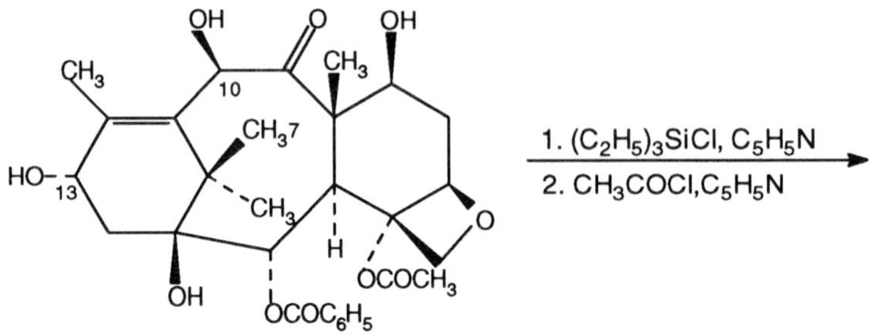

4

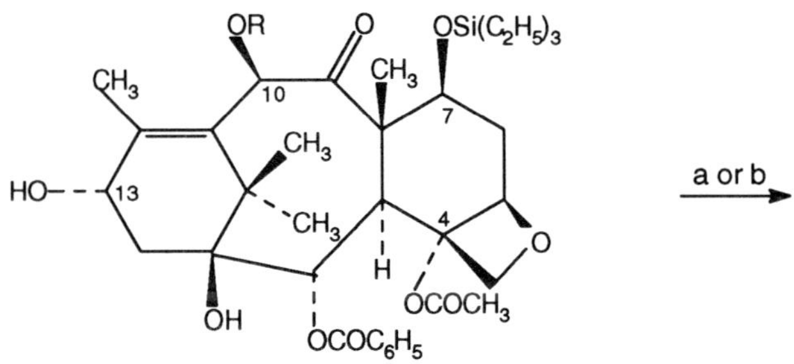

11a, R = H
11b, R = $COCH_3$

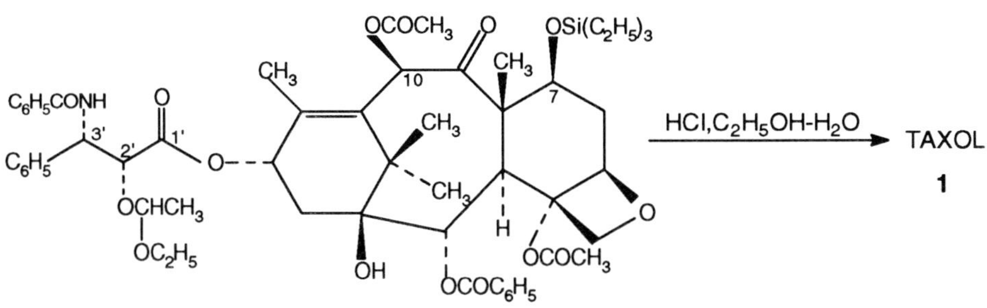

12

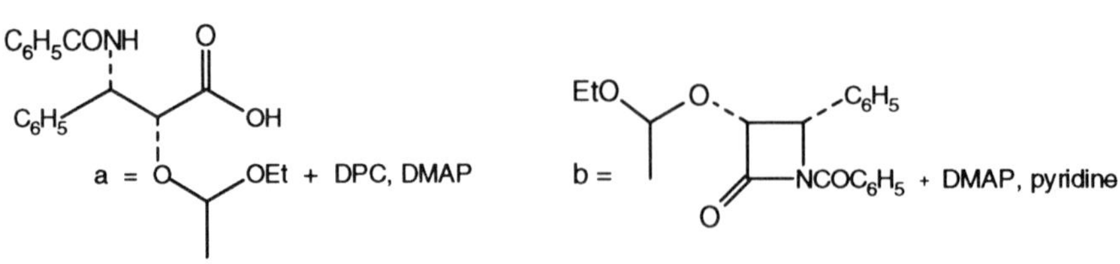

20 **21**

7-triethylsilylbaccatin III (**11b**) in 86% yield. Coupling of the protected baccatin III derivative (**11b**) with previously synthesized optically pure (2*R*, 3*S*)-*N*-benzoyl-O-(1-ethoxyethyl)-3-phenylisoserine (**20**, vide infra) in the presence of di-2-pyridyl carbonate and 4-dimethylaminopyridine (DMAP) gave the protected taxol derivative **12** (80% yield based on 50% conversion) which, upon treatment with dilute HCl, gave taxol in 38% overall yield from **4**.

The second high yielding semisynthesis of taxol from **11b** has been recently reported by Holton [27] (Chart 3). In this approach, a suitably-protected *N*-benzoylated β-lactam **21** (vide infra) was coupled with **11b** in the presence of DMAP in pyridine to give **12** in excellent yield which, upon acidic hydrolysis as described above, gave taxol (**1**):

Syntheses of C-13 Ester Side Chain. Ojima and coworkers [28] have published a highly enantioselective and practical method for the synthesis of the C-13 ester side chain of taxol. The procedure known as the β-lactam synthon method is based on the chiral ester enolate-imine condensation strategy giving 3-hydroxy-4-aryl-β-lactam as the key intermediate. It is claimed that with this approach the phenylisoserine side chain can be obtained in three steps in good yields with almost 100% enantiomeric excess (ee).

Thus, as shown in Chart 4, deprotonation of the triisopropylsilyl ether of the (-)-trans-2-phenyl-1-cyclohexyl ester **13** of glycolic acid with lithium diisopropylamide (LDA) gave the lithium enolate **14** which was condensed with N-(trimethylsilyl)benzaldimine **15** to give the protected β-lactam **16** in >96% ee. Deprotection with tetra-*n*-butylammonium fluoride gave the 3-hydroxy-β-lactam **17**. Acid hydrolysis of **17** gave the aminoacid **18**, which gave **19** upon *N*-benzoylation. Ethoxyethylation of the hydroxyl group of **19** yielded the protected acid **20**. Ethoxyethylation of the hydroxy lactam **17** followed by benzoylation yielded the key lactam **21** required for the taxol semisynthesis by the Holton procedure (*vide supra*).

Another synthesis of the hydroxyl protected side chain by the French group [29] started with the inexpensive (*s*)(+)-phenylglycine (**22**) and led to the final product in four steps with an overall yield of 30% (Chart 5).

The amino acid **22** was reduced with lithium aluminum hydride and benzoylated *in situ* to give **23**. The benzoylation of the intermediate amino alcohol facilitated its isolation. It was recognized that *N*-protected α-amino aldehydes are susceptible to racemization. Therefore, in order to prevent racemization during the conversion of **23** to **24**, the aldehyde obtained by the Swern oxidation of **23** was treated *in situ* with vinylmagnesium bromide by inverse addition to give the desired product **24** with good diastereoselection (9:1) and ee. Ethoxyethylation of the vinyl alcohol **24** with ethyl vinyl ether in the presence of pyridinium *p*-toluenesulfonate followed by oxidation with sodium periodate and catalytic amount of ruthenium chloride in the presence of sodium bicarbonate completed the synthesis to give the protected acid **20**.

A highly efficient synthesis of the taxol C-13 side chain reported by Deng and Jacobsen [30] involves enantioselective epoxidation reaction catalyzed by the readily accessible catalyst (salen)Mn(III)complex **25** (Chart 6). It is claimed that this may be the most practical, commercially feasible route to **19** to date because of the low cost of all reagents and no chromatographic separations.

Chart 4

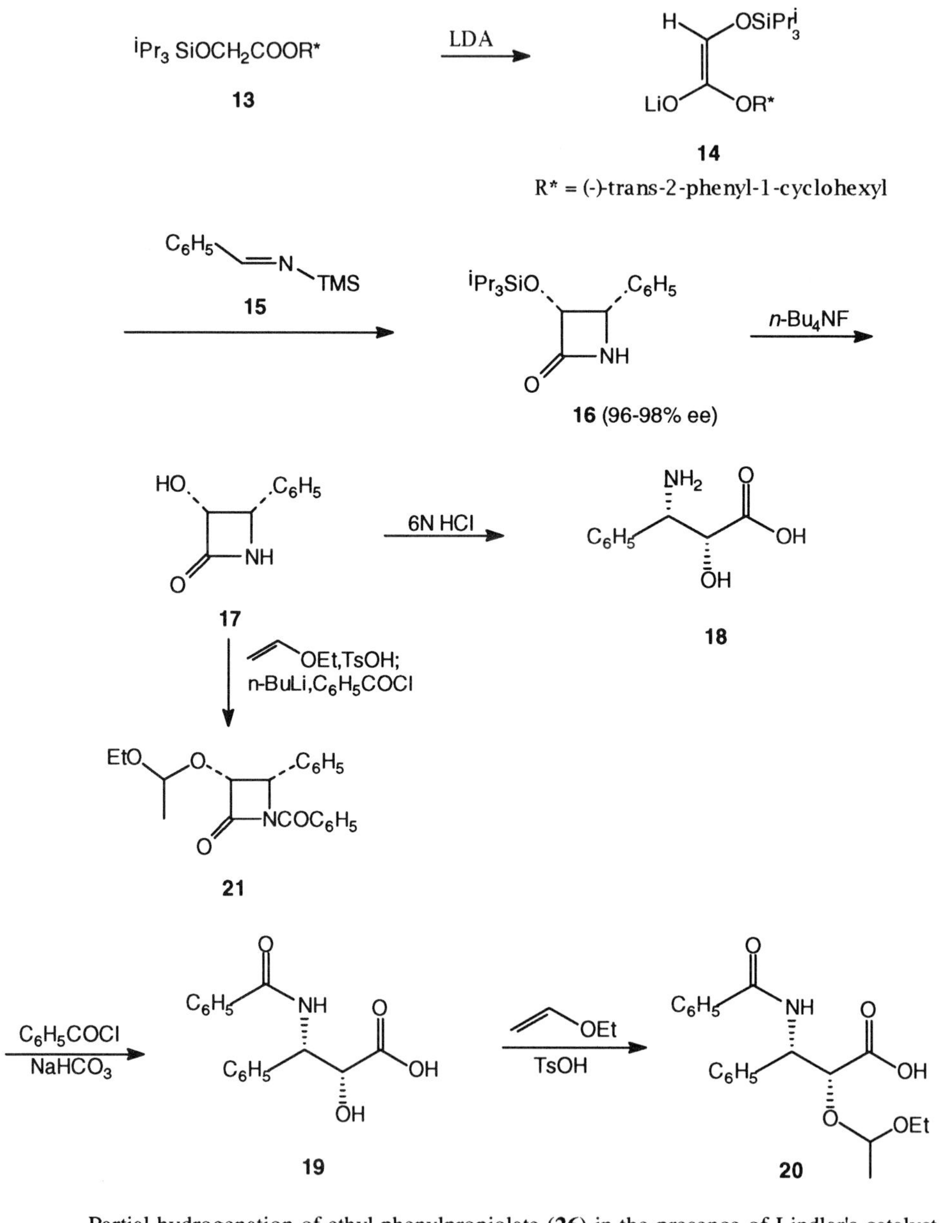

Partial hydrogenation of ethyl phenylpropiolate (**26**) in the presence of Lindlar's catalyst gave predominantly the desired cis-ethyl cinnamate (**27**) along with small amounts of overreduced material and trans alkene. However, these side products did not interfere with subsequent steps.

Epoxidation of **27** with sodium hypochlorite in the presence of 6 mol % (R,R)**25** afforded the cis epoxide **28** in 95-97% ee. The enantioselectivity of the reaction was sensitive to the nature of alkyl group. For example, epoxidation of cis-methyl cinnamate under similar conditions gave the epoxide with decreased ee (87-89%). Addition of a hydrophobic pyridine *N*-oxide such as 4-phenlypyridine *N*-oxide (4-PPNO) was also essential to the success of the reaction. Trans epoxide was also formed in significant amounts (cis:trans ~3.5:1) in this reaction. However, the mixture could be carried forward in the synthetic sequence. Thus, treatment of the mixture of cis and trans epoxides with ammonia in ethanol gave 3-phenylisoserine amide **29** regioselectively. The diastereomeric product from the trans epoxide could be removed by crystallization of the crude product from methanol. Hydrolysis of amide **29** was accomplished with aqueous barium hydroxide without racemization. The reaction was acidified with H_2SO_4, the precipitated $BaSO_4$ was removed by filtration and (+)-(2*R*,3*S*)-3-phenylisoserine (**18**) was obtained by crystallization from the aqueous solution. Benzoylation of **18** gave the taxol side chain **19**.

Taxotere. Utilizing the naturally occurring 10-deacetylbaccatin III (**4**) and an appropriately protected side chain, the French group [31] has reported an efficient semisynthesis of a very active taxol analog named Taxotere (**30**). The analog **30** differs from taxol (**1**) in that the side chain benzoyl moiety is replaced by the *t*-butoxycarbonyl moiety.

Chart 5

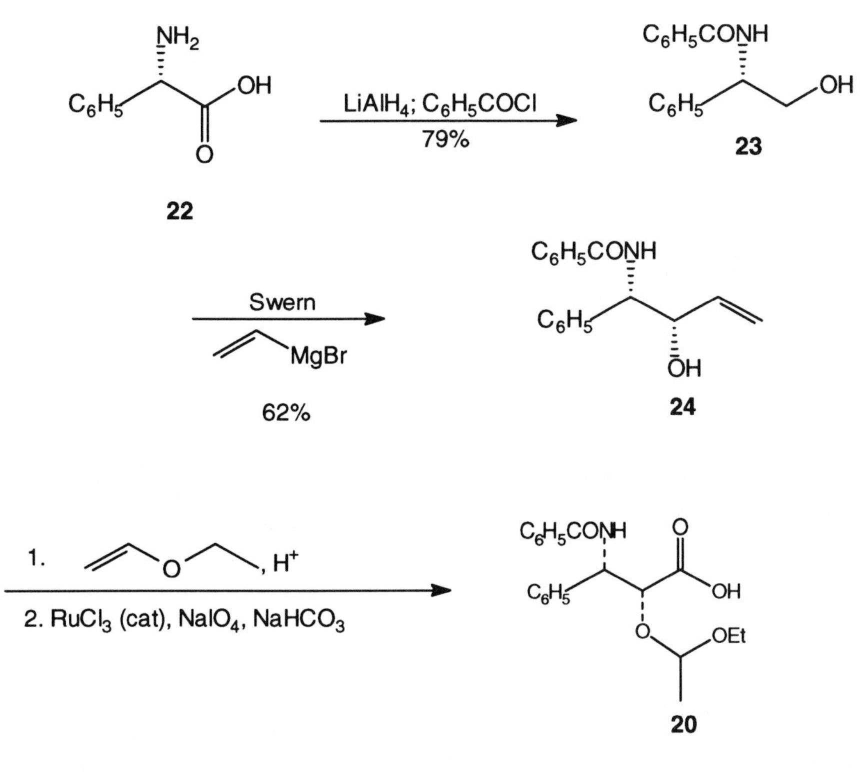

Chart 6

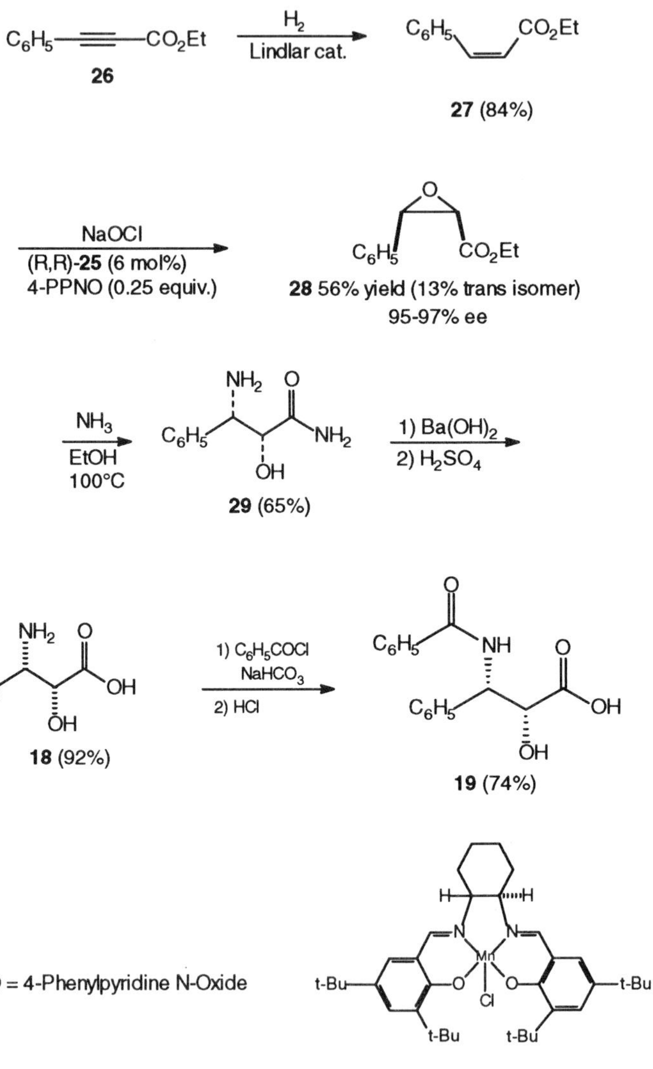

5. MECHANISM OF ACTION OF TAXOL

The mechanism of action of taxol has recently been reviewed in considerable detail [25,32-34]. Therefore, this section will review only the major features of the mechanism of action of this now widely-used antitumor drug. The first mechanistic study reported that taxol was a mitotic spindle poison [35]. Shortly afterward the surprising observation was reported from Horwitz's laboratory that unlike all of the mitotic inhibitors known at that time, taxol did not bind tubulin dimers and thereby block microtubule assembly, but rather accelerated the polymerization of tubulin and stabilized the polymerized microtubules [36].

It has been found that taxol alters the normal equilibrium between soluble tubulin dimers and polymerized microtubules [32]. Taxol increases the polymerization of tubulin into stable microtubules even in the absence of GTP, a cofactor normally required for microtubule polymerization *in vitro* [37,39]. Taxol binds specifically and reversibly to microtubules, in particular to the β subunit of tubulin [40]. It binds to cells in a specific, saturable manner [41].

As stated by Horwitz (32), "In addition to being an essential component of the mitotic spindle and required for the maintenance of cell shape, microtubules are involved in a wide variety of cellular activities, such as cell motility and transport between organelles within the cell. Any disruption of the equilibrium within the microtubule system would be expected to disrupt cell division and normal cellular activities in which microtubules are involved."

Horwitz has stated that, "The ability of taxol to polymerize tubulin into stable microtubules in the absence of any cofactors and to induce the formation of stable microtubule bundles in cells are the unique properties of this drug. Taxol is thus a prototype of a new class of antitumor drugs and has focused attention on microtubules as a worthy target for cancer chemotherapeutic drugs. At the molecular level, the mechanism by which taxol interacts with microtubules and blocks cells in mitosis is poorly understood. Essentially, there is no information on the site(s) at which taxol binds to microtubules or on the specificity that taxol displays toward certain malignant tumors [32]."

6. STRUCTURE ACTIVITY RELATIONSHIPS (SAR) OF TAXOL AND ANALOGS

The SAR of Taxol and analogs has been of great interest for a number of years. Major advances in the synthesis of the taxol side chain and in the semisyntheses of taxol (**1**), taxotere (**30**), and analogs from synthetic side chains and naturally occurring taxanes have been made during the last decade [25-31]. As a consequence, SAR data is available for a large number of taxol analogs modified either in the side chain or nucleus. Much of this data was assembled from the outstanding SAR studies of Kingston [25], Guérrite-Voegelein et al. [42] and Swindell et al. [43].

Excellent reviews by Kingston and Suffness give data for many taxol and taxotere analogs modified either in the side chain or in the nucleus [25,34]. The purpose of the various synthetic studies and associated tubulin or cytotoxicity assays has been to (a) determine what substituents on the side chain or nucleus could be modified or eliminated without loss of activity in order to simplify further synthetic studies and (b) to determine if more active analogs could be prepared. With a few notable exceptions, the majority of analogs modified in the side chain or nucleus have been less active than taxol or taxotere. Because reviews are available which present in great detail the effects of modification of the side chain or taxane ring of **1** or **30** on tubulin disassembly, tubulin assembly, cytotoxicity, and *in vivo* acitivity, this section will deal primarily with SAR studies based on tubulin disassembly data, but will also include some tubulin assembly data. This data was assembled mainly from a review by Kingston [25].

6.1 The C-13 Side Chain

The taxol sidechain may be regarded as a derivitive of 3-phenylisoserine with the 2*R*,3*S* configuration as shown below.

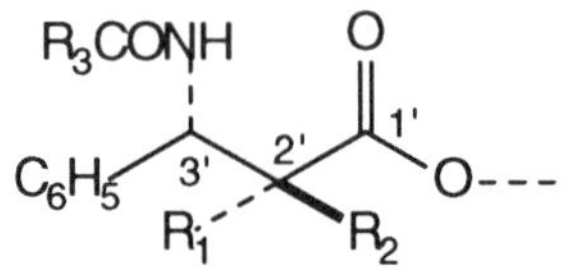

The first SAR study showing that both the side chain and the nucleus (taxane ring) in combination were required for activity was carried out more than twenty years ago. During the course of our studies on the structure of taxol, the side chain was cleaved from the nucleus by low temperature methanolysis. Although taxol demonstrated considerable cytotoxicity, both the cleaved side chain and nucleus were essentially inactive [4,5]. It is of interest to note that a number of cytotoxic agents isolated from plants have ester side chains and a ring nucleus [44]. These include compounds such as harringtonine [45], maytansine [46], and a number of quassinoids [47]. The specific functions of the side chains of taxol and other cytoxic agents is unknown but undoubtedly involves binding to a specific receptor, enzyme, or protein. In every case hydrolysis of the ester group resulted in complete inactivation as judged by cytotoxicity.

Effects of Substituent Changes at C-2' Position. The configuration at C-2' position found in taxol (**1**) and taxotere (**30**) is 2*R* (R_1 = OH, R_2 = H). Acetylation of the 2'-hydroxyl moiety to give the acetate **31** results in considerable loss of activity [48,50]. The 2'-desoxy analog **32** ($R_1 = R_2$ = H) also suffers major loss of activity [42]. Reversal of the configuration at C-2' to give compound **33** (R_1 = H, R_2 = OH) also considerably reduced tubulin disassembly activity [43].

Effects of Substituent Changes at 3' Position. Two substituents are present at the 3'-position of the taxol (**1**) side chain, a phenyl (C_6H_5) group and an *N*-carboxamide moiety ($R_3 = C_6H_5$). The 3'-phenyl group is essential for activity [25] but considerable modification of the 3'-carboxamide moiety can be made without loss of major tubulin disassembly activity.

Replacement of the N̲-carboxamide group ($R_3 = C_6H_5$) by the *N*-t-butoxy-carbonyl moiety ($R_3 = (CH_3)_3CO$) gives taxotere (**30**) which has greater activity than **1**. The naturally-occurring taxane cephalomannine [49] (**33**) ($R_3 = CH_3CH{=}C(CH_3)$) is slightly less active than **1** [53,54]. Replacement of both phenyl and R_3CONH groups leads to great loss of activity [43].

6.2. Taxane Ring

Because of the complexity of the taxane ring and the difficulty of conducting reactions which affect only one ring substituent at a time, less SAR information is available than is present for sidechain analogs. Most of the taxane ring studies have been carried out on ring C-7 or C-10 substituents. Although a number of studies have been conducted on baccatin III (**7**) derivatives [25,34], only ring analogs of **1** or **30** will be considered in this section (cf. structure below).

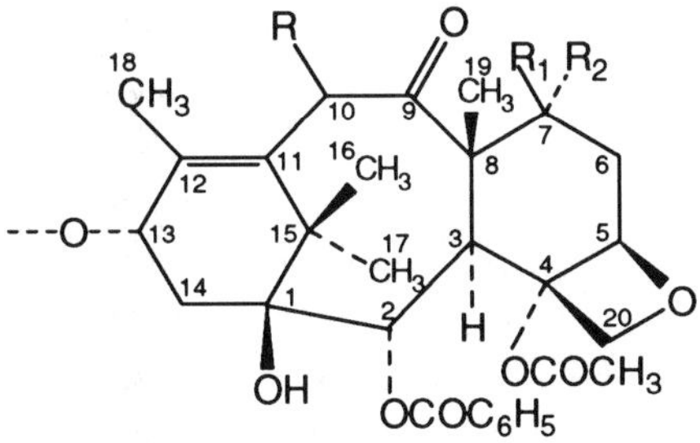

34, R = β-OAc, R_1 = OAc, R_2 = H
35, R = β-OAc, R_1 = O-Xylosyl, R_2 = H
36, R = β-OAc, R_1 = H, R_2 = OH
37, R = β-OAc, R_1, R_2 = O
38, R = β-OH, R_1 = OH, R_2 = H
39, R = O, R_1 = OH, R_2 = H

Effects of Changes at C-7 Position. The 7β -hydroxyl moiety is found in **1**, **30**, and **33**. Acylation to give the 7β-acetate (**34**) does not seriously diminish tubulin disassembly activity [50]. Polar sugar residues such as the 7β-xylose derivative of **1** (**35**) increase activity to some extent [50]. The 7α-hydroxyl epimer, 7-epitaxol (**36**) exhibits a modest reduction in activity [51]. The cytotoxicity (KB) of 7-oxotaxol (**37**) is much less than that of **1** [52].

Effects of Changes at C-10 Position. Removal of the C-10 acetyl group of **1** yields 10-deacetyl taxol (**38**) which is slightly less active than **1** [53,54]. It should be noted, however, that in the case of taxotere (**30**) in which deacetylation at C-10 is also accompanied by a change in the C-3' substituent, a two fold increase in tubulin disassembly activity was found [42]. Oxidation of the 10-hydroxyl moiety of 10-deacetyltaxol (**38**) gives the corresponding 10-oxo-analog (**39**) with greatly reduced cytotoxicity [52].

Changes in Other Taxane Ring Substituents. Only a few studies have been conducted involving changes at positions other than C-7 or C-10 in the taxane ring. Opening the oxetane ring yielding 20,O-secotaxol (**40**) resulted in marked loss of KB cytotoxicity [52]. Ring contraction yielding 11(15→1)abeotaxol (**41**) with about 1/3 the tubulin disassembly action but with almost complete loss of KB cytotoxicity [52]. Removal of the 4-acetyl moiety in cephalomannine resulted in considerable loss of cytotoxicity [53].

7. CLINICAL RESULTS

Phase I clinical trials of taxol with different schedules commenced in 1984 [55]. Numerous obstacles and difficulties ensued, including hypersensitivity to the Cremophor EL (polyethoxylated castor oil) formulation, neutropenia, the principle dose-limiting toxicity, peripheral neurotoxicity, and cardiac rhythm disturbances [33].

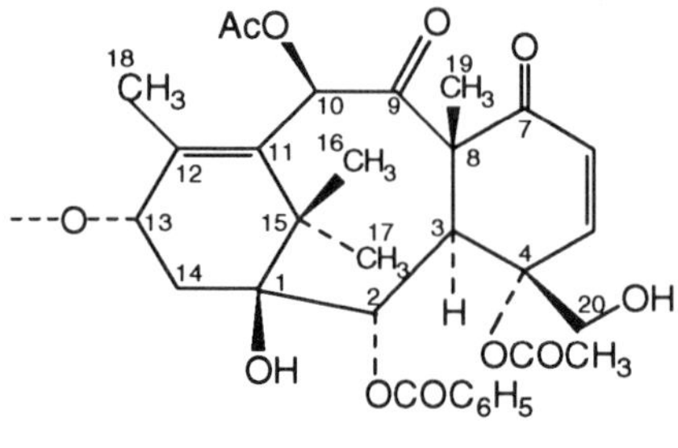

40

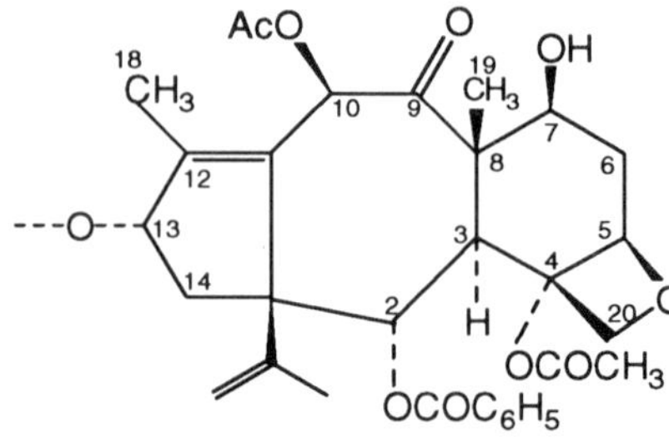

41

A major problem, which was in effect until almost the end of 1992, was the extremely limited supply of taxol. Previously, patients could not receive taxol therapy without having had three prior chemotherapy regimens. As a consequence, most of the previous clinical trials have involved subjects with chronic, usually metastatic disease. The situation in regard to taxol supply is rapidly improving (cf. Section 4). Moreover, in December 1992, Bristol-Myers/Squibb received FDA approval to market taxol. In addition, a promising taxol analog named taxotere (RP56976) has been prepared by semisynthesis from 10-deacetylbaccatin (**4**) by the French pharmaceutical company, Rhone-Poulenc and is now in clinical trial. Taxotere appears to be somewhat more potent and water soluble than taxol. Since there are excellent comprehensive reviews of previous clinical studies with taxol [33,34,55], and abstracts of several recent symposia are available [56,57], this section will review only the major aspects of the current clinical situation. With the advent of ample supplies of both taxol and the taxol analog, taxotere, it is likely that, in the future, clinical utilization of taxol will involve extensive use of combination therapy with other anticancer drugs at earlier stages of the disease [33].

7.1. Formulation

Cremophor EL. Taxol is a lipophilic substance insoluble in water. It is currently formulated as a 50% mixture of Cremophor EL (polyethoxylated castor oil) and 50% alcohol [34]. The drug is then diluted with saline and administered at doses of 135–250 mg/m^2 as a 24 hour infusion [33,58]. The major adverse effects ascribed to the presence of Cremophor EL in the formulation are hypersensitivity reactions mediated by direct release of histamine [33,34]. This problem has largely been overcome by premedication with steroids, histidine antagonists, and the utilization of 24 hour infusion rather than shorter periods [33].

Liposomes. Considerable research is underway to prepare formulations which will have fewer side effects than Cremophor EL. Currently, liposomes are under active investigation. Liposomes represent a mature, versatile technology which has been applied to lipophilic drugs such as adriamycin, cyclosporine, and amphotericin B [59,60].

Water Soluble Taxol Analogs. Several water soluble esters of the 2'-hydroxyl group of taxol (cf. Section 6) have been prepared, including succinate and other diacid derivatives [61], taurine and related analogs [48], and amino acid derivatives [62,63]. Unfortunately, these 2'-hydroxyl esters are very reactive and undergo rapid hydrolysis yielding water insoluble taxol. However, recently Nicolaou et al. [64] have designed and synthesized a series of water-soluble taxol-releasing derivatives (protaxols) with improved pharmacological properties. These prodrugs, which contain the water-solubilizing moiety at the C-2' hydroxyl group, release taxol under basic or physiological conditions.

7.2. Current Clinical Usage of Taxol and Taxotere

Ovarian Cancer. Taxol is in extensive use for ovarian cancer [33]. A 1989 Phase II study showed a 30% response rate, which was particularly encouraging because the patients had been heavily pretreated with chemotherapy and radiation [65]. Subsequently, somewhat higher response rates were observed. Trials with combination therapy, taxol plus cisplatin, are in an early stage [33].

Breast Cancer. After treatment of patients with breast cancer with taxol, a 50% response rate was noted in a 1991 MD Anderson Cancer Center Study [66]. A later study with patients with zero prior stage IV regimens showed a 60% response. Those with one or two prior treatments gave a 35% response [67]. Even more promising results have been observed with taxotere when used as a first line chemotherapy with responses ranging from 55–75%, or as second line therapy for anthracycline resistant diseases with responses averaging 63% [68].

Future Prospects. As supplies of both taxol and taxotere increase, combination therapy will become more frequent. Patients with other solid tumors will be treated. It is evident that early treatment prior to the development of resistance will be invaluable.

From our perspective as the discoverers of taxol some twenty-five years ago, the evolvement of this compound from an interesting natural product with a complex structure to a chemotherapeutic agent which has aroused great excitement on the part of scientists and physicians can only be most gratifying. Hopefully, interest and support for natural product scientists will continue on a high level.

REFERENCES

1. LA Mitscher, S Drake, SR Gollapudi, JA Harris, and DM Shankel, In Antimutagensis and Anticarcinogenesis Mechanisms, DM Shankel, Ed., Plenum Press, New York, 1986, p. 153.
2. DH Williams, JJ Stone, PR Hauck and SK Rahman, *J. Nat. Prod.* 52:1189 (1989).
3. AM Woodbury, ME Wall, and JJ Willaman, *J. Econ. Bot.* 15:79 (1961).
4. MC Wani, HL Taylor, ME Wall, P Coggin,and AT McPhail, *J. Am. Chem. Soc.* 93:2325 (1971).
5. ME Wall, in: Chronicles of Drug Discovery, Vol. 3, D Lednicer, Ed., *Am. Chem. Soc.*, in Press (1993).
6. DD Von Hoff, Private Communication to M.E. Wall, 1989.
7. S Broder, Private Communication to M.E. Wall, 1991.
8. EK Rowinsky, MP McGuire, and RS Donehower, in: Workshop on Taxol and Taxus: Current and Future Perspectives, Bethesda, MD, June 1990.
9. L Lenaz, and MD De Furia, Fitoterapia 64, Suppl. 1 (1993).
10. ME Wall, MC Wani, and HL Taylor, Cancer Chemother. Repts. 60:1011 (1976).
11. MC Wani, HL Taylor, and ME Wall, Chem. Comm. 390 (1973).
12. MC Wani, HL Taylor, ME Wall, AT McPhail, and KD Onan, J. Am. Chem. Soc. 97:5955 (1975).
13. MC Wani, HL Taylor, JB Thompson, and ME Wall, *J. Nat. Prod. 41:*578 (1978).
14. ME Wall, Letter to JL Hartwell, April 1966.
15. ME Wall, RTI Progress Report #18 May, 1966.
16. ME Wall and MC Wani, Paper M-006, 153rd National Meeting, American Chemical Society Miami Beach, FL, 1967.

17. MC Woods, K Nakanishi, and NS Bhacca, Tetrahedron 22:243 (1966).
18. IW Harrison, RM Scrowston, and BJ Lythgoe, Chem. Soc. 1933 (1966).
19. DP Della Casa de Marcano and TG Halsall, Chem. Commun. 1382 (1970).
20. G Zemplen,A Gerecs, and I Hadacsy, Ber. 69:1827 (1936).
21. M Kurono, Y Nakadaira, S Onuma, S., Y Sasaki, and K Nakanishi, Tetrahedr. Lettr. 2153 (1963).
22. IM Goldman, J. Org. Chem. 34:1979 (1969).
23. A Stierle, G Strobel, and D Stierle, Science 260:214 (1993).
24. P Wender and TP Mucciaro, J. Am. Chem. Soc. 114:5878 (1992).
25. DGI Kingston, *Pharmac. Ther.* 52:1 (1991).
26. JN Denis, AE Greene, D Guenard, F Gueritte-Voegelein, L Mangatal, and PJ Potier, J. Am. Chem. Soc. 110:5917 (1988).
27. RA Holton, Eur. Pat. Appl. 0400971 (1990).
28. I Ojima, I Habus, and M Zhao, J. Org. Chem. 56:1681 (1991).
29. JN Denis, A Correa, and A Greene, J. Org. Chem. 56:6939 (1991).
30. L Deng and EN Jacobsen, J. Org. Chem. 57:4320 (1992).
31. L Mangatal, MT Adeline, D Guenard, F Gueritta-Voegelein, and P Potier, Tetrahedron 45:4177 (1989).
32. SB Horwitz, Trends Pharmacol. Sci. 13:134 (1992).
33. EK Rowinsky, N Onetto, RM Cametto, and SG Arbuck, in: Seminars on Oncology 19:646, 1992.
34. M Suffness, Ann. Reps. Med. Chem . In press, 1993.
35. DA Fuchs and RK Johnson, Cancer Treat. Repts. 62:1219 (1978).
36. PB Schiff, J Fant, and SB Horwitz, Nature 277:665 (1979).
37. PB Schiff and SB Horwitz, Biochemistry. 20:3247 (1981).
38. W Kumar, J. Biol. Chem. 256:10435 (1981).
39. J Parness and SB Horwitz, J. Cell Biol. 91:479 (1981).
40. S Rao, SB Horwitz, and I Ringel, J. Natl. Cancer Inst. 84:789 (1992).
41. JJ Manfredi, J Parness, and SB Horwitz, J. Cell Biol. 94:688 (1982).
42. F Gúeritte-Voegelein, D Guénard, F Cavelle, M-T LeGaff, L Mangatal, and P Potier, J. Med. Chem. 34:992 (1991).
43. CS Swindell, NF Krauss, SB Horwitz, and I Ringel, J. Med. Chem. 34:1176 (1991).
44. JM Cassidy and JD Douros, Anticancer Agents Based on Natural Product Models, Academic Press, New York, 1980.
45. RG Powell, D Weisleder, and CR Smith, J. Pharm. Sci. 61:1227 (1972).
46. S Kupchan, Y Komoda, WA Court, GJ Thomas, RM Smith, A Karim, CJ Gilmore, RC Haltiwanger, and RP Bryan, J. Am. Chem. Soc. 95:1354 (1972).
47. ME Wall and MC Wani, in: Ann. Rev. Pharmacol. and Toxicol., Vol. 17, HW Elliott, R George, R Okun, Eds., Ann. Reviews Inc., Palo Alto, 1977, p. 42.
48. NF Magri and DG Kingston, J. Nat. Prod. 51:298 (1988).
49. RG Powell, RW Miller, and CS Smith, Jr., J. Chem. Soc. 102 (1979).
50. H Lataste, V Sénilh, M Wright, D Guenard and P Potier, Proc. Nat. Acad. Sci. USA 81:4090 (1984).
51. I Ringel and SB Horwitz, J. Pharmacol. Exp. Therap. 692 (1987).
52. DGI Kingston, G Samaranayake and GA Ivey, J. Nat. Prod.53:1 (1990).
53. G Chauviere, D Guénard, F Picot, V Senihl and P Potier, C.R. Acad. Sc. Paris, Series II, 193:501 (1981).
54. J Parness, DGI Kingston, RG Powell, C Harracksingh and SB Horwitz, Biochem. Biophys. Res. Commun. 105:1082 (1982).
55. WJ Slichenmyer and DD Von Hoff, Anticancer Drugs 2:519 (1991).
56. Second National Cancer Institute Workshop on Taxol and Taxus, September 23-24, 1992, Alexandria, VA.
57. Stonybrook Symposium on Taxol and Taxotere—New Hope for Breast Cancer Chemotherapy, Stonybrook, NY, May 14-15, 1993.
58. EK Rowinsky, LA Casenave, and RC Donehower, J. Natl. Cancer Inst. 82:1247 (1990).
59. RM Straubinger, A Sharma, and E Mayhew,. Second National Cancer Institute Workshop on Taxol and Taxus, September 23-24, 1992, Alexandria, VA.
60. RM Straubinger, Stonybrook Symposium on Taxol and Taxotere—New Hope for Breast Cancer Chemotherapy, Stonybrook, NY, May 14-15, 1993, p. 9.
61. HM Deutsch, JA Glinski, RD Hernandez, RD Haugwitz, VL Narayanan, M Suffness, and LH Zalkow, J. Med. Chem. 32:788 (1989).

62. AI Mathew, MR Mejillano, JP Nath, RH Himes, and VJ Stella, J. Med. Chem. 35:145 (1992).
63. Z Zhou, DGI Kingston, and AR Crosswell, J. Nat. Prod. 54:1607 (1991).
64. KC Nicolaou, C Reimer, MA Kerr, D Rideout, and W. Wrasidlo, Nature 364:464 (1993).
65. WP McGuire, EK Rowinsky, WB Rosenheim, FC Grumbine, DS Ettinger, DK Armstrong, and RC Donehower, Ann. Intern. Med. 111:273 (1989).
66. FA Holmes, RS Walters, and RL Theriault J. Natl. Cancer Inst. 83:1797 (1991).
67. AD Seidman, Stonybrook Symposium on Taxol and Taxotere, Stonybrook, NY, May 14-15, 1993, p. 14.
68. PM Ravdin, Stonybrook Symposium on Taxol and Taxotere, Stonybrook, NY, May 14-15, 1993, p. 18.

Chapter Two

The Synthesis of Macroline Related Sarpagine Alkaloids

Linda K. Hamaker and James M. Cook
Department of Chemistry,
University of Wisconsin Milwaukee,
P.O. Box 413, Milwaukee,
Wisconsin 53201-0413,
U.S.A.

CONTENTS

1. INTRODUCTION

Indole alkaloids have long held a prominent position in the history of natural products chemistry because of the structural similarity to the essential amino acid tryptophan and related metabolites of tryptophan, such as the neurotransmitter serotonin. New alkaloids have been isolated from a variety of sources with increasing frequency and characterized *via* the latest spectroscopic techniques; moreover, thousands of alkaloids have been obtained from plant sources worldwide (1-6). The medicinal properties of these natural products remain of great interest, as well as the nature of their structure and stereochemistry. The construction of these structurally complex molecules remains of paramount importance to the synthetic chemist. This challenge has grown considerably in an enantiospecific sense to permit comparison of the biological properties of the unnatural antipodes to those of the natural alkaloids (7-12). It is now important to design synthetic routes to provide entry into either antipode in high optical purity. For the above reasons, studies which involve indole alkaloids continue to play a prominent role in organic chemistry. Synthetic routes are designed and revised continually in order to provide material for testing and for commercial uses. However, many alkaloids with potential medicinal importance have not been evaluated, to date, due to the paucity of isolable material. Consequently, an important goal of the synthetic chemist continues to center on the preparation of these alkaloids on a gram scale to permit biological screening.

Interest in macroline-related plant alkaloids from the species *Alstonia* originated as a result of folk tales describing the medicinal properties of these plants (13, 14). The macroline/sarpagine alkaloids bear important structural similarities to the ajmaline alkaloids, the latter of which are well known for their biological activity (15-28). The most prominent action of ajmaline is an antiarrhythmic effect on the heart that is less pronounced than that of propranolol (29, 30), but is superior in terms of the ratio of prolongation of the refractory phase over reduced conduction to that of procaine amide and quinidine (31). For this reason (+)-ajmaline has been extensively used in Europe to treat arrhythmias (29-32). Stöckigt has recently shown experimentally with the aid of the enzyme *vinorine synthase* that a biogenetic link exists between the sarpagine and ajmaline alkaloids (33). In a structural sense, macroline can be envisioned to originate *via* bond cleavage between the N_b nitrogen atom and the carbon atom at C(21) of the sarpagine alkaloids (34, 35). A discussion of the aforementioned three classes of alkaloids will be presented with emphasis on the macroline/sarpagine series. The preparation of the indole alkaloid ajmaline and a few other syntheses of related interest will also be briefly described. A description of the newly isolated alkaloids in the macroline/sarpagine series will be described first, followed by the syntheses of the macroline/sarpagine/ajmaline indole alkaloids.

1.1. Classification of Alkaloids

1.1.1. Sarpagine

The sarpagine alkaloids (see Figure 1) are the largest class of natural products related to the macroline bases, and both series originate from common biogenetic intermediates. The two classes can be related in a synthetic sense (Figure 2) by a Michael addition of the nitrogen atom of N(4) of macroline **2** to the α,β-unsaturated carbonyl system at C(21), or by direct 1,2 addition of N(4) to the ketone at C(19) (36). Described in this review are the recent isolation and synthesis of sarpagine alkaloids which contain a hydrogen atom in the β-position at C(16). Many other alkaloids which contain two functional groups at C(16) exist in this series and provide a direct relationship to the ajmaline alkaloids (33); however, they will not be covered extensively here.

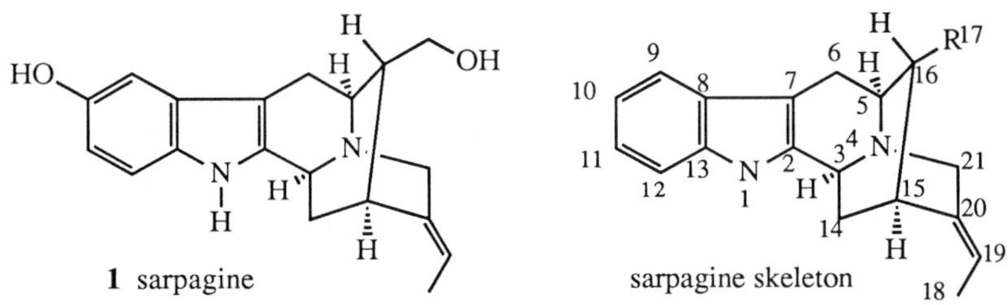

Figure 1. The numbering system of the sarpagine alkaloids.

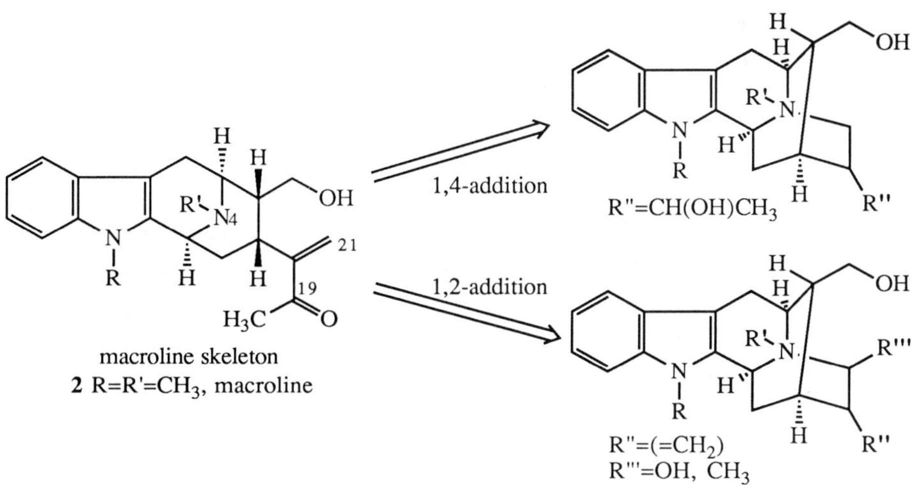

Figure 2. Retrosynthetic relationship of macroline to the sarpagine alkaloids.

1.1.2. Macroline

Macroline **2** has not yet been isolated as a natural product but is believed to be a biomimetic precursor to many *Alstonia* alkaloids (37). During the elucidation of the structures of *Alstonia* bisindole alkaloids by Schmid *et al.*, macroline **2** was obtained as a degradation product from villalstonine (37-42). Depicted in Figure 3 is the numbering system employed for macroline **2**. Note the four stereocenters common to this important biogenetic intermediate at C(3), C(5), C(15), and C(16). The β-hydrogen atom at C(15) of this group, as well as the chiral centers at C(3), C(5), and C(16), are the same as those in the sarpagine series (Figure 1).

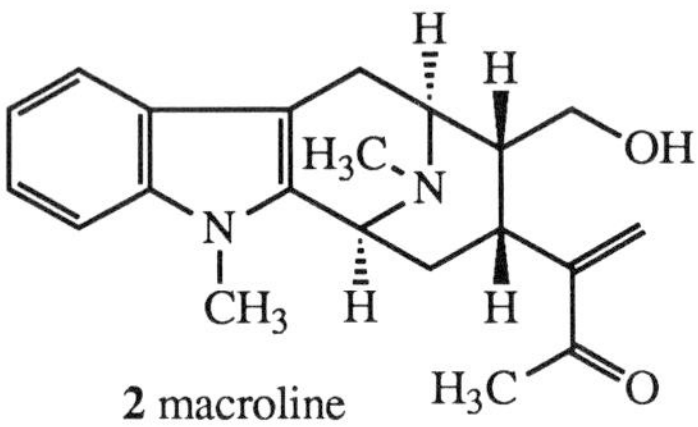

2 macroline

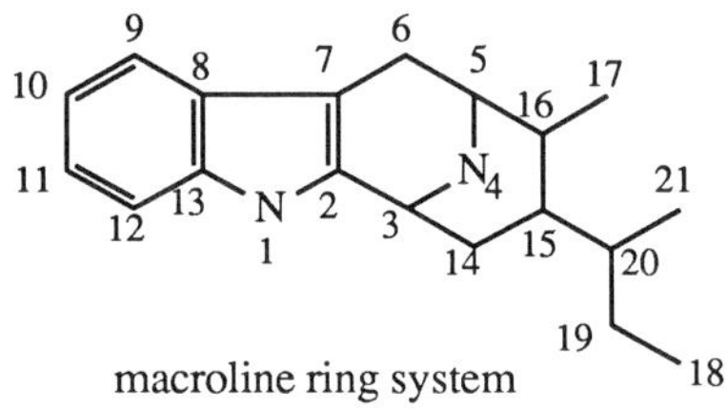

macroline ring system

Figure 3. The numbering system of the macroline skeleton.

1.1.3. Suaveoline

Suaveoline **3** and the two related alkaloids norsuaveoline **4** and macrophylline **5** (43-48) are unique because rings A-D are identical to those in macroline **2**; however, ring E in this series is comprised of an aromatic pyridine moiety devoid of stereochemistry at C(15) and C(16). The suaveoline alkaloids will be considered "macroline related" here, and the reasons for this designation will become apparent.

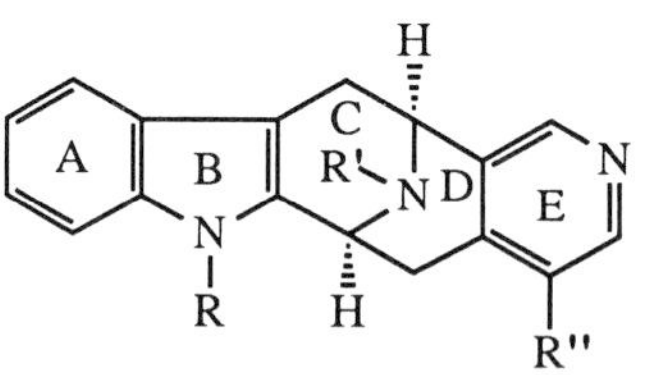

3 R=CH_3, R'=H, R"=C_2H_5, suaveoline
4 R=H, R'=CH_3, R"=C_2H_5, norsuaveoline
5 R=H, R'=CH_3, R"=CH_2CH_2OH, macrophylline

Figure 4. The suaveoline alkaloids.

1.1.4. Ajmaline

As mentioned above, Stöckigt has investigated the biogenetic link between the sarpagine and ajmaline alkaloids (33). Prior to Stöckigt's work, Woodward (49) had suggested that the sarpagine alkaloids (**6**) which bear an *endo* aldehyde function at C(16) might act as a biogenetic precursor of the ajmaline skeleton (**8**). Taylor then performed this conversion chemically under strongly acidic conditions (50), as shown in Scheme 1. A short time later Masamune and coworkers carried out a similar transformation in their synthesis of (±)-ajmaline (51). Bartlett and Taylor's reductive alkylation was also utilized in van Tamelen's biogenetic synthesis of ajmaline (52). In 1983 Stöckigt confirmed the biogenetic link between sarpagine and the ajmaline alkaloids by conversion of 16-epi-vellosimine **7** into deacetylvinorine **9** in the presence of the acetyl-CoA dependent *vinorine synthase* (Scheme 1). Attempts were made to isolate the intermediate deacetylvinorine **9** but were not successful. Based on the specificity of *vinorine synthase*, however, Stöckigt clearly demonstrated that 16-epi-vellosimine **7** was a biogenetic precursor of vinorine **10**, and only alkaloids that carried an *endo* aldehyde function served as substrates for the enzyme (33). Recently, interest in the well known biological properties of ajmaline **11** (15-28) has led to the search for new antiarrhythmic drugs derived from indole alkaloids which are related to this natural product (53).

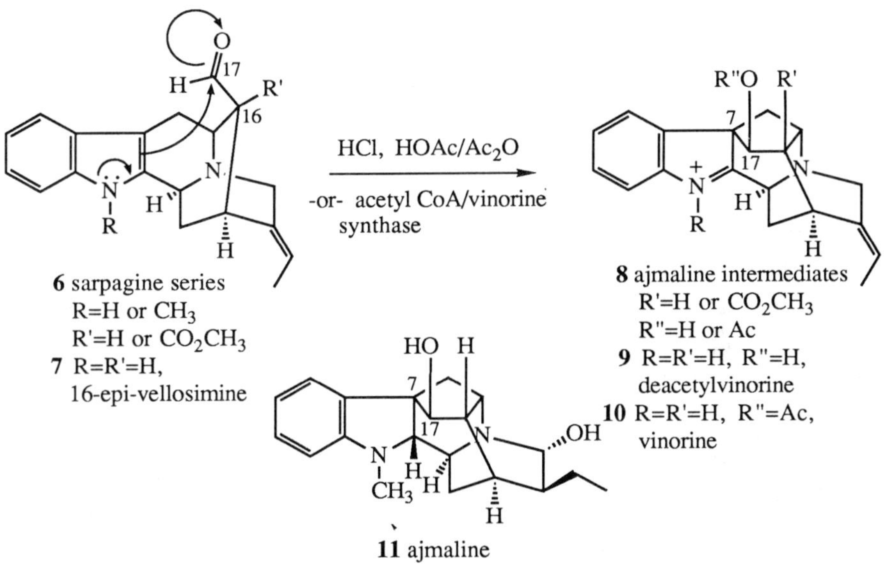

Scheme 1.

1.1.5. Oxindoles

Macroline related oxindoles are a rather small class of alkaloids of the general structure illustrated in Figure 5. The structures of the recently isolated oxindoles along with references on their isolation will be presented in the new alkaloids section. For oxindole alkaloids in the

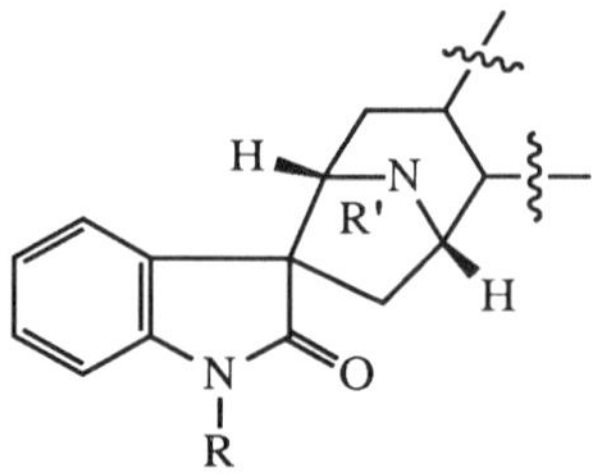

Figure 5. The general structure of macroline related oxindoles.

macroline/sarpagine series, other than those listed in the new alkaloids section, one is referred to the last review in this series (36). Interest in the synthesis of this class of indole alkaloids has been stimulated by the desire to determine the biogenetic pathway to these bases (54). Synthetic studies on oxindole alkaloids outside of the macroline/sarpagine class have also been reported (55, 56). However, studies on the synthesis of the macroline/sarpagine related oxindoles has been rather limited to date (54, 57-59).

1.1.6. Degradation Products

Illustrated below in Figure 6 are a number of bases not yet isolated as natural products

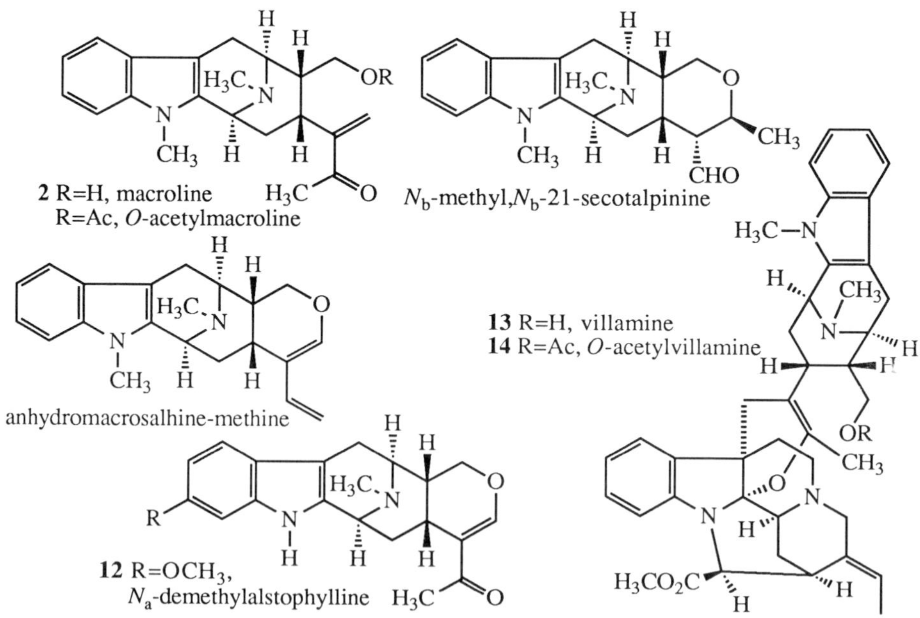

Figure 6. Intermediates from degradation of *Alstonia* bisindole alkaloids.

but obtained from the degradation of *Alstonia* bisindoles (37-39, 42, 60-63). The biomimetic synthesis of a number of bisindole alkaloids from macroline **2** was reported originally by LeQuesne *et al.* and recently reviewed (36). The reader is referred to the synthesis section for a more detailed discussion of bisindoles.

1.2. Recently Isolated Alkaloids

The alkaloids illustrated in the section below are macroline/sarpagine alkaloids that have been isolated since these bases were last reviewed (36). Although emphasis in this report is placed on the alkaloids of the macroline/sarpagine substructure, the closely related ajmaline alkaloids along with a few sarpagine bases with the α-(H)-configuration at C(16) are also

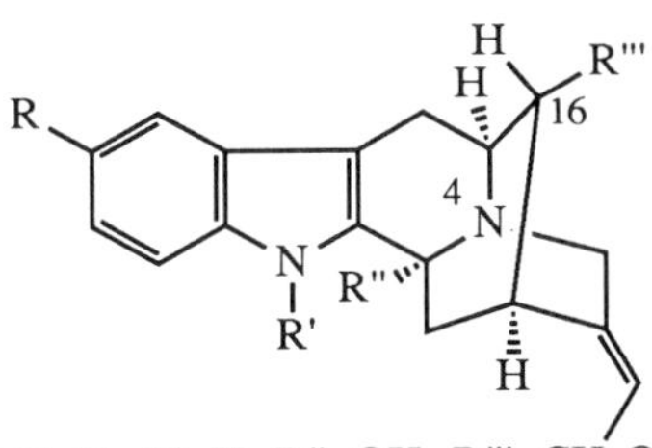

15 R=H, R'=H, R"=OH, R'''=CH_2OH, N_a-demethylaccedine (64)
16 R=H, R'=H, R"=H, R'''=CH=CHCOCH_3, difforine (65)
17 R=H, R'=H, R"=OH, R'''=CH_2OH, amerovolfine (66)
18 R=H, R'=CH_3, R"=H, R'''=CO_2CH_3, N_a-methyl-16-epipericyclivine (67)
19 R=H, R'=CH_3, R"=H, R'''=CH_2OH, N(4) → O affinisine N_4-oxide (68)

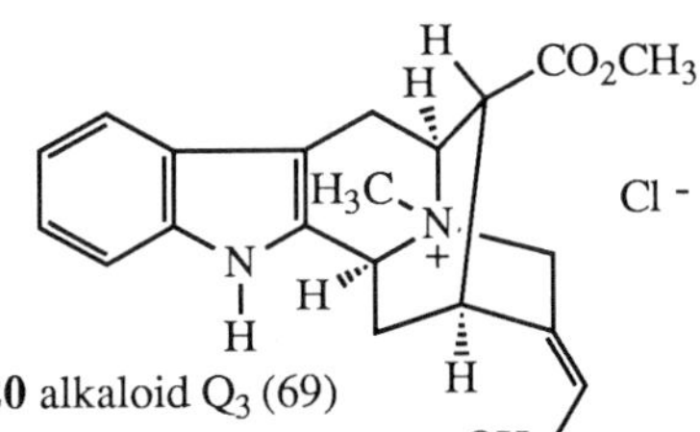

20 alkaloid Q_3 (69)

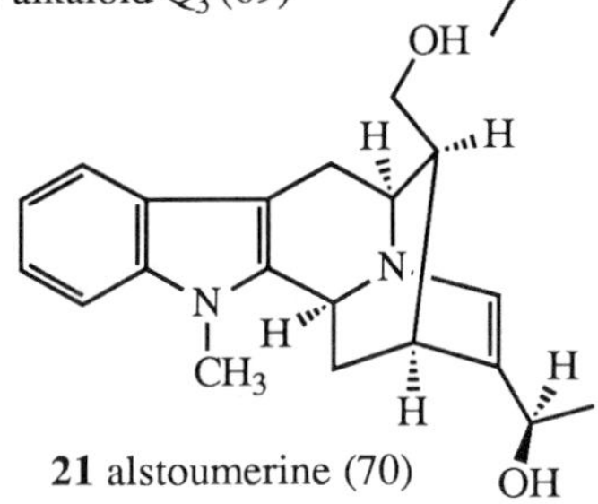

21 alstoumerine (70)

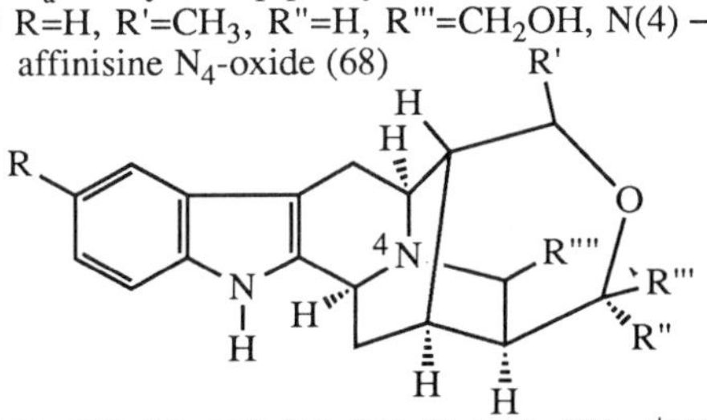

22 R=OH, R'=OH, R"=R'''=H, R''''=CH_3, $^+$N(4)-CH_3, Cl^-, verticillatine (71)
23 R=H, R'=OH, R"=R'''=H, R''''=CH_3, peraksine (72)
24 R= R'=R"=R'''=H, R''''=CH_3, deoxyperaksine (72)
25 R=R'=H, R"=CH_3, R'''=OH, R''''=H, trinervine (73)

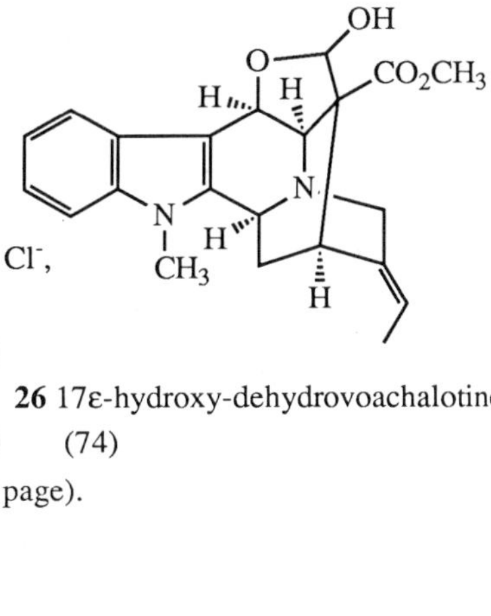

26 17ε-hydroxy-dehydrovoachalotine (74)

Figure 7. Recently isolated alkaloids (continued on next page).

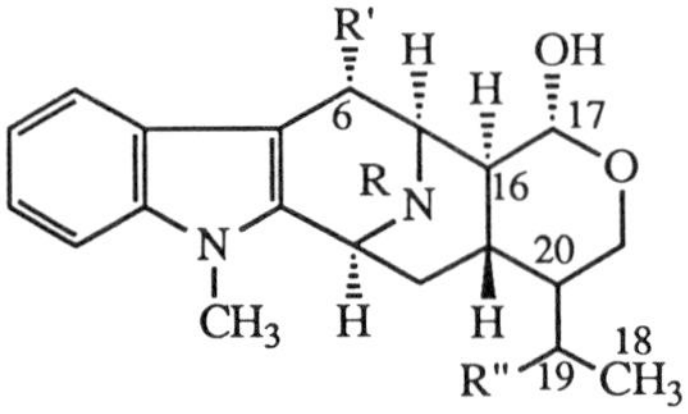

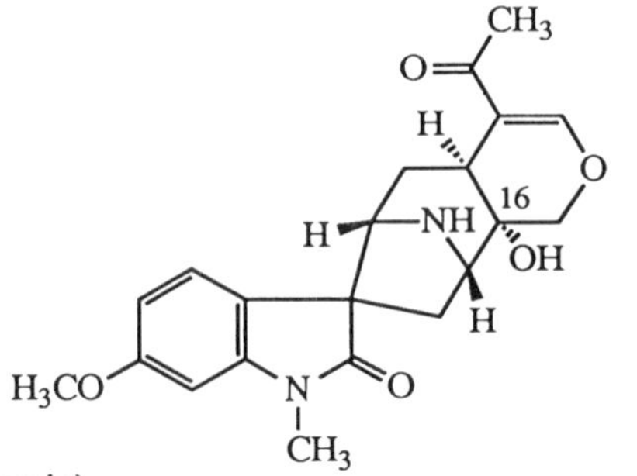

27 R=R'=R"=H, 20(*S*), raumacline (75 and refs. therein)
28 R=CH_3, R'=R"=H, 20(*S*), N_b-methylraumacline (75)
29 R=H, R'=OH, R"=H, 20(*S*), 6α-hydroxyraumacline (75)
30 R=H, R'=OCH_3, R"=H, 20(*S*), 6α-methoxyraumacline (75)
31 R=CH_3, R'=H, R"=OH, 19(*S*), 20(*S*), 19-hydroxy-N_b-methylraumacline (76)
32 R=R'=R"=H, 20(*R*), isoraumacline (75)

33 16-hydroxy-N_b-demethyl alstophylline oxindole (59)

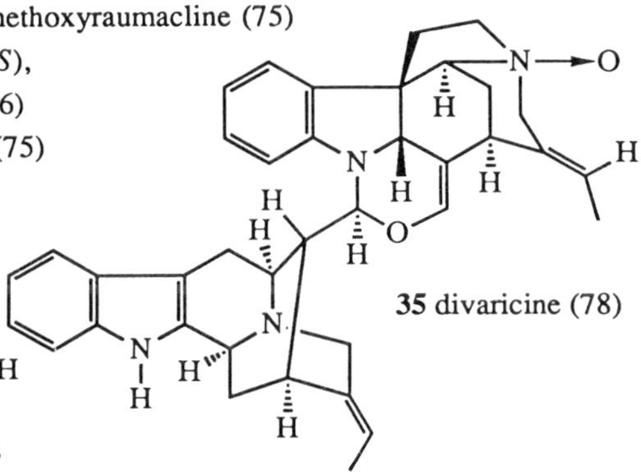

35 divaricine (78)

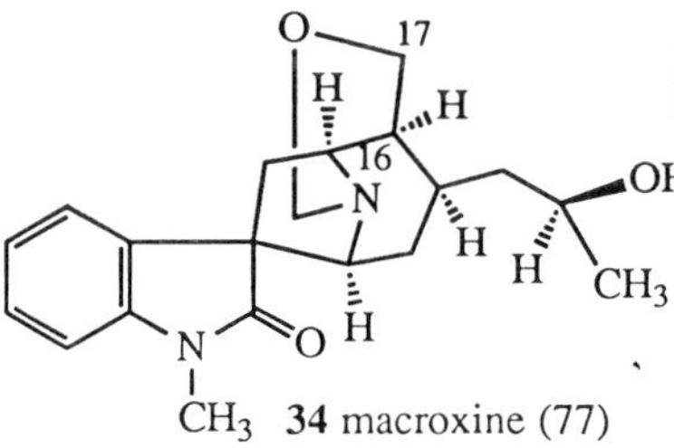

34 macroxine (77)

Figure 7. Recently isolated alkaolids.

described. These indole alkaloids (59, 64-78) are depicted in Figure 7, accompanied by the references to their isolation and structural determination. N_a-demethylaccedine **15** isolated from *Rauwolfia tetraphylla* and *Rauwolfia cubana* appears to have the exact same structure as that of amerovolfine **17** (isolated from *R. cubana*). The reader is referred to references 64 and 66 for details of this contradiction.

2. SYNTHESIS

2.1. The Tetracyclic Ketone

The macroline/sarpagine/ajmaline alkaloids have very similar skeletons, as shown in Figures 1-4 and Scheme 1. Consequently, an ideal approach to these bases might rest on the multigram synthesis of a common, optically active intermediate that could be employed for the synthesis of many related natural products. This common intermediate would at the very least contain the requisite tetracyclic ring system which could be readily functionalized for further transformations. The (-)-tetracyclic ketone **36a** (Figure 8) was synthesized in 1988 with these goals in mind (79-83), while the racemic compound had been prepared on kilogram scale in the late 1970's (84).

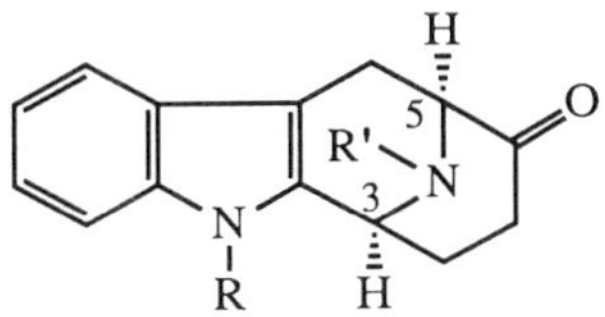

36a R=CH_3, R'=CH_2Ph
36b R=CH_3, R'=CH_3
36c R=H, R'=CH_2Ph
36d R=H, R'=CH_3
36e R=R'=CH_2Ph

Figure 8. The tetracyclic ketone.

2.1.1. Enantiospecific Synthesis of the Tetracyclic Ketone

The synthesis of (±)-5-methyl-9-oxo-12-benzyl-6,7,8,9,10,11-hexahydro-6,10-imino-5H-cyclooct[b]indole **36a** was first reported by Yoneda (79) and was improved by Soerens (80). The enantiospecific preparation of tetracyclic ketone **36a** in optically active form was developed by Zhang (82, 83) and is illustrated in Scheme 2. The synthesis of **36a** began with D(+)-tryptophan since Zhang had found earlier that the Pictet-Spengler reaction of aldehydes with N_b-benzyl substituted tryptophan methyl esters exhibited a strong preference for the enantiomerically pure *trans* diester. The 1,3-transfer of chirality from position-3 to position-1 of **36a** would impart the correct configuration at C(3) of macroline to the tetrahydro β-carboline intermediate **40**. Methylation of D(+)-tryptophan **37** was accomplished with sodium in liquid ammonia and methyl iodide in 92% yield. Fischer esterification of the methylated D(+)-tryptophan gave N_a-methyltryptophan methyl ester **38** (87%). The benzylation of the N_b-nitrogen function was carried out without racemization if care was taken to keep the imine intermediate cold during the reduction and to limit the reaction time (three hours). The tryptophan methyl ester **38** was treated with benzaldehyde at 22 °C, and the imine which resulted was reduced with sodium borohydride (at -5 °C) to provide N_a-methyl, N_b-benzyltryptophan methyl ester **39** (greater than 98% ee) in 88% yield. The Pictet-Spengler condensation of **39** with α-ketoglutaric acid in benzene/dioxane with the removal of water *via* a Dean-Stark trap was followed by esterification in 1% methanolic HCl to afford the required *trans* diester **40a** enantiospecifically. The details of the important conversion of the *cis* diastereomer into the *trans* isomer are described in the literature (82, 83). Dieckmann cyclization of the *trans* diester **40a** afforded the β-ketoester **41** (92%). After acid-mediated decarboxylation of β-ketoester **41**, the (-)-tetracyclic ketone **36a** was obtained in 91% yield. The enantiomeric purity of this ketone (-)-**36a** was shown to be greater than 98% ee by use of both ^1H NMR spectroscopy with the chiral shift reagent (85) tris-[3-(heptafluoropropyl-hydroxymethylene)-(+)-camphorato], europium (III) and by HPLC on a diastereomeric urea

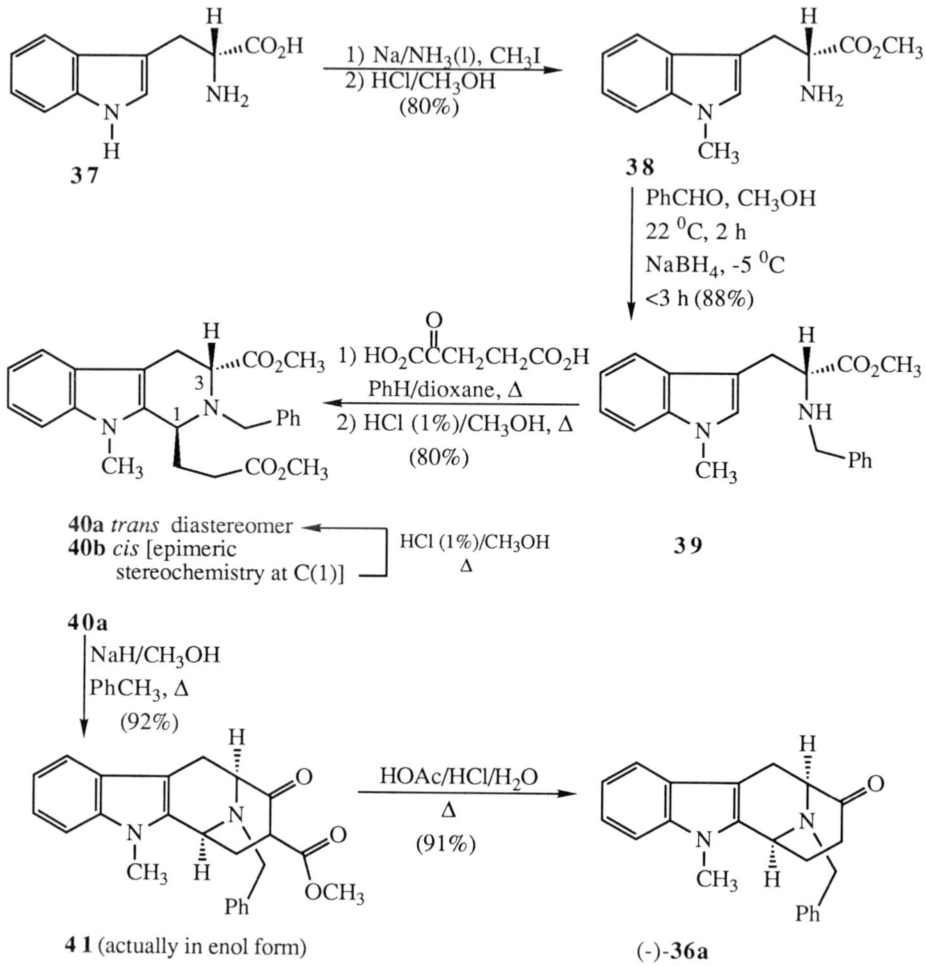

Scheme 2.

derivative of **36a** (86). The utility of this enantiospecific sequence rests on the fact that these reactions can be run on multigram scale to provide the (-)-tetracyclic ketone **36a**, which can now be considered a readily available starting material for the synthesis of optically pure macroline/sarpagine/ajmaline alkaloids. In addition, both D(+)-tryptophan and L(-)-tryptophan are readily available from commercial sources permitting entry into both antipodes of the natural products for biological screening.

2.1.2. Studies on the Stereospecific Pictet-Spengler Reaction for the Synthesis of Macroline Related Indole Alkaloids

The Pictet-Spengler cyclization has been utilized for many years for the synthesis of indole alkaloids (79, 80, 82, 87-96). With the increasing interest in the enantiospecific synthesis of alkaloids, many improvements have been made toward stereochemical control of this important condensation. In the synthesis of the optically active tetracyclic ketone **36a**, the Pictet-Spengler reaction was employed to set the stereochemistry at C(1) and C(3) of the tetrahydro β-carboline ring system in stereospecific fashion. Yoneda (79) had earlier reported the synthesis of a mixture of the racemic diesters **40a** and **40b** *via* the Pictet-Spengler reaction. The *cis* isomer **40b** was originally reported to be the main constituent (79), but this was later corrected (95) to consist of a *trans* **40a** to *cis* **40b** ratio of 5:4 (89.1% yield). Meanwhile, Ungemach demonstrated the utility of the Pictet-Spengler reaction by reporting the 100% stereoselective formation of *trans*-1,3-disubstituted-1,2,3,4-tetrahydro-β-carbolines in aprotic media when various aldehydes were heated with N_b-benzyltryptophan methyl ester (92, 97). This was the first report of a 100% stereoselective cyclization in the Pictet-Spengler condensation and is largely due to steric constraints placed upon the transition state by the N_b-benzyl and C(3) carbomethoxy groups. In the optically active series, Sakai and coworkers extended the study of the Pictet-Spengler reaction to include the synthesis of (-)-trypargine. Although the synthesis by Sakai *et al.* was in the N_a-H series, use of the method developed by Ungemach (97) with an N_b-benzyl group, provided a remarkable *trans* to *cis* preference (96). In the N_a-methyl series, however, Zhang (82) observed a 72:28 ratio of *trans* **40a** to *cis* **40b** diastereomers when N_a-methyl, N_b-benzyltryptophan was treated with methyl-3-formyl propionate under aprotic conditions (90% yield). More importantly, there was no racemization at C(3) (Scheme 3). Under the protic conditions involving α-ketoglutaric acid, Zhang observed

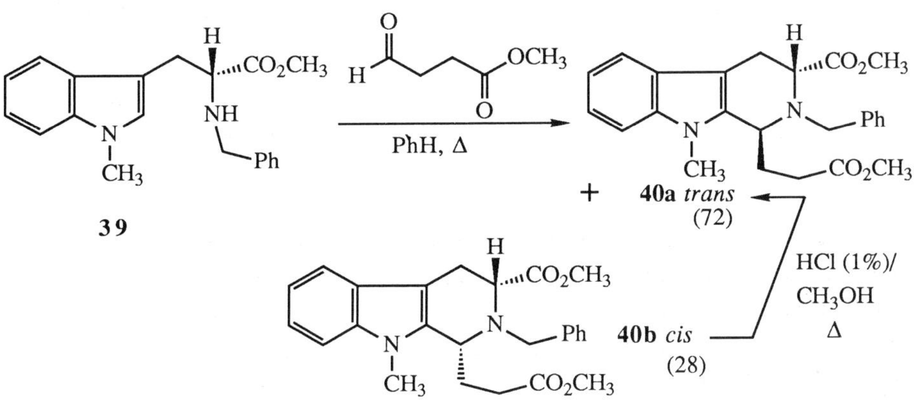

Scheme 3.

almost complete *trans* stereospecificity after esterification. The remaining small amount of *cis* isomer had been converted, with no loss of optical activity, into the *trans* diastereomer upon heating in 1% methanolic HCl. Hence, a sequence had been developed to provide the *trans* isomer in high enantiomeric purity even in the N_a-methyl series in the absence of time consuming separations.

In 1987 Ottenheijm (98) observed a different stereochemical outcome in the Pictet-Spengler reaction when the N_b-hydroxyl- and N_b-(benzyloxy)tryptophan ethyl esters (**42a** and **42b**, respectively) were treated separately with acetals **43a** or **43b** (Scheme 4). The results from these reactions are summarized in Table 1. Because the *cis* and *trans* ratios differed from those previously observed (97), Sandrin *et al.* repeated the above experiments under the stereoselective conditions of Ungemach (80, 93). By performing these experiments in aprotic media (80), Sandrin was able to rationalize the effects of an N_b-oxygen substituent on the stereoselectivity of the reaction. The pK_a values of the oxygenated amines are significantly lower than their corresponding tertiary amines; consequently, the imine intermediates are more

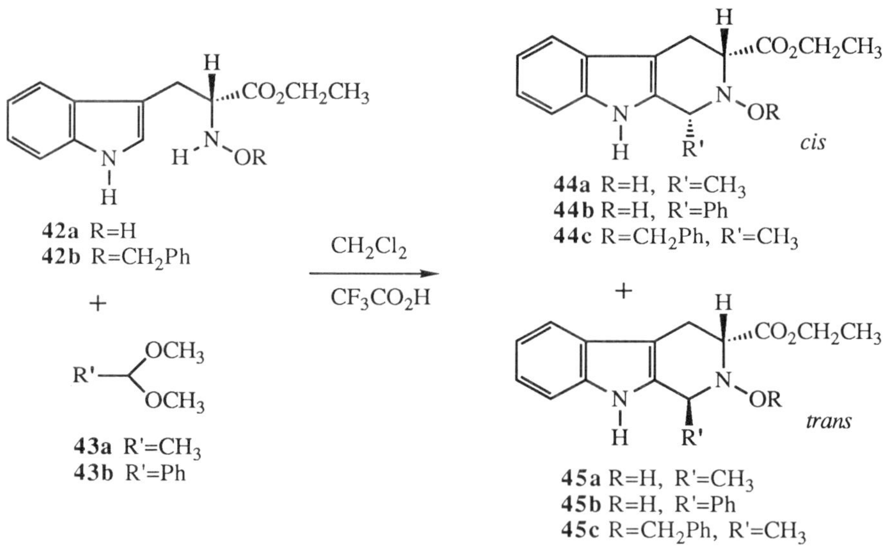

Scheme 4.

Table 1.

reactants	R	R'	products (%) *cis*	products (%) *trans*	time, h	yield
42a + **43a**	H	CH_3	**44a**, (67)	**45a**, (33)	72	95
42a + **43b**	H	Ph	**44b**, (40)	**45b**, (60)	6	77
42b + **43a**	CH_2Ph	CH_3	**44c**, (50)	**45c**, (50)	3	96

reactive, and the cyclizations less stereoselective. This explains the higher proportion of *cis* diastereomer in the Ottenheijm report in comparison with the 100% *trans* stereoselectivity observed in the N_b-benzyl series by Ungemach (91). With these results in hand, Sandrin continued to look at the effect of the N_b-benzyl group on stereoselectivity of the Pictet-Spengler reaction. He repeated the reactions originally reported in benzene (80) under the acidic conditions of Ottenheijm (98). The results, as expected, confirmed that the amount of the *trans* diastereomer produced in the Pictet-Spengler reaction of N_b-alkyltryptophan methyl esters increased in relation to the size of the N_b-alkyl substituent (Table 2). This trend was also observed in the oxygenated substrates of Ottenheijm *et al.*, as shown in Table 1.

Table 2.

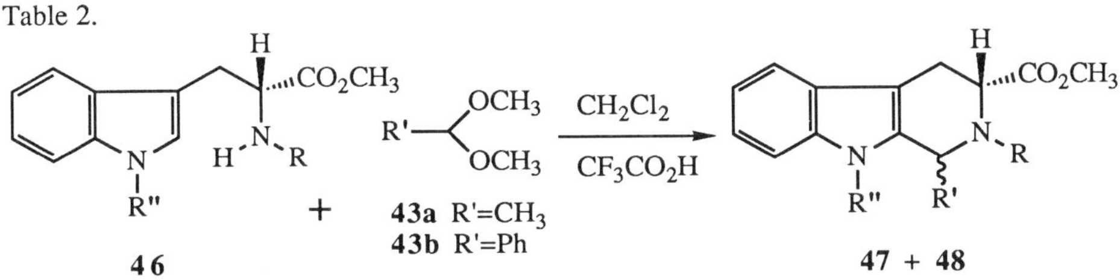

	R"	R	R'	*cis* (%)	*trans* (%)	time, h
46a	H	H	CH_3	**47a** (75)	**48a** (25)	48
46b	H	CH_3	CH_3	**47b** (66)	**48b** (34)	24
46c	H	CH_2Ph	CH_3	**47c** (16)	**48c** (84)	72
46d	H	CH_2CH_2Ph	CH_3	**47d** (16)	**48d** (84)	168
46e	H	CH_2Ph	Ph	**47e** (0)	**48e** (100)	48
46f	CH_3	CH_3	CH_3	**47f** (16)	**48f** (84)	288

Recently, Czerwinski *et al.* have achieved complete *trans* stereoselectivity in the Pictet-Spengler cyclization even when aldehydes as small as acetaldehyde are employed in the condensation (99). Czerwinski rationalized this 100% diastereoselectivity in the Pictet-Spengler cyclization of N_b-diphenylmethyltryptophan isopropyl ester with a variety of smaller aldehydes, by examination of the steric stabilities of the E and Z isomers of the imine intermediates as well as the spiroindolenine intermediates (99). Represented in Figure 9 are the two possible imine intermediates. In agreement with the report of Ungemach, attack is favored from the face opposite the ester function of the more stable imine intermediate **49**.

Czerwinski went on to note that the *trans*-favored diastereoselectivity correlated extremely well with the energy difference between the two possible spiroindolenine intermediates (Figure 10). MacroModel version 2.5-MM2 force field calculations revealed that the *anti* spiroindolenine intermediate **51** was 2.1 kcal/mole more stable than the corresponding

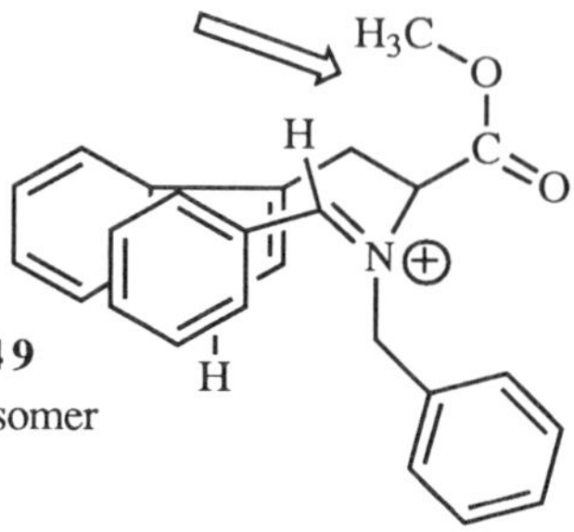

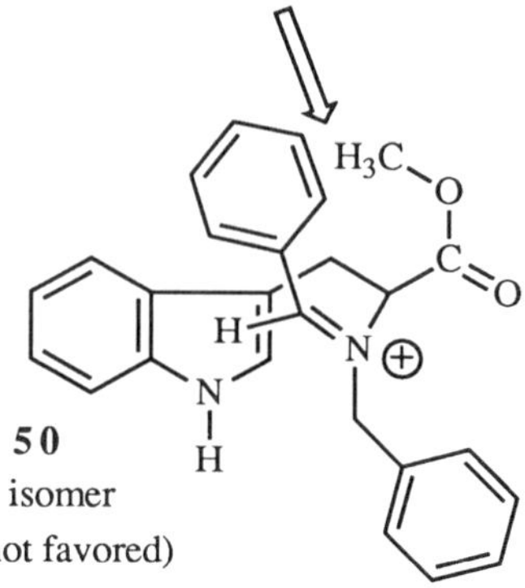

49
E isomer

50
Z isomer
(not favored)

Figure 9. Steric interaction between the ester function and the aldehyde substituent.

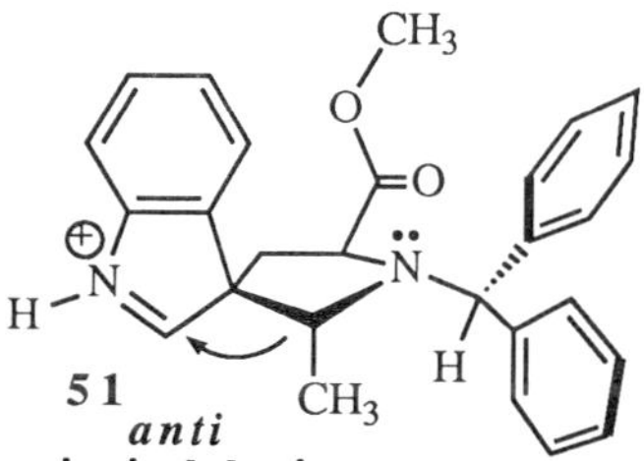

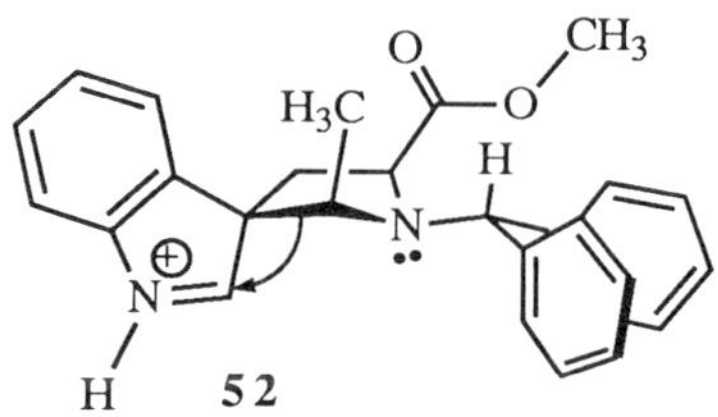

51
***anti*
spiroindolenine**

52
***syn*
spiroindolenine**

Figure 10. *Anti* and *syn* spiroindolenine intermediates.

Table 3.

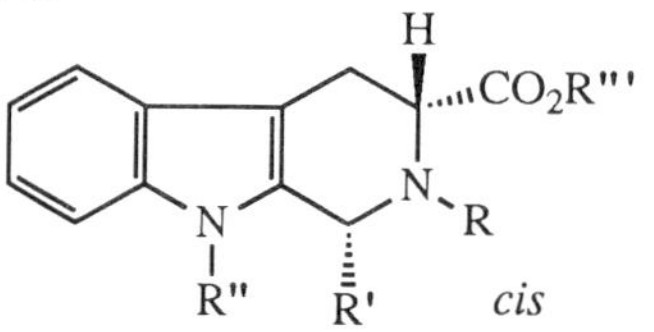

cis

racemic series

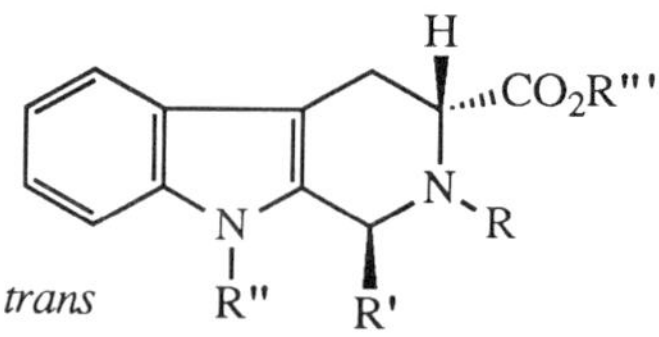

trans

	R	R'	R"	R'''	*cis:trans* aprotic	*cis:trans* in TFA
53a	**$CHPh_2$**	**CH_3**	**H**	**CH_3**	**10:90**	**0:100**
53b	$CHPh_2$	$CH_2CH_2CH_3$	H	CH_3	0:100	0:100
53c	$CHPh_2$	C_6H_{11}	H	CH_3	N.R.	0:100
54a	**$CHPh_2$**	**CH_3**	**H**	**$CH(CH_3)_2$**	**0:100**	**0:100**
54b	$CHPh_2$	$CH_2CH_2CH_3$	H	$CH(CH_3)_2$	N.R.	0:100
54c	$CHPh_2$	C_6H_{11}	H	$CH(CH_3)_2$	N.R.	0:100

syn isomer **52**. This complete *trans* stereoselectivity of the Pictet-Spengler cyclization in both the *N*b-diphenylmethyltryptophan methyl and isopropyl ester series is clearly illustrated in Table 3. Note that a similar *trans* stereoselectivity was also achieved in the tryptophan methyl ester cases. Pictet-Spengler reactions with bulky aldehydes were conducted in the presence of trifluoroacetic acid. The strong acid facilitated the cyclizations that would not otherwise occur due to the extreme steric bulk of the aldehydes. Hence, the results in TFA mirror the thermodynamic ratio of tetrahydro-β-carbolines while the cyclizations in benzene at reflux represent the ratio from a kinetic trapping experiment. In either case, the versatility and complete *trans* stereoselectivity of the Pictet-Spengler reaction can now be utilized in the enantiospecific total synthesis of the macroline/sarpagine/ajmaline alkaloids.

2.1.3 Studies on the Dieckmann Cyclization

There are contrasting reports on the Dieckmann cyclization (**40** to **41**) with regard to time of reaction. Yoneda reported that the *cis* diastereomer of diester **40b** (*N*a-methyl series) could not be coerced to undergo the Dieckmann reaction to provide the antipode of tetracyclic β-ketoester **41** under conditions employed for cyclization of *trans* diester **40a**. Zhang in 1988 demonstrated that the *trans* diastereomer **40a** (under these Dieckmann conditions) in the *N*a-methyl series was converted into the β-ketoester **41** more rapidly than the corresponding *cis* diastereomer underwent the cyclization under conditions similar to those reported by Yoneda (95). Zhang was, however, able to effect cyclization of the *cis* diastereomer to produce the tetracyclic system by employing additional quantities of base and longer reaction times (82). In contrast, Magnus *et al.* in 1990 reported that the *cis* diester **56** (Figure 11) in the related *N*a-benzyl series underwent the Dieckmann reaction faster than the corresponding *trans* diester **55** (52). Because these results were obtained in two different cases (*N*a-methyl vs. *N*a-benzyl), a detailed study of the Dieckmann cyclization in both series was carried out in our laboratories (82, 86).

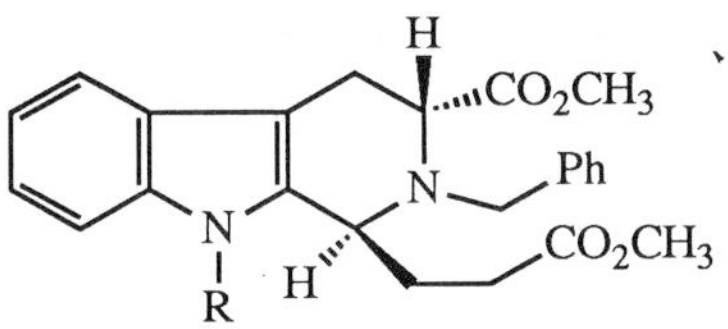

trans

40a R=CH_3
55 R=CH_2Ph

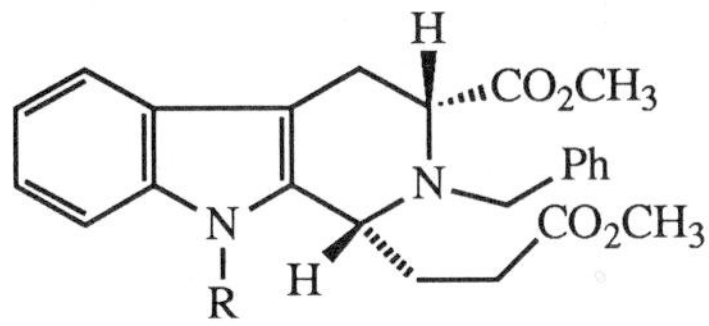

cis

40b R=CH_3
56 R=CH_2Ph

Figure 11. *Cis* and *trans* diastereomers.

Examination of the experimental results clearly indicated that the *trans* diastereomer in both series is the thermodynamically more stable isomer. The *cis* diastereomer was completely converted into the corresponding *trans* diastereomer, respectively, under either acidic or alkaline conditions (82, 86). Execution of the Dieckmann reaction in the N_a-methyl series under the conditions of Zhang always resulted in complete conversion of the *trans* diastereomer to provide **41** more rapidly than the corresponding *cis* diastereomer was converted. Evidence indicated that the *cis* diastereomer epimerized to the *trans* diastereomer and then underwent the Dieckmann reaction, but with *cis* stereochemistry (86). Because the rate of these cyclizations is highly dependent on the amount of sodium hydride and methanol present, an equimolar mixture of *cis* (**56**) and *trans* (**55**, R=CH_2Ph) diastereomers was subjected to the Dieckmann reaction. It was found that both the *cis* (**56**) and *trans* (**55**) diastereomers in the mixture in the N_a-benzyl, N_b-benzyl series underwent the Dieckmann reaction at the same rate in contrast to the earlier reports of Magnus (100, 101). Although the *trans* diastereomer **40a** in the N_a-methyl, N_b-benzyl series cyclized to completion (see **41**) consistently faster than the corresponding *cis* diastereomer, under the conditions of Zhang (86), an equimolar mixture of the two diastereomers (*cis/trans*) yielded both the (+) and (-) β-ketoesters (**41**) at the same rate (86). This is in agreement with the results of the same experiment in the N_a-benzyl, N_b-benzyl series and is in contrast to the earlier reports of Magnus *et al.* (100, 101). It should be pointed out, however, in agreement with Magnus, the rate of this cyclization is very sensitive to the amount of methanol present in the solution (102). In 1993 Bailey reported that in the N_a-methyl series both diastereomers cyclized at comparable rates under the conditions of the Dieckmann reaction (103). Although Bailey *et al.* claimed that these results were in agreement with Magnus (101), in fact this statement is in error. Magnus reported that in the N_a-benzyl series the *cis* isomer cyclized to completion in 5 hours; whereas, the *trans* isomer required 27 hours to provide the β-ketoester (101). The report of Bailey in 1993 confirms the report from our laboratory in 1992. The report of Bailey *et al.* (1993) on this Dieckmann reaction contains several inaccurate statements (103) with regard to previous reports in 1988 (82), 1990 (104), and 1992 (86) on this process. The original papers in this series should be consulted for the details of this cyclization in the optically active series.

The importance of the discrepancies about the rates of the Dieckmann cyclization between laboratories is not significant from an experimental point of view but is extremely important from a stereochemical point of view. Enantiospecific synthesis of indole alkaloids in the macroline/sarpagine/ajmaline series rests on the accurate identification of the stereogenic centers at C(1) and C(3) in the 1,3-disubstituted tetrahydro-β-carbolines. Although Bailey *et al.* have reported a ^{13}C NMR method to differentiate between the *cis* and *trans* diastereomers in the N_a-H, N_b-benzyl series, this method is not 100% effective as noted by the authors (105, 106). Moreover, Toth *et al.* have examined this method for stereochemical assignments and also found exceptions (107). Consequently, if the rates of the Dieckmann cyclization were ever taken as evidence of *cis* stereochemistry at C(1) and C(3), the rate differences encountered in the various laboratories could be problematic. Rate differences between laboratories in this series generally stem from the scale of the reaction and the amount of sodium hydride and methanol employed in the cyclization (79, 82). To date, accurate stereochemical assignments for the *cis* and *trans* 1,3-disubstituted N_b-benzyltetrahydro β-carbolines in this series can only be made in 100% of these cases by removal of the N_b-benzyl group (catalytic transfer hydrogenation) followed by identification of the diastereomers by the ^{13}C NMR method developed earlier by Sandrin (108) and Ungemach (97, 109) in these laboratories.

Other routes to the tetracyclic ketone **36** have been published and are described in brief below. In 1989 Magnus (100, 101) reported the synthesis of the (-)-N_a-benzyl, N_b-benzyl tetracyclic ketone from (L)-tryptophan. However, Magnus utilized the Pictet-Spengler reaction in the N_a-benzyl, N_b-benzyl series to maximize *cis* isomer selectivity. The cyclization provided the desired *cis* isomer in a ratio of 1:2 (*cis:trans*). The *cis* isomer was then separated from the *trans* isomer and carried through to the (-)-tetracyclic ketone. Magnus then elegantly converted this ketone into the indole alkaloids (+)-koumine, (+)-taberpsychine and (+)-koumidine (100, 101). This is the same ketone that was prepared stereospecifically in 1988 by Zhang from D(+)-tryptophan (82). In 1991 and 1993, Bailey (103, 110) reported the Pictet-Spengler cyclization with N_a-H, N_b-H tryptophan methyl ester in the desire to achieve complete *cis* stereoselectivity. Bailey observed *cis* selectivity in a ratio of 4:1 over the *trans* diastereomer. However, this selectivity fell short of that reported by Zhang and Czerwinski (82, 99); from D(+)-tryptophan excellent stereoselectivity was observed to provide the desired enantiomer of the tetracyclic ketone required for the preparation of the macroline/sarpagine/ajmaline alkaloids. Bailey also reported the cyclization of the *cis* diastereomer under Dieckmann conditions to afford the (-)-N_a-benzyl, N_b-H tetracyclic ketone which had earlier been prepared by Magnus (101). Kluge *et al.* (111) and Hobson *et al.* (112) employed the Fischer-indole cyclization several years ago to provide a route to the [3.3.1]system in the racemic series.

2.2. Synthesis of Macroline/Sarpagine Alkaloids

2.2.1. Suaveoline

In 1972 Potier *et al.* isolated suaveoline **3** and eight other alkaloids from the alkaloidal components of the trunk bark of *Rauwolfia suaveolens* S. The structure of suaveoline was elucidated on the basis of mass and proton spectroscopy, as well as a partial synthesis of N_b-methylsuaveoline from ajmaline (43, 44). In 1989 Trudell reported the total synthesis of the macroline related alkaloid (±)-suaveoline **3** (113). This synthesis employed the Pictet-Spengler cyclization, the Dieckmann condensation and an ortho-ester Claisen rearrangement, all of which occurred with high stereoselectivity. In 1992 Fu (114, 115) completed the first enantiospecific total synthesis of (-)-suaveoline **3**. An optical rotation of $[\alpha]^{25}_D$ = -9.3° (c = 0.30, CHCl3) was determined for pure **3** in contrast to earlier reports of a 0° ± 2° reported

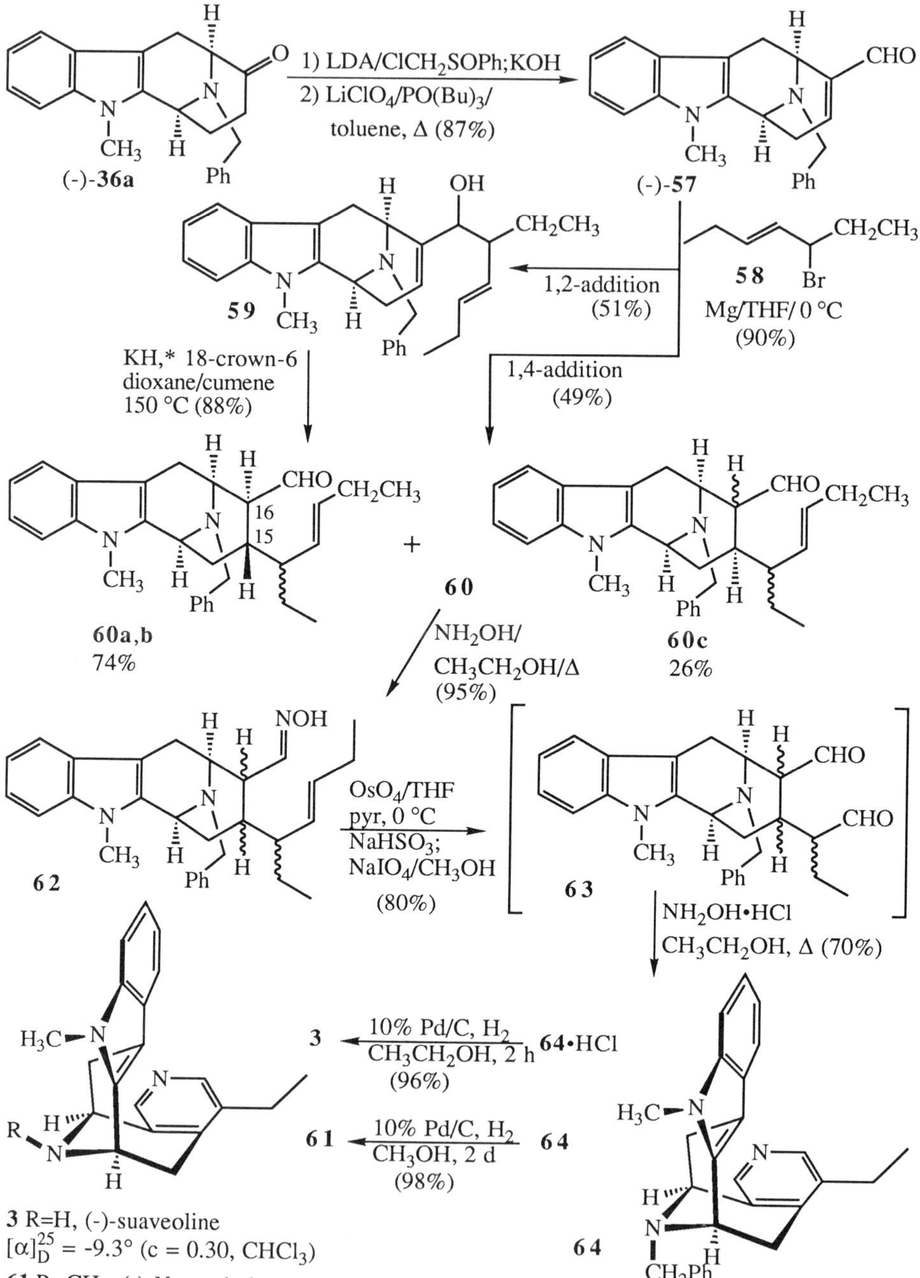

3 R=H, (-)-suaveoline
$[\alpha]_D^{25}$ = -9.3° (c = 0.30, $CHCl_3$)
61 R=CH_3, (-)-N_b-methylsuaveoline

*Ratio of **60a,b** to **60c** from the anionic oxy-Cope rearrangement was 3:2.

Scheme 5. The synthesis of (-)-suaveoline.

(43, 44) for this base. The total synthesis of (-)-suaveoline (Scheme 5) will be described beginning from (-)-N_a-methyl, N_b-benzyltetracyclic ketone **36a**, the synthesis of which was illustrated in Scheme 2. Conversion of the carbonyl function of (-)-**36a** into the α,β-unsaturated aldehyde **57** *via* the spiro-oxiranophenylsulfoxide was accomplished in 87% yield by the method of Trudell (113, 116) in the racemic series and of Zhang (104) in the (-)-N_b-methyl series. The pseudosymmetric Grignard reagent, available from 5-bromo-3-heptene **58**, was added to the α,β-unsaturated aldehyde **57** at low temperature to provide the products of 1,2- (**59**) and 1,4-addition (**60a-c**) in a combined yield of 90% [ratio 51(**59**):49(**60**)]. When this sequence was repeated at room temperature, only the product of 1,2-addition (**59**) was isolated and in high yield. The alcohol **59** was purified and subjected to conditions that promote an oxyanion Cope rearrangement (150 °C) to furnish the same C-15 functionalized tetracyclic systems **60a,b** and **60c** obtained from the 1,4-addition in a ratio of 3:2. Although the stereoselectivity in the oxyanion-Cope process was only 3:2 with preferred attack from the desired bottom face of the C(15)-C(16) olefinic bond, the mixture of **60a-c** could be employed in the synthesis of (-)-suaveoline **3**. The 1,4-addition of **58** to **57** was unprecedented in these systems and provided the diastereomers **60a** and **60b** with the ajmaline configuration at C(15) and C(16) in a ratio of 3(**60a,b**):1(**60c**). Previous attempts (84, 105, 117) to effect 1,4-addition to **57** proved unsuccessful; therefore, this example serves as the first case of such an addition in this hindered N_b-benzyl-azabicyclo[3.3.1]nonane system. Since the configurations of the newly formed stereocenters in **60a-c** will eventually be destroyed, the aldehyde functions of the mixture of **60a-c** were protected by treatment with hydroxylamine hydrochloride in refluxing ethanol. A diastereomeric mixture of oximes represented by **62** was obtained in 95% yield. The mixture of oximes was osmylated and subsequently hydrolyzed reductively with $NaHSO_3$ to provide the desired diol which was subjected directly to the oxidative cleavage sequence ($NaIO_4$). The desired dialdehyde was obtained in 80% overall yield based on recovered starting oxime **62**. The mixture of dialdehydes **63** was cyclized *in situ* with hydroxylamine hydrochloride to provide N_b-benzylsuaveoline **64** in 70% yield. When **64** was subjected to the conditions of catalytic debenzylation with excess 10% Pd/C (1.5:1 w/w) and hydrogen in methanol, a 98% yield of (-)-N_b-methylsuaveoline **61** ($[\alpha]^{25}_D$ = -89.5°, c = 0.35, $CHCl_3$) was realized in greater than 98% ee. Although the mechanism of the benzyl/methyl transformation was not clear, it provided a simple manner in which to execute a benzyl/methyl transfer in the latter stages of the synthesis. This process can be employed in the preparation of a number of macroline/sarpagine/ajmaline alkaloids (36, 118). Catalytic debenzylation of the hydrochloride salt of (-)-N_b-benzylsuaveoline **64** with 10% Pd/C (0.7:1.0 w/w) and hydrogen in ethanol provided a 96% yield of (-)-suaveoline **3** (114, 115). This sequence represents the first enantiospecific total synthesis of (-)-suaveoline and provides material upon which an accurate optical rotation could be obtained. Since the intermediates in this route are closely related to those previously reported in the synthesis of (±)-ajmaline, the strategy employed in the macroline series can be extended to alkaloids of the ajmaline family. Later in 1993, Bailey

(103) described a formal synthesis of (-)-suaveoline (from L-tryptophan) which rested on the preparation of the optically active N_a-methyl, N_b-benzyltetracyclic ketone **36a**. This (-)-ketone is identical with that reported earlier by Zhang (104) in 1988 and Fu (114) in 1992.

2.2.2. Alstonerine

Alstonerine **65** was first isolated from *Alstonia muelleriana* Domin by Elderfield and Gilman (119, 120), and its structure was elucidated by LeQuesne *et al.* (121). The indole alkaloid alstonerine **65** is closely related to the oxindole alkaloid alstonisine. Zhang (104) reported the synthesis of (-)-alstonerine in 1990 starting from the (-)-tetracyclic ketone **36a** which was prepared earlier in enantiospecific fashion (82, 86). The N_b-benzyltetracyclic ketone **36a** was methylated with methyl trifluoromethanesulfonate followed by catalytic debenzylation with Pd/C and hydrogen to afford the N_b-methyltetracyclic ketone **36b** in high yield (Scheme 6). The ketone (**36b**) was converted into the α,β-unsaturated aldehyde **66** in

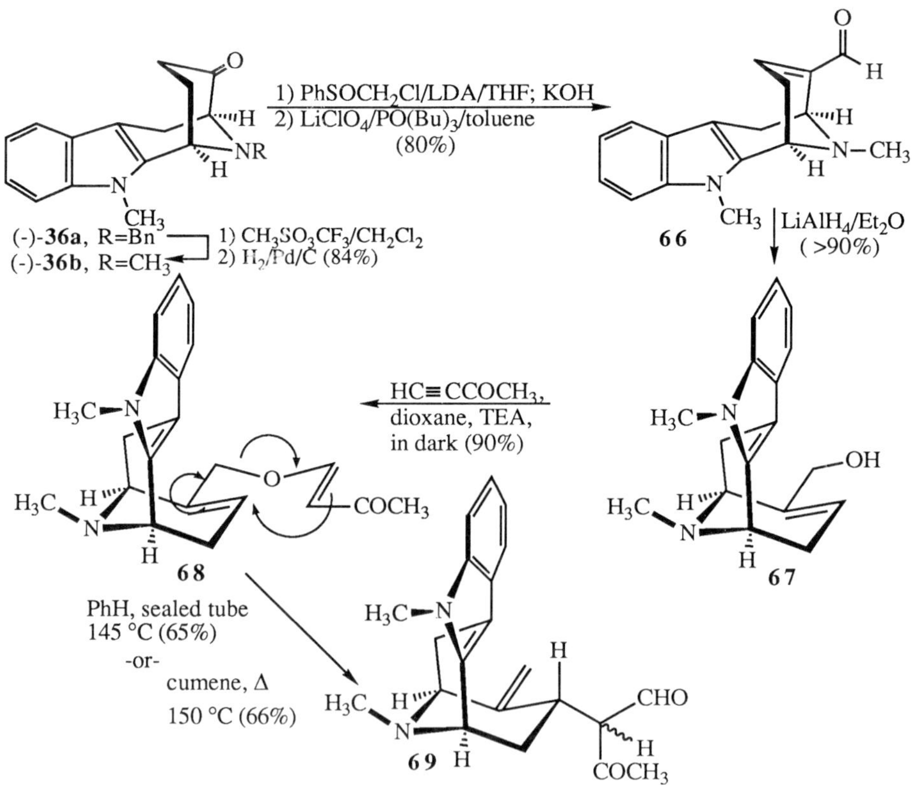

Scheme 6.

80% overall yield using conditions analogous to those reported by Trudell (113). The α,β-unsaturated aldehyde **66** was then transformed into the allylic alcohol **67** with lithium aluminum hydride in ether at -20 °C. Michael addition of 3-butyne-2-one in the absence of light gave the desired enone **68** in excellent yield. The Claisen rearrangement (**68** to **69**) proceeded *via* the preferred chair transition state primarily from the bottom face of the double bond to afford the desired β-dicarbonyl compound **69** (104, 122) with a diastereoselectivity in cumene (150 °C), of 4:1 in 82% yield (104, 123). The β-dicarbonyl compound **69** was reduced with sodium borohydride to the diol **70**, as illustrated in Scheme 7. Hydroboration of the

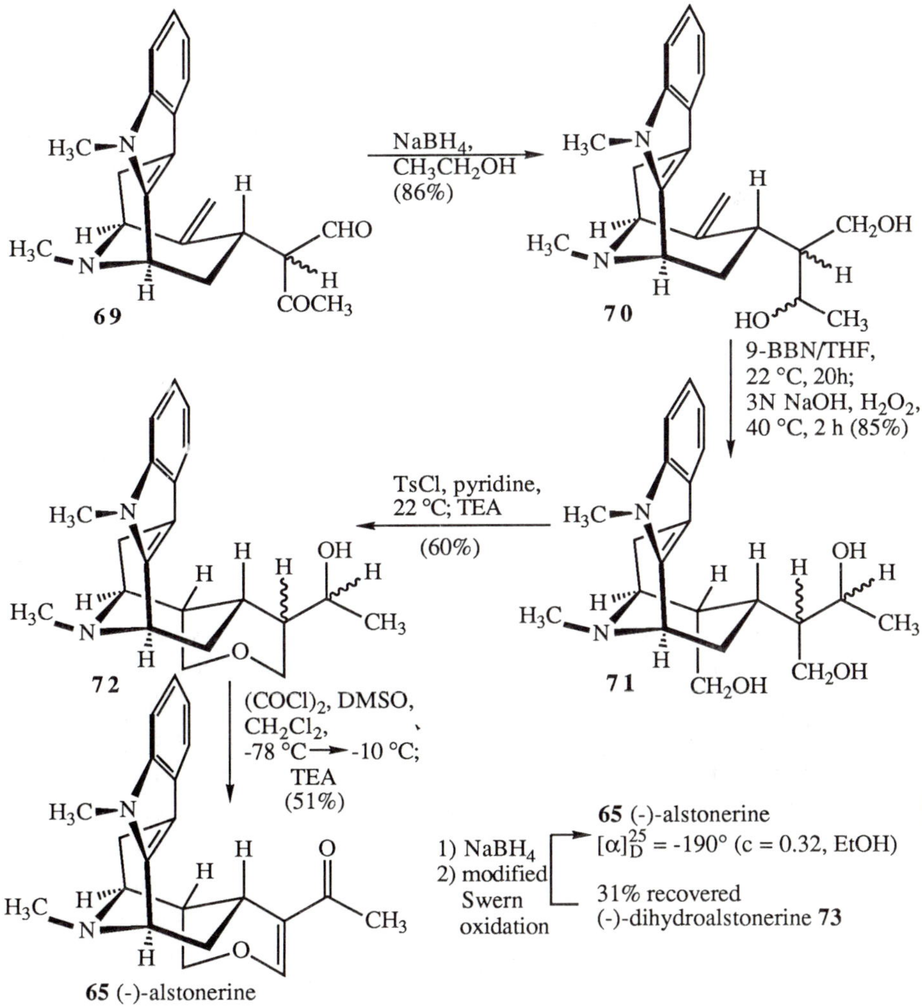

Scheme 7. Synthesis of (-)-alstonerine.

exocyclic methylene function of **70** with an excess of 9-BBN occurred stereospecifically from the β-face of the double bond and after oxidative work up provided the triol **71**. As planned, one equivalent of 9-BBN complexes to the N_b-nitrogen function which hinders attack from the bottom face of the double bond and results in exclusive hydroboration from the β-face of the exocyclic methylene function. Upon stirring with tosyl chloride (1 equiv) in pyridine followed by treatment with triethylamine, the triol **71** was regioselectively cyclized to the desired monol **72** in 60% yield, accompanied by recovered starting triol **71** (33%). Additional quantities of alcohol **72** could be obtained by subjecting the recovered triol **71** to the same tosylation process. The alcohol **72** underwent a modified Swern oxidation to provide (-)-alstonerine **65** in 51% yield, accompanied by dihydroalstonerine **73** (31%). The proposed mechanism of this transformation has been reported (36). The dihydroalstonerine could be recycled to provide additional quantities of **65** by sodium borohydride reduction; the monol of which was subjected to the conditions of the modified Swern oxidation. This procedure may provide a general method for the conversion of hydroxy substituted tetrahydropyrans into enones which are commonly found in other *Alstonia* alkaloids such as alstophylline **78** and alstonisine. The enantiospecific synthesis of the tetracyclic ketone (-)-**36a** coupled with the Claisen rearrangement (C15) and the hydroboration process (C16) provided a route of high diastereoselectivity for the enantiospecific synthesis of the macroline/sarpagine alkaloid, (-)-alstonerine **65**.

2.2.3. Macroline

The synthesis of (+)-macroline **2** has recently been completed (Scheme 8) in enantiospecific fashion starting with D(+)-tryptophan (123). The significance of this synthesis becomes apparent when one considers that over sixty macroline related alkaloids have been isolated (36). Macroline is not stable in a vial for long periods of time; therefore, the synthesis of (+)-macroline **2** is presented in Scheme 8 along with the preparation of a stable macroline equivalent that can be employed for the synthesis of *Alstonia* bisindoles. The tetracyclic ketone (-)-**36a** was prepared *via* the synthetic route (Scheme 2) employed for (-)-alstonerine (104); moreover, the required allylic alcohol **67** was also prepared *via* the route shown in Scheme 6. The synthesis of macroline **2**, as illustrated in Scheme 8, began with the intermediate alcohol **67**. Michael addition of 3-butyne-2-one to the allylic alcohol **67** in the absence of light provided the enone **68** in excellent yield. The Claisen rearrangement of enone **68** took place stereoselectively from the desired α-face (4:1) in cumene at reflux to afford the same dicarbonyl compound **69** employed for the synthesis of (-)-alstonerine. Although the stereoselectivity was reported to be 4:1, it may be much higher because the three byproducts formed in this pericyclic event are inseparable, rendering their structure determination difficult

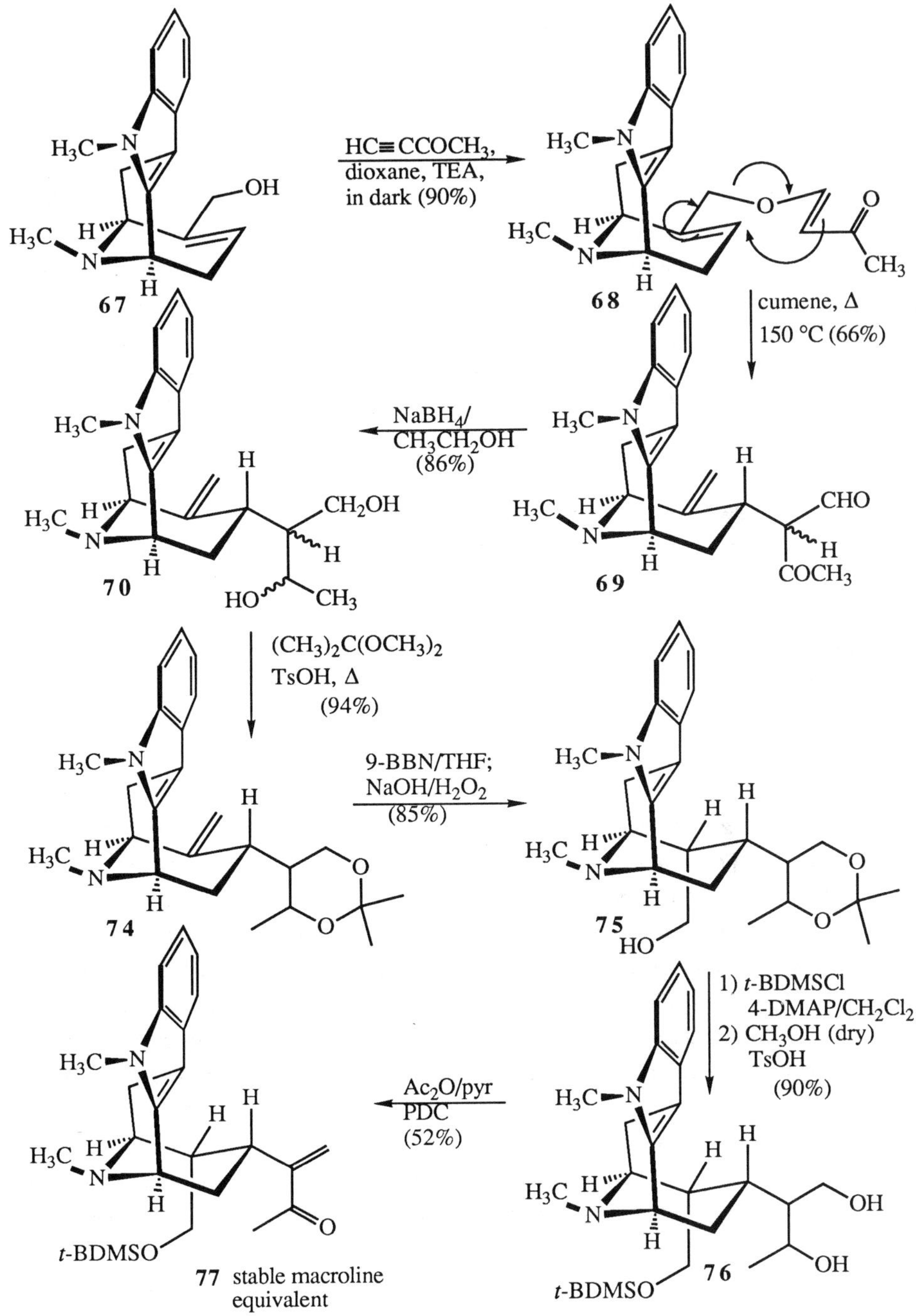

Scheme 8. The synthesis of (+)-macroline (continued on next page).

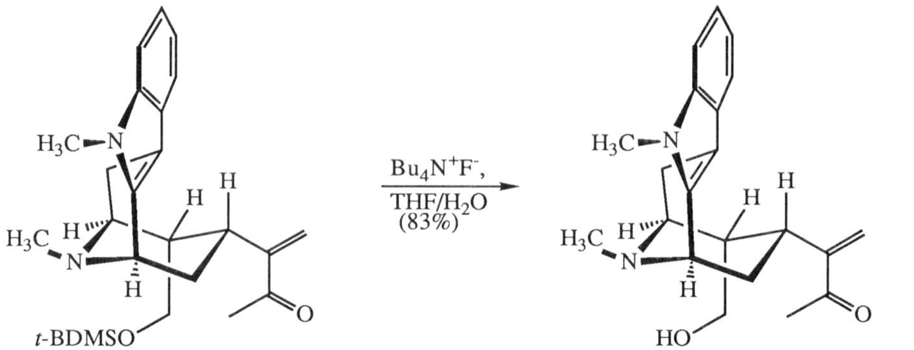

77
stable macroline
equivalent

(+)-2
macroline

Scheme 8. Synthesis of (+)-macroline.

at this juncture. Reduction of dicarbonyl compound **69** produced the diol **70**. The diol **70** was converted into the triol previously described in Scheme 7; however, attempts to utilize the triol **71** for the synthesis of macroline proved impractical. Consequently, diol **70** was protected as the acetonide **74** before the hydroboration/oxidation process with 9-BBN/OH^-/H_2O_2 was carried out. Hydroboration of acetonide **74** occurred exclusively from the β-face of the C(16-17) olefinic bond, as planned, to provide the desired primary alcohol **75**. The primary hydroxyl moiety of **75** was converted into the *t*-butyldimethylsilyl ether; afterwhich, the acetonide was selectively removed upon stirring this compound with *p*-toluenesulfonic acid in dry methanol. Acetic anhydride was then used to protect the primary alcohol in diol **76**, and the acetate which resulted served as the desired leaving group. This protection of the primary hydroxyl group of **76** was followed by oxidation with pyridinium dichromate (PDC) to provide the stable macroline derivative **77** in a one-pot process. After the oxidation of the secondary alcohol of **76** to the corresponding ketone had occurred, the pyridine present in solution promoted the loss of the β-ketoacetate function to provide the stable macroline enone **77**. When enone **77** was stirred in THF with tetrabutylammonium fluoride, (+)-macroline **2** was obtained. Macroline **2** is known to cyclize to dihydroalstonerine **73** when exposed to base; therefore, the synthesis of the macroline equivalent **77** was designed to facilitate its use in the synthesis of bisindole alkaloids.

2.2.4. Alstophylline

Alstophylline **78** was first found in the bark rind of the species *Alstonia macrophylla* Wall (124) and has since been isolated from other *Alstonia* species (125). Alstophylline may be more notably recognized as the lower half of the structure of the hypotensive bisindole macralstonine **79** (38-40, 126, 127). In addition, N_a-demethylalstophylline **12** comprises the lower portion of the bisindole alkaloid H **80** (Figure 12) obtained by LeQuesne and Lazar from *Alstonia muelleriana* Domin (128, 129).

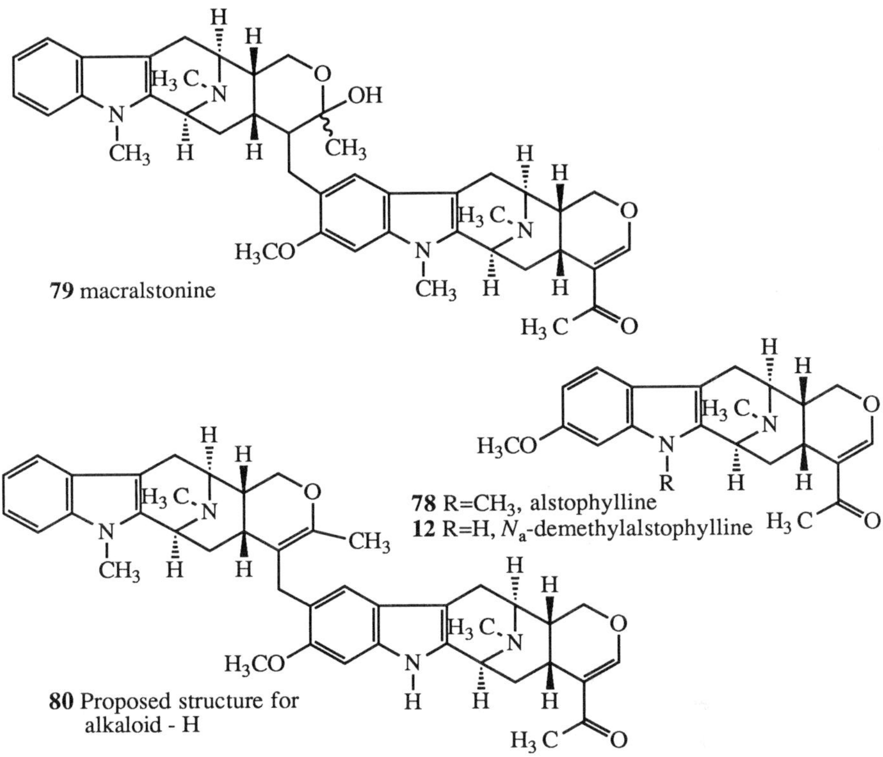

79 macralstonine

78 R=CH_3, alstophylline
12 R=H, N_a-demethylalstophylline

80 Proposed structure for alkaloid - H

Figure 12.

The biogenetic Michael condensation pioneered by LeQuesne (37, 40, 41) of the above-mentioned two monomeric bases alstophylline and **12** to macroline to form bisindoles **79** and **80** will be discussed in the section on bisindoles. The total synthesis of macralstonine **79** and alkaloid H **80** both then revert to the total synthesis of alstophylline **78** and N_a-demethylalstophylline **12**, as well as (+)-macroline **2**. A potential route to alstophylline **78** can be envisaged to proceed in an analogous fashion to the synthesis of alstonerine **65** developed by Zhang (104); however, D(+)-6-methoxytryptophan would be required in place of D(+)-tryptophan. In addition, if the synthon required for the construction of D(+)-6-methoxytryptophan could be modified, a potential route to N_a-demethylalstophylline **12** could also be developed. Outlined in Schemes 9 and 10 is a potential route to alstophylline **78** based on the chiral amino ester synthon, 1-benzenesulfonyl-6-methoxy-D(+)-tryptophan ethyl ester **81a**. This amino ester **81a** was stirred with benzaldehyde followed by reduction at -5 °C to furnish the *N*b-benzyl analog **82**, as illustrated in Scheme 9. Pictet-Spengler reaction of the *N*b-benzyl analog **82** with α-ketoglutaric acid in aprotic media, equipped with a Dean-Stark trap (DST), provided an excellent yield of a mixture of *trans* and *cis* diastereomers (51:49) **83**, as expected. When the Pictet-Spengler reaction medium was allowed to stir at 78 °C for greater than 90 hours only the *trans* acid was observed, albeit in poor yield. Presumably, this is due to the decomposition of the *cis* diastereomer. The mixture of *cis* and *trans* acids **83** was esterified in 2% ethanolic hydrogen chloride to provide a mixture of the *cis* and *trans* diesters **84**. However, when the mixture of esters **84** was heated further in 2% ethanolic hydrogen

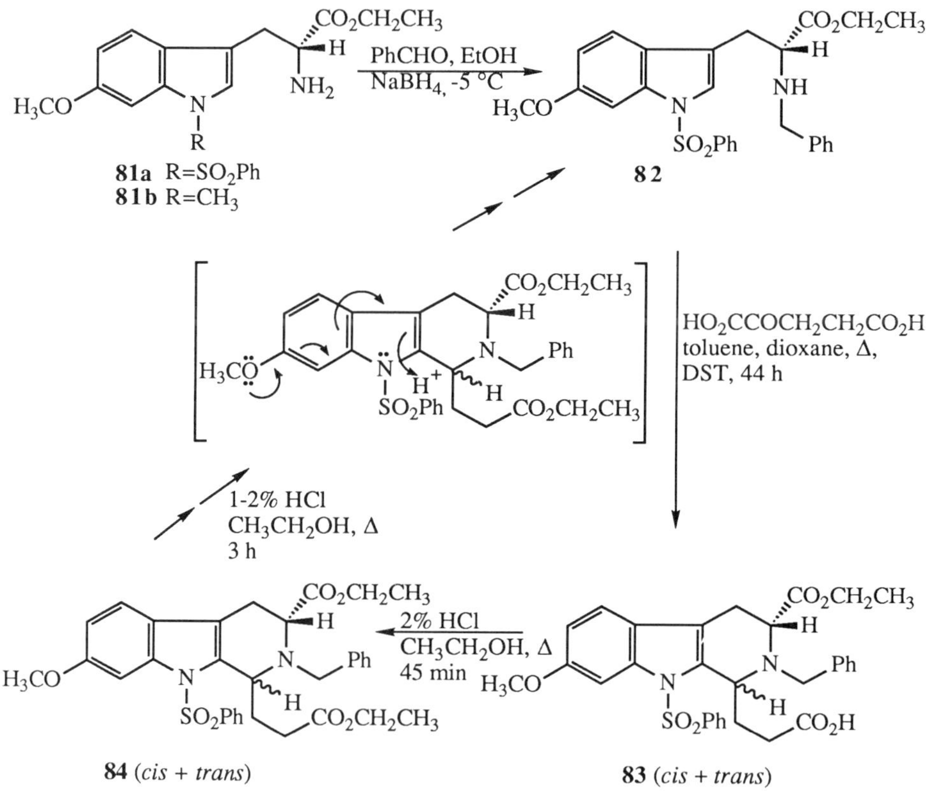

Scheme 9.

chloride to effect *cis* to *trans* epimerization, a 6-methoxy-mediated side reaction to provide the starting amino ester **82** was observed (130). Presumably, the *N*a-substituted sulfonamide retarded cleavage across the C(1)-N(2) single bond (86) of **84** and permitted competitive protonation of the indole double bond, as illustrated. Participation of the lone pair of electrons on the *N*b-nitrogen atom followed by imine formation and hydrolysis would generate the starting *N*b-benzyl analog **82**. Since it is felt that the presence of the sulfonamide is responsible for this side reaction, the amino ester **85** has been prepared and converted into the *N*a-methyl analog **86** (Scheme 10).

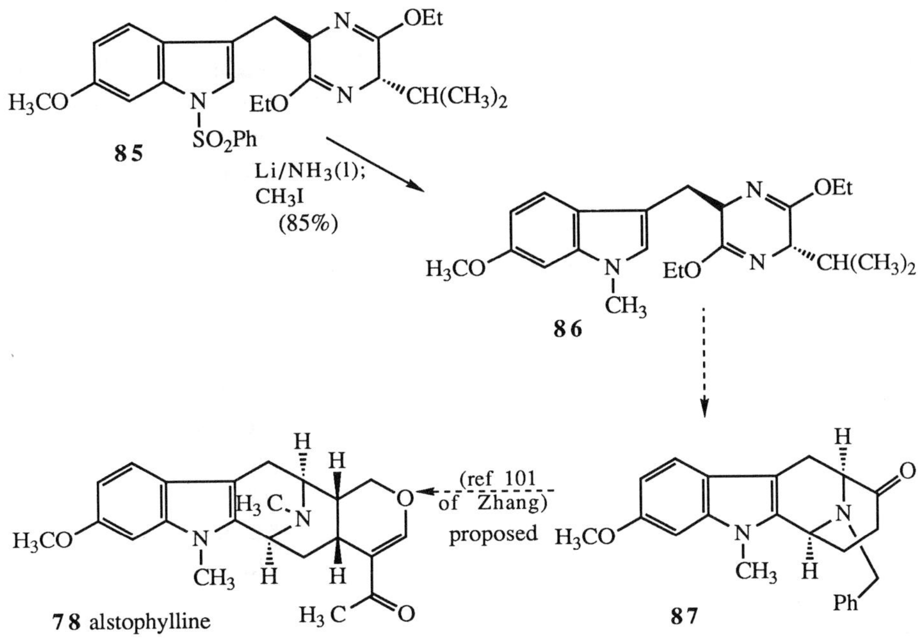

Scheme 10.

The sulfonamide **85**, available from the Schöllkopf chiral auxiliary of L-valine, was stirred with lithium and ammonia in the presence of methyl iodide to provide the pyrazine **86** required for the synthesis of *N*a-methyl-6-methoxy-D(+)-tryptophan. Currently, efforts are underway to convert the pyrazine **86** into the required 6-methoxy substituted analog of the (-)-tetracyclic ketone (see **87**) needed for the synthesis of alstophylline **78** (see Scheme 10). The chemistry to convert ketone **87** into alstophylline **78** should follow that previously reported by Zhang (104).

2.2.5. Entry into Optically Active 6-Methoxytryptophans

Alstophylline **78** and alkaloid H **80** represent just two of the many ring-A oxygenated indole alkaloids isolated from natural sources. To illustrate the importance of a multigram synthetic route to optically active 6-methoxytryptophans, a number of biologically important ring-A oxygenated alkaloids are presented in Figure 13 (see **88** to **99**). Most notable among these are the *Catharanthus* antitumor alkaloids vinblastine **90**, vincristine **91** and leurosine **92**.

The pharmacology, structure activity relationships and therapeutic uses of these clinically important bisindoles have been reviewed extensively (131). Certainly, 6-methoxytryptophan is an important intermediate for the construction of the well known antihypertensive agent reserpine **98** (82, 132-135), as well as for the related heteroyohimbinoid alkaloid, 11-methoxytetrahydroalstonine (136). In addition, optically active 6-methoxytryptophan would provide a route to the active series of mycotoxins (see **99**), the fumitremorgins. In the series of

macroline/sarpagine class

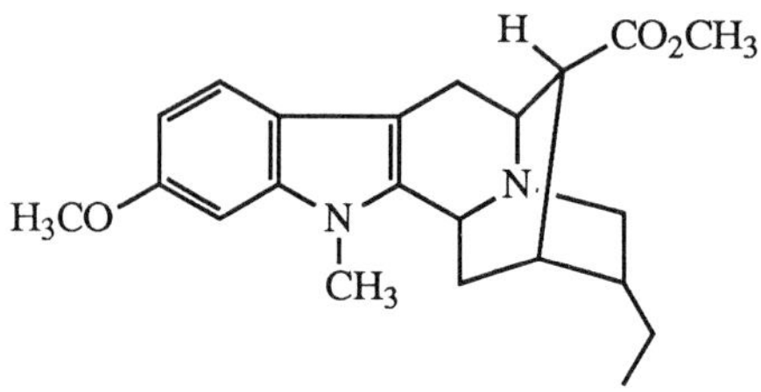

88 11-methoxy-*N*-methyl dihydropericyclivine
(stereochemistry not shown in isolation reference)

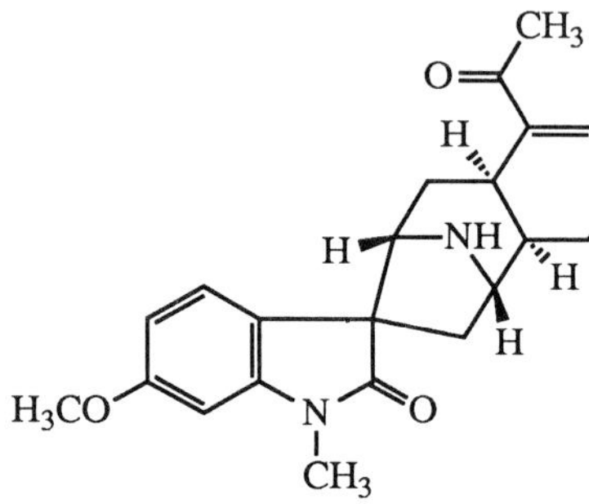

89 N_b-demethylalstophylline oxindole

Catharanthus alkaloids

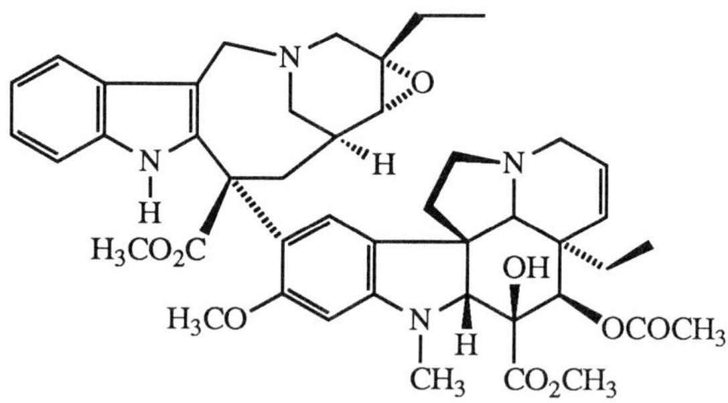

92 leurosine

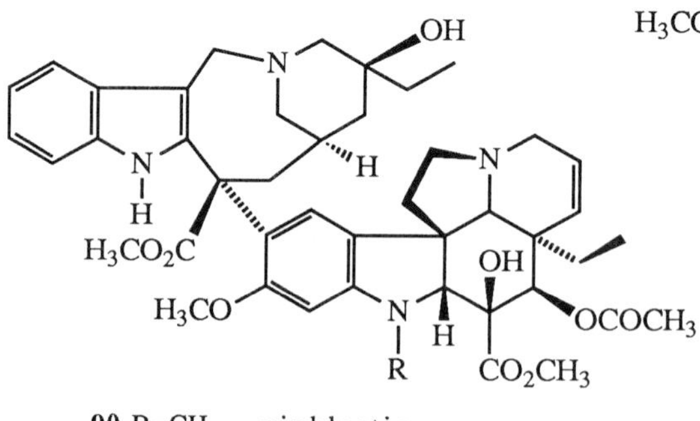

90 R=CH_3, vinblastine
91 R=CHO, vincristine

Figure 13. (continued on next page)

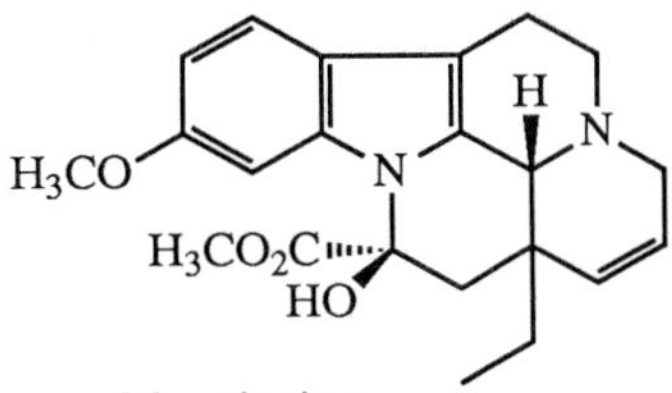

93 vincine

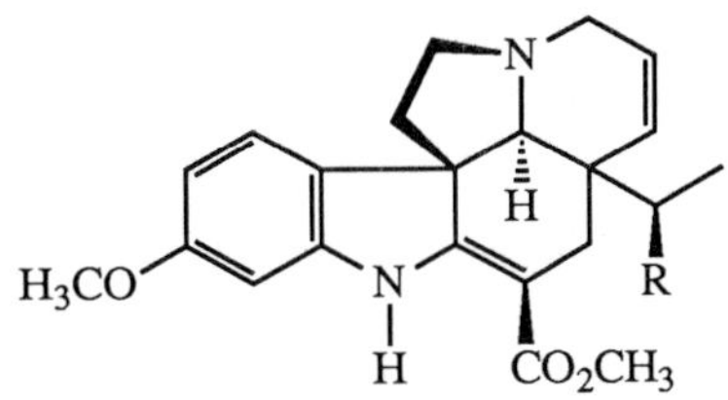

94 R=H, 11-methoxytabersonine
95 R=OH, 11-methoxy-19*R*-hydroxytabersonine

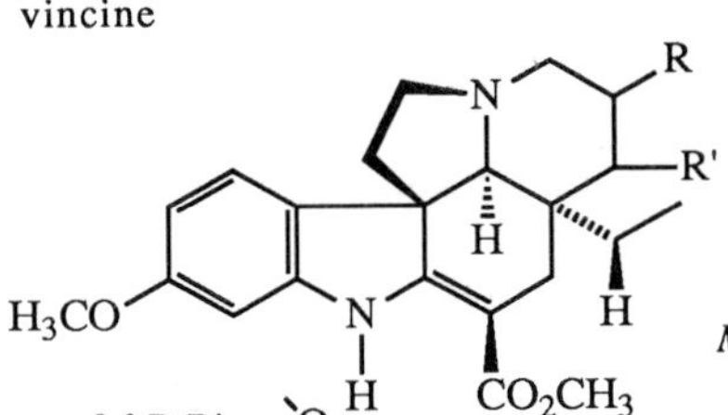

***Melodinus suaveolens* alkaloids**

96 R,R'= O
hazuntine
97 R=R'=H,
11-methoxyvincadifformine

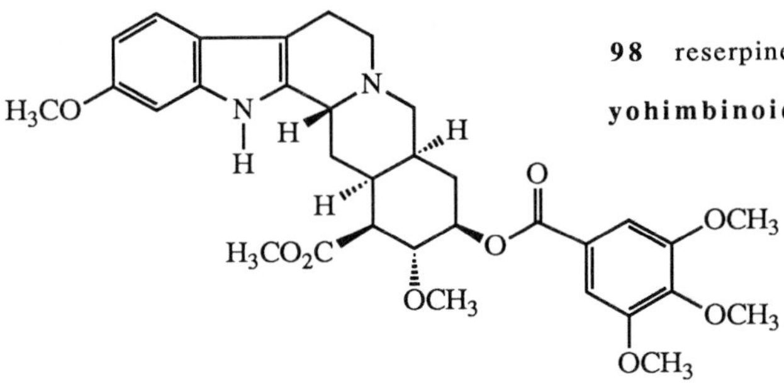

98 reserpine

yohimbinoid alkaloids

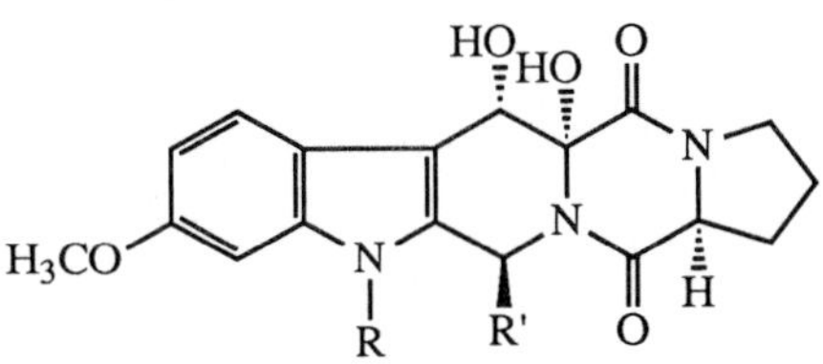

99 R = $CH_2C=C(CH_3)_2$,
R' = $CH=C(CH_3)_2$,
fumitremorgin B

fumitremorgins

Figure 13.

indole alkaloids isolated from *Alstonia* species, 6-methoxy-D-(+)-tryptophan would provide a route to gram quantities of the bisindoles **79** and **80**. The biological activity of these alkaloids, to date, has been explored only sparingly due to the paucity of isolable material (13, 41, 128, 129, 137, 138). Any route developed to provide optically active 6-methoxytryptophan for alkaloid synthesis must be capable of scale up and be relatively easy to perform.

Recently Allen *et al.* reported a strategy for the synthesis of ring-A methoxylated indole alkaloids utilizing a Moody azide/Schöllkopf chiral auxiliary protocol (139). In this method illustrated in Scheme 11, 4-methoxybenzaldehyde was condensed with methyl azidoacetate **100** in the presence of sodium methoxide at low temperature. The azidocinnamate **101** that resulted was then dissolved in xylenes and added slowly (rate monitored by observing nitrogen gas evolution *via* a gas bubbler) to a solution of xylenes at reflux. This sequence provided the methyl-6-methoxyindole-2-carboxylate **102** in high yield. Hydrolysis of the ester function of indole **102** and Cu/quinoline mediated decarboxylation of the resulting acid **103** afforded 6-methoxyindole **104**. The attractive feature of this sequence was the ability to prepare over one hundred grams of 6-methoxyindole in a very short time (several days) without time consuming chromatographic separations.

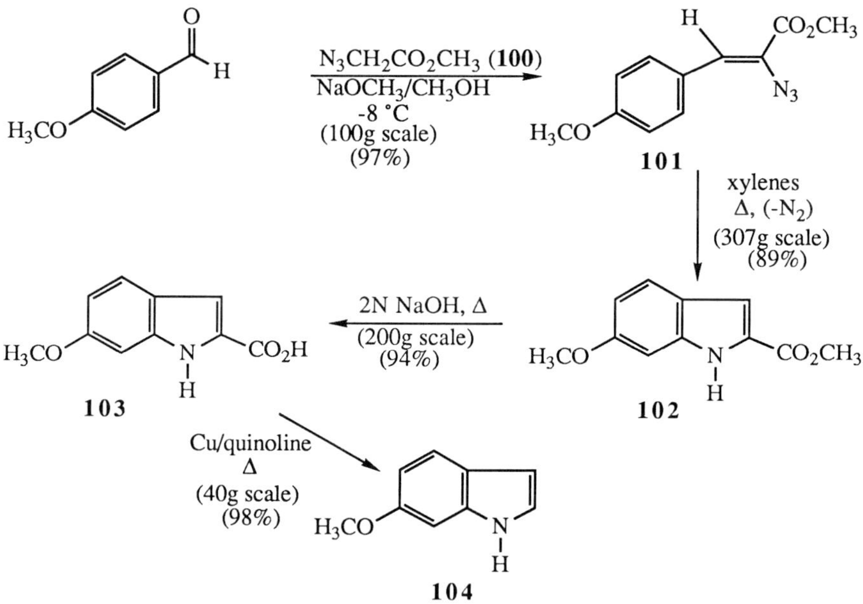

Scheme 11. Preparation of 6-methoxyindole.

A classical Vilsmeier-Haack reaction was employed to formylate the 6-methoxyindole **104** to give the 6-methoxyindole-3-carboxaldehyde **105** in 95% yield (Scheme 12). The indole N(H) functionality was then protected as the sulfonamide by treating indole **105** with sodium hydride and benzenesulfonyl chloride to provide 1-benzenesulfonyl-6-methoxyindole-3-carboxaldehyde **106**. The importance of the sulfonamide moiety was to provide a route both to N_a-demethylalstophylline **12** and alstophylline **78** *via* the same D(+)-6-methoxytryptophan intermediate. Reduction of aldehyde **106** with sodium borohydride furnished the hydroxymethyl analog **107**, which was then treated with dibromotriphenylphosphorane in

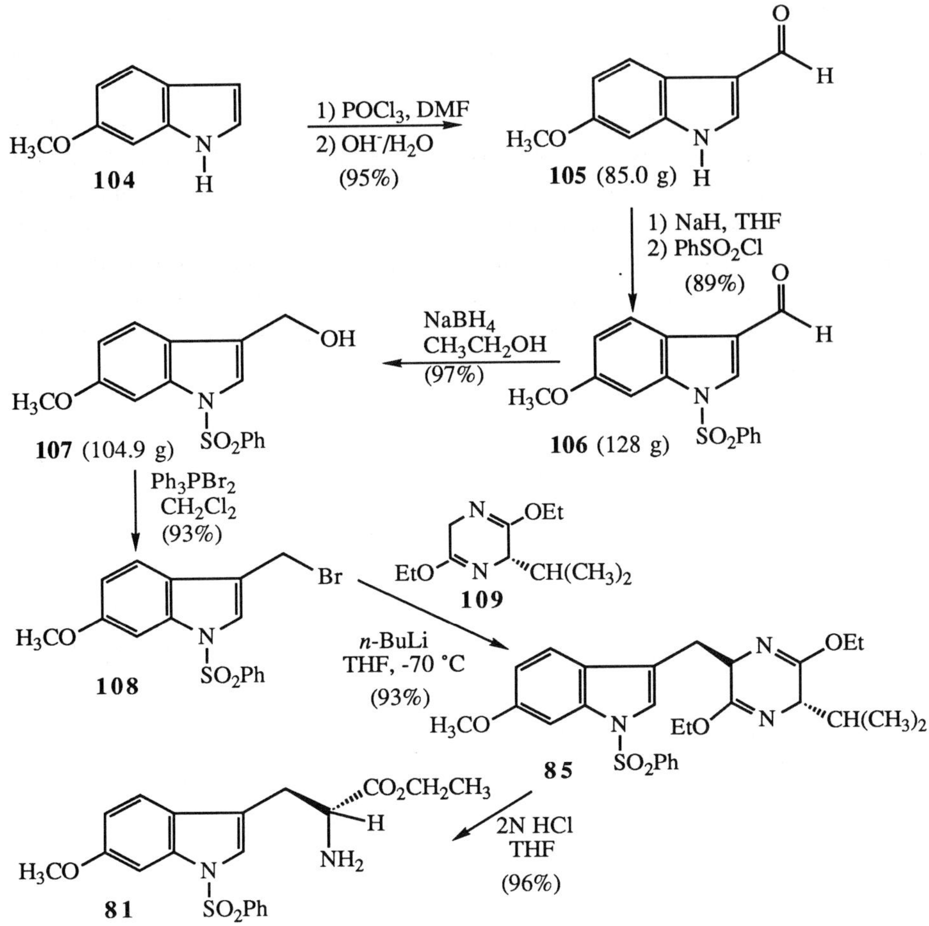

Scheme 12.

methylene chloride to afford the bromomethyl derivative **108**. To incorporate the correct stereochemistry at the α-amino ester position for the synthesis of D(+)-tryptophans, the Schöllkopf (140) chiral auxiliary [3*S*-isopropyl-2,5-diethoxypyrazine **109** (from L-valine)] was chosen as the chiral building block. Bromomethylindole **108** was treated with the anion derived from the chiral auxiliary **109** to provide the (3*R*,6*S*)-3-[1-(benzenesulfonyl-6-methoxy)-3-indolyl]-methyl-3,6-dihydro-6-isopropyl-2,5-diethoxypyrazine **85**. Acid mediated hydrolysis of the pyrazine **85** afforded the optically pure 1-benzenesulfonyl-6-methoxy-D(+)-tryptophan ethyl ester **81** in 96% yield. Allen extended this approach to include the preparation of 4-methoxy and 4,6-dimethoxyindoles as well (139). These indoles should serve as intermediates for the synthesis of many natural cytotoxic canthin-6-one alkaloids (139) as well as other macroline/sarpagine/ajmaline alkaloids.

2.2.6. 19,20-Dehydrotalcarpine

The indole base 19,20-dehydrotalcarpine **110** has not yet been isolated as a natural product, but its 10-methoxy derivative **111** was isolated from the leaves of *Alstonia angustifolia* Wall by Ghedira *et al.* in 1988. Given the proper methoxylated starting material, the natural product **111** could be synthesized *via* the same route employed to prepare 19,20-dehydrotalcarpine **110**. With this strategy in mind, the synthesis of the desmethoxy analog **110** was pursued. Illustrated in Scheme 13 is the synthesis of the desmethoxy analog **110** starting

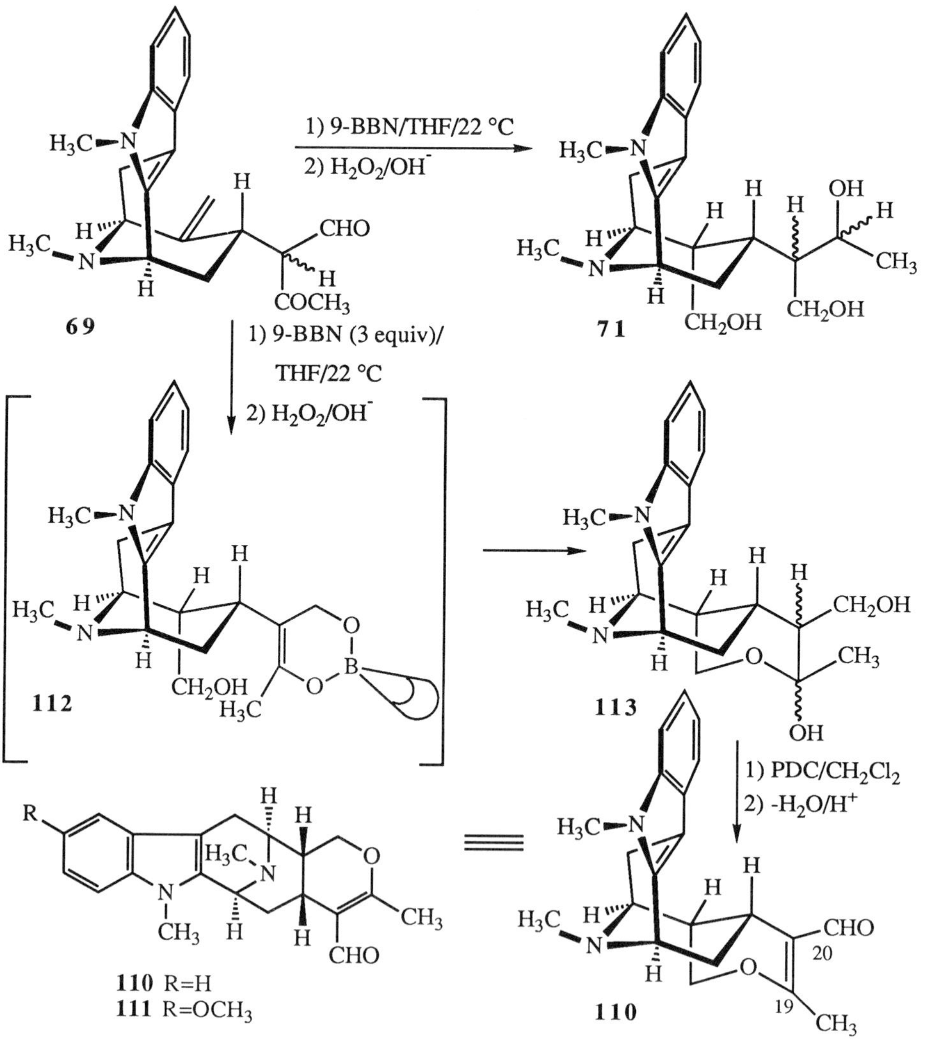

Scheme 13. Synthesis of 19,20-dehydrotalcarpine **110**.

with the dicarbonyl compound **69** available from the Claisen rearrangement employed earlier in the synthesis of alstonerine **65** (Scheme 7). Treatment of the β-dicarbonyl compound **69** with three equivalents of 9-BBN afforded the hemiketal **113**, accompanied by a small amount of the triol **71** earlier employed in the synthesis of alstonerine **65**. Oxidation of the primary alcohol function of **113** with PDC to furnish an aldehyde was followed by an acid mediated dehydration sequence to provide 19,20-dehydrotalcarpine **110** in 71% yield. The total synthesis of the natural analog **111** will now rest on the substitution of 5-methoxy-D(+)-tryptophan for D(+)-tryptophan.

2.2.7. Talcarpine/Talpinine

Talcarpine **114** and talpinine **115** are macroline-related indole alkaloids both of which were isolated (61) from *Pleiocarpa talbotii* Wernham by Schmid *et al.* Talpinine **115** was converted into talcarpine **114** upon N_b-methylation with methyl iodide followed by cleavage of the N_b-C(21) bond to afford N_b-methyl,N_b-21-secotalpinine **116**. Epimerization of the aldehyde function at C(20) in the secotalpinine **116** proceeded to give talcarpine **114** with the correct stereochemistry at C(20). This epimerization presumably (61) goes through a retro-Michael reaction under thermal conditions (Scheme 14). Talcarpine **114** was then converted into N_b-methyl,N_b-21-secotalpinine **116** upon treatment with potassium carbonate in methanol. The transformation of both talpinine **115** and talcarpine **114** into N_b-methyl,N_b-21-secotalpinine **116** occurred without alteration of the configuration of the bases at C(19); consequently, these three alkaloids possess the same absolute configuration at C(19).

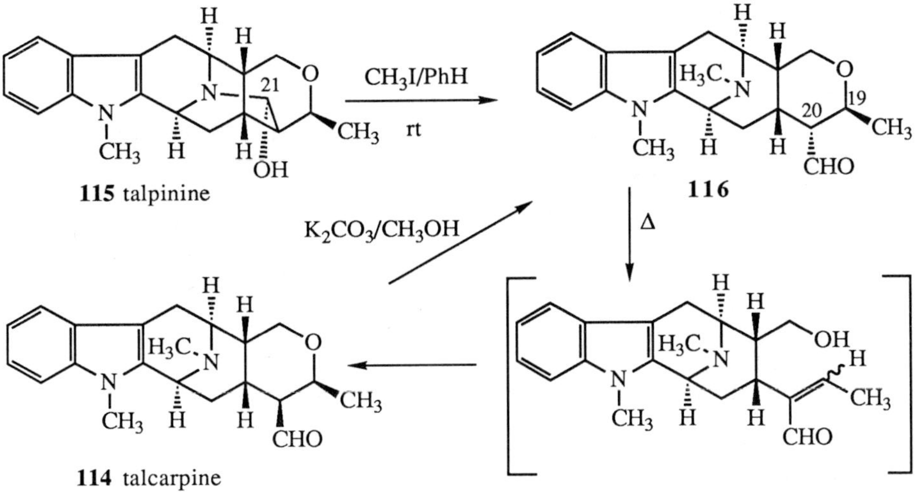

Scheme 14.

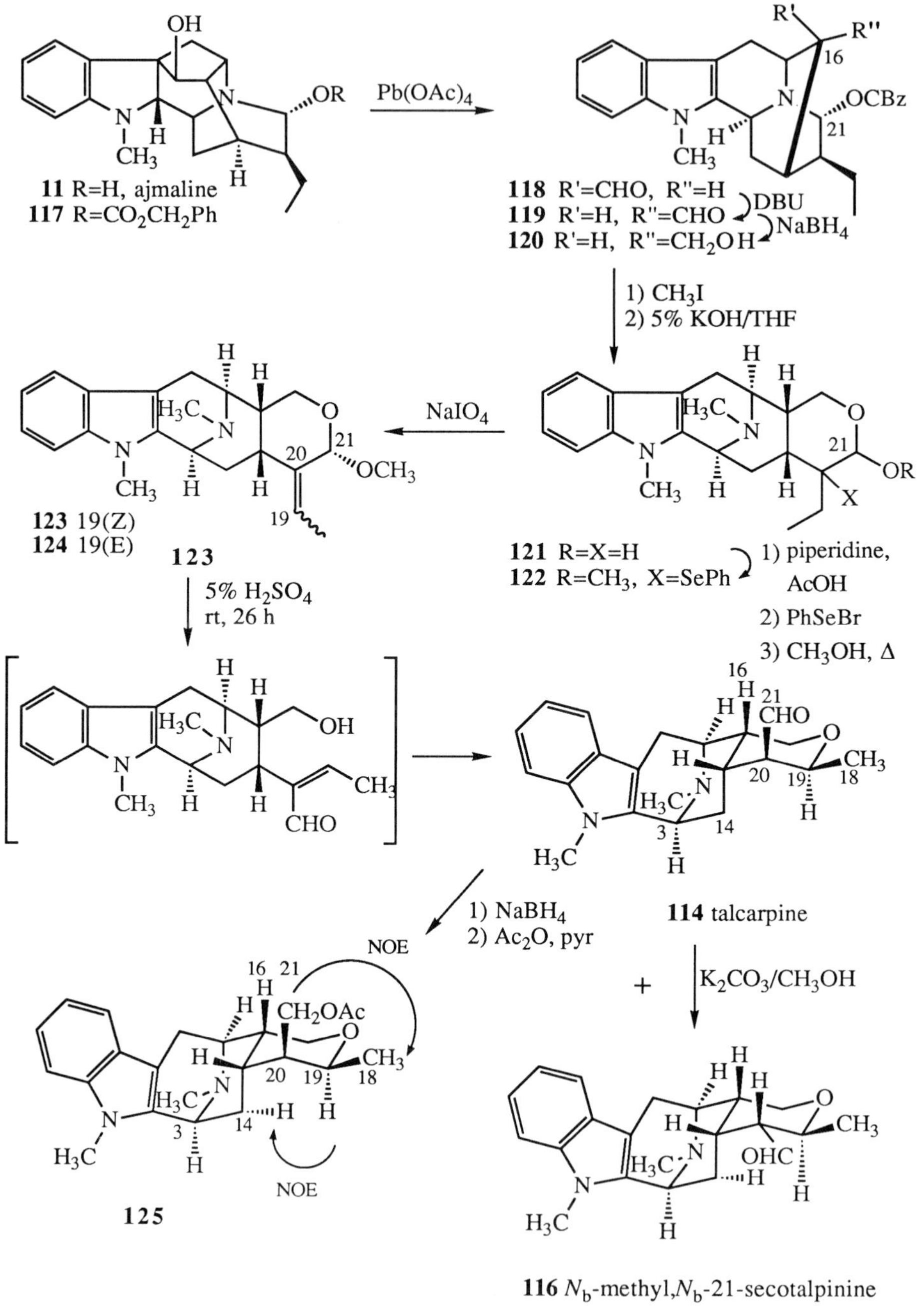

Scheme 15. Synthesis of talcarpine **114** from ajmaline **11**.

The configuration at C(19) in talcarpine **114** had not been established, unequivocally, until Sakai *et al.* performed NOESY NMR experiments on an acetyl derivative of talcarpine (60). Sakai converted ajmaline **11** into talcarpine **114** *via* the series of transformations illustrated in Scheme 15. Ajmaline was transformed into its carbobenzyloxy derivative **117** and oxidized with lead tetraacetate to afford the aldehyde **118**, analogous to the previous work of Taylor and Bartlett (141). The aldehyde function at C(16) of **118** was epimerized to provide epimeric aldehyde **119** upon treatment with diazabicyclo[5.4.0]undec-7-ene (DBU). Reduction of aldehyde **119** with sodium borohydride followed by methylation and base mediated hydrolysis afforded the macroline related hemiacetal **121**. A series of standard transformations followed by sodium periodate oxidation of the intermediate gave olefinic isomers **123** and **124** in a 2:1 ratio in 42% overall yield from **121**. The Z-olefin **123** was separated and stirred with 5% aqueous sulfuric acid for 26 hours to furnish talcarpine **114** and *N*b-methyl,*N*b-21-secotalpinine **116** in 30% and 59% yield, respectively. Since a number of signals in the NOE spectrum of talcarpine overlapped the acetyl derivative of talcarpine **125** was prepared. Clear interactions between the protons labelled H(19) and αH(14) and between C(18)H3 and one of the protons on C(21) of **125** were observed, which confirmed the configuration (*S*) at C(19).

A total synthesis of a diastereomer of talcarpine is underway in these laboratories beginning from an intermediate employed in the synthesis of (-)-alstonerine (Scheme 16). The

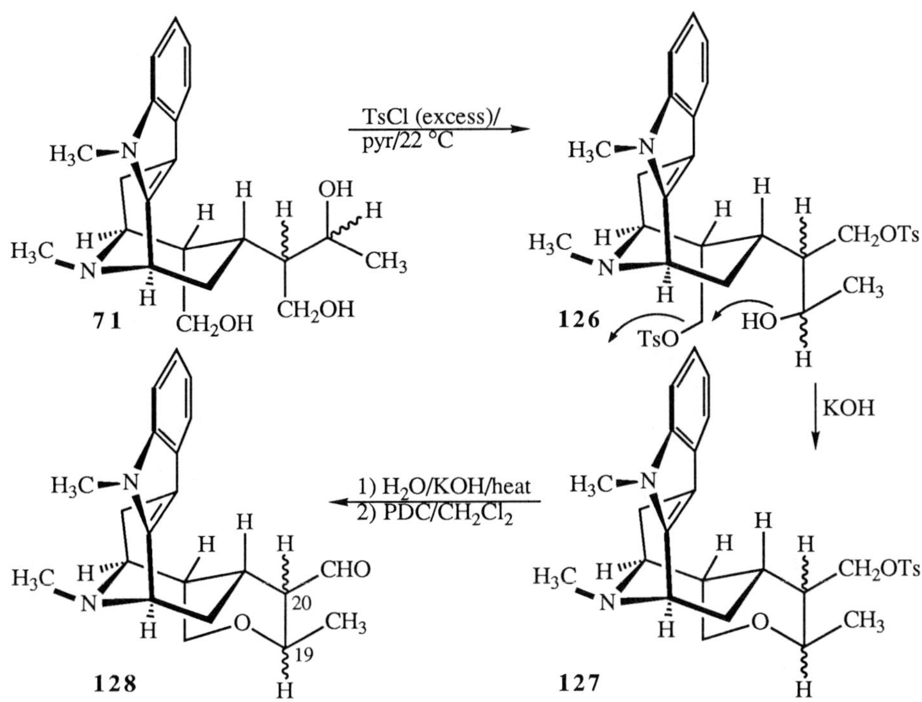

Scheme 16.

triol **71** previously synthesized from D-(+)-tryptophan was treated with excess tosyl chloride to provide the bistosylate **126**. Upon treatment with base, the secondary hydroxyl group at C(19) displaced the tosyl group at C(17). The tosyl substituted tetrahydropyran **127** that resulted was hydrolyzed to the alcohol, which upon oxidation with pyridinium dichromate afforded the diastereomer of talcarpine **128**. The nature of the stereochemistry at C(19) and C(20) is currently under study *via* NMR spectroscopy. Initial studies indicate that the stereochemistry at C(19) and C(20) is different from that of talcarpine **114** and N_b-methyl,N_b-21-secotalpinine **116**.

2.2.8. Bisindoles

Bisindole alkaloids comprise a major portion of the macroline/sarpagine alkaloids. However bisindoles in general, whether macroline related or not, are a class of alkaloids that present a significant synthetic challenge for the natural products chemist. Interest in bisindole alkaloids arises from the nature of their complex structure and their important biological activity. Bisindole alkaloids as a class have been reviewed by Lounasmaa, Cordell and Kutney (142-144); consequently, this section will focus only on the macroline related bisindoles. Many of these alkaloids are comprised of two units directly related to macroline, while others originate from the condensation of macroline with another alkaloid. The isolation and structure determination of these alkaloids have recently been reviewed (36). The biomimetic synthesis of these bisindoles pioneered by LeQuesne (40) will be described here in brief. The biomimetic synthesis of *Alstonia* alkaloids involves the Michael addition of a monomeric alkaloid to C(21) of the α,β-unsaturated enone moiety of macroline **2** (40). The formation of villalstonine **129** from macroline **2** and pleiocarpamine **130** is represented in Scheme 17. The C(7) carbon atom of pleiocarpamine was activated by the lone pair of electrons on the indole nitrogen atom. This activation facilitated the Michael addition to the C(21) enone of macroline **2**. The iminium ion which formed in this process was then attacked nucleophilically by the oxygen atom of the developing hemiacetal to provide villalstonine **129** in a stereospecific coupling process (40). Since the total synthesis of (+)-macroline **2** has been completed and coupled to natural pleiocarpamine **130** a partial total synthesis of villalstonine **129** has recently been achieved (145).

Macralstonine **79**, a ring-A oxygenated bisindole, was formed biomimetically by a similar process; however, the Michael addition took place between the C(10) carbon atom of alstophylline **78** and C(21) of macroline. Again, hemiketal formation followed the Michael addition and macralstonine **79** resulted (Scheme 18). Macralstonidine **131**, isolated (146) from *Alstonia macrophylla* Wall, was formed by Michael addition of the hydroxyl activated carbon atom C(9) of the sarpagine alkaloid N_a-methylsarpagine **132** to the enone system C(21) of macroline **2**. The carbonyl group which remained then reacted with the two hydroxyl functions

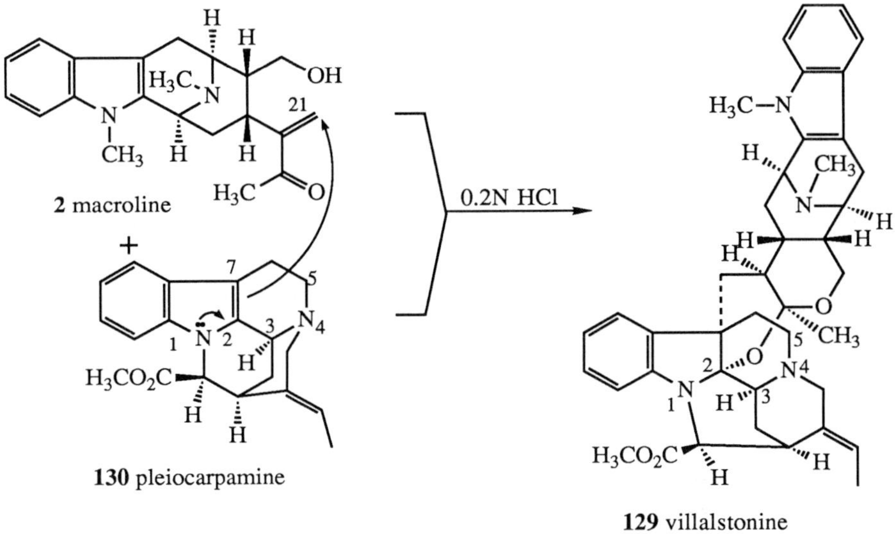

Scheme 17. Formation of villalstonine.

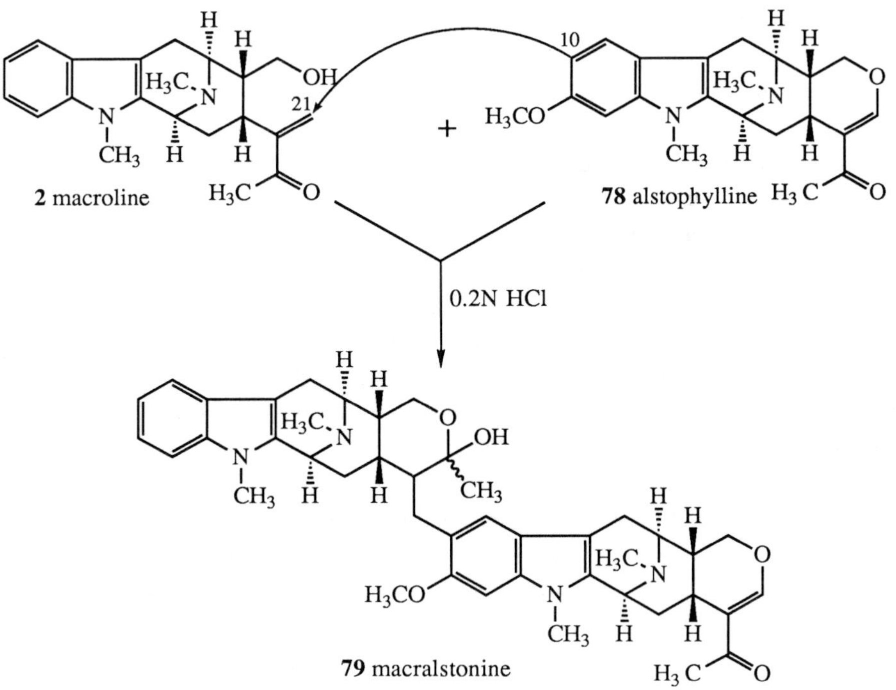

Scheme 18. Biomimetic condensation to form macralstonine.

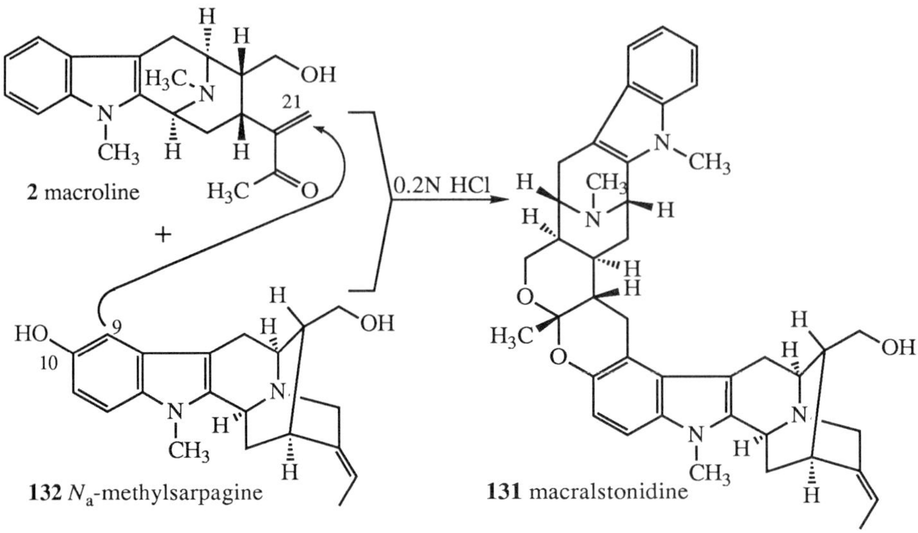

Scheme 19. Biomimetic condensation to form macralstonidine.

to provide the ketal moiety of macralstonidine **131**. Again this sequence occurred stereospecifically (Scheme 19).

Illustrated in Scheme 20 is the stereospecific coupling of macroline **2** to quebrachidine **134** to furnish alstonisidine **133**. Michael addition of the indoline nitrogen atom of the ajmaline related alkaloid quebrachidine **134** to C(21) of macroline **2** provided an intermediate hemiacetal **135**, which upon treatment with boron trifluoride etherate gave alstonisidine **133**. All twelve of the stereogenic centers in synthetic **133** were identical with those of the natural product (40).

Alkaloid H (**80**) is a macroline related bisindole that has been formed biomimetically from a unit of macroline **2** and N_a-demethylalstophylline **12**. However, not enough material has been isolated or synthesized to provide a thorough spectroscopic investigation of the structure of alkaloid H. Represented in Scheme 21 is the proposed structure of alkaloid H from the condensation of monomers **2** and **12** (128, 129), although Garnick has proposed two possible alternatives to **12** for alkaloid H (42).

Recent isolation of the novel bisindole undulatine **136** (147) has promoted interest in the origin of the unusual C(10) to C(6') bond present in this base. Undulatine **136** possesses the opposite stereochemistry at C(16) to that of sarpagine but is worth mentioning due to the unique bond joining the two monomeric units. Massiot *et al.* attempted to couple the two monomeric indoles cabucraline **137** and the pericyclivine derivative **138** *via* a DDQ mediated process (148). However, DDQ was found to be too reactive with both partners under standard conditions of oxidation. Treatment of **138** under milder conditions (1 mol DDQ/THF/rt, 30 min) did, however, provide the 6-hydroxy derivative **139** (Scheme 22) whose structure was

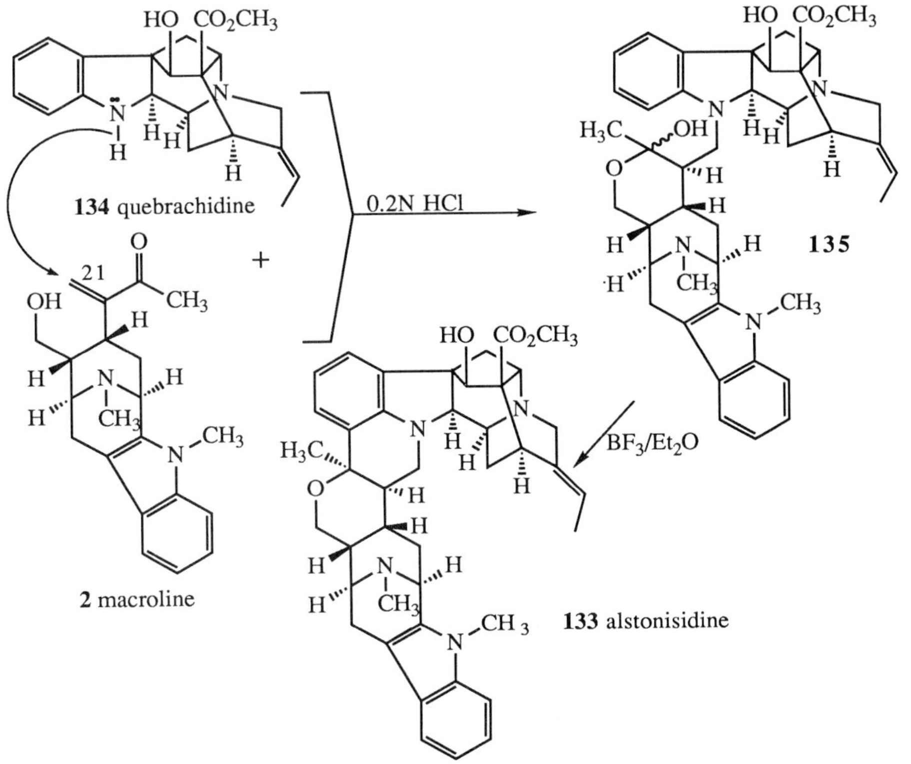

Scheme 20. Formation of alstonisidine.

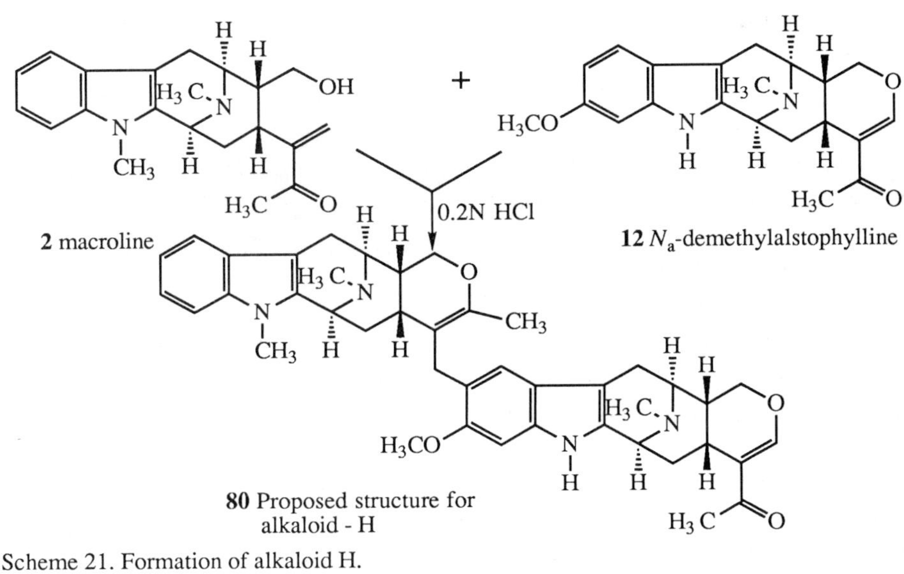

Scheme 21. Formation of alkaloid H.

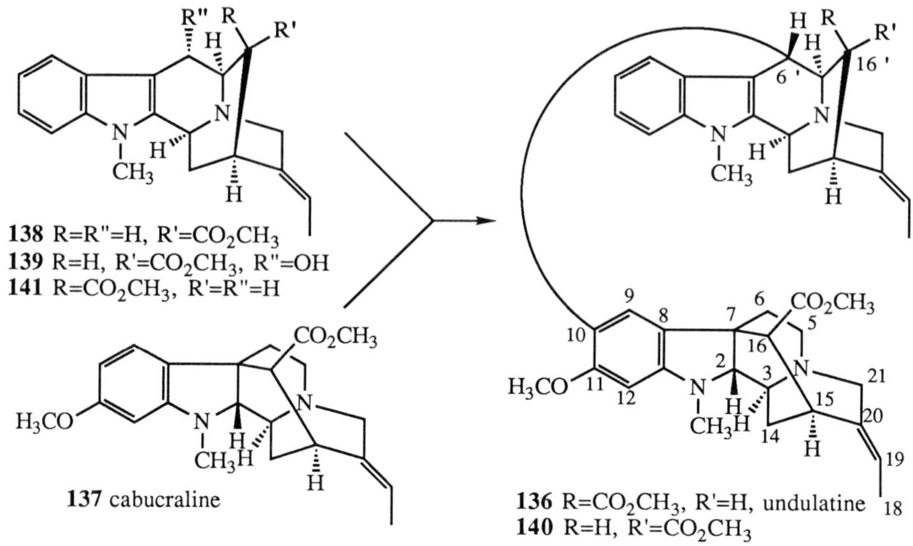

Scheme 22.

determined by NMR and mass spectroscopy. When the hydroxy derivative **139** was heated with cabucraline **137** in ethanolic hydrogen chloride the bisindole **140** was obtained in 38% isolated yield. The synthesis of the bisindole undulatine **136** now rests on the isolation or preparation of the monomeric indole N_a-methyl pericyclivine **141** or its 6'-hydroxy derivative.

2.3. Synthesis of Ajmaline Alkaloids

Ajmaline related alkaloids, including ajmaline itself, make up a "quebrachidine group" which includes nearly forty bases (149). Therefore, the main focus of this section will be to present the syntheses, to date, of ajmaline and emphasize their importance. Also included will be recent synthetic studies directed toward a newer subclass of ajmaline alkaloids, the raumaclines. For further information on specific properties, chemical conversions and syntheses of other alkaloids in this ajmaline class, the reader is referred to the excellent reviews by Lounasmaa *et al.*, (118) Joule (149) and Creasey (29).

2.3.1. Synthesis of Ajmaline and Ajmaline Related Alkaloids

In 1967, Masamune *et al.* (51) reported the first total synthesis of (±)-ajmaline **11** (Scheme 23). Condensation of the magnesium chelate of the cyclopentene derivative **143** with

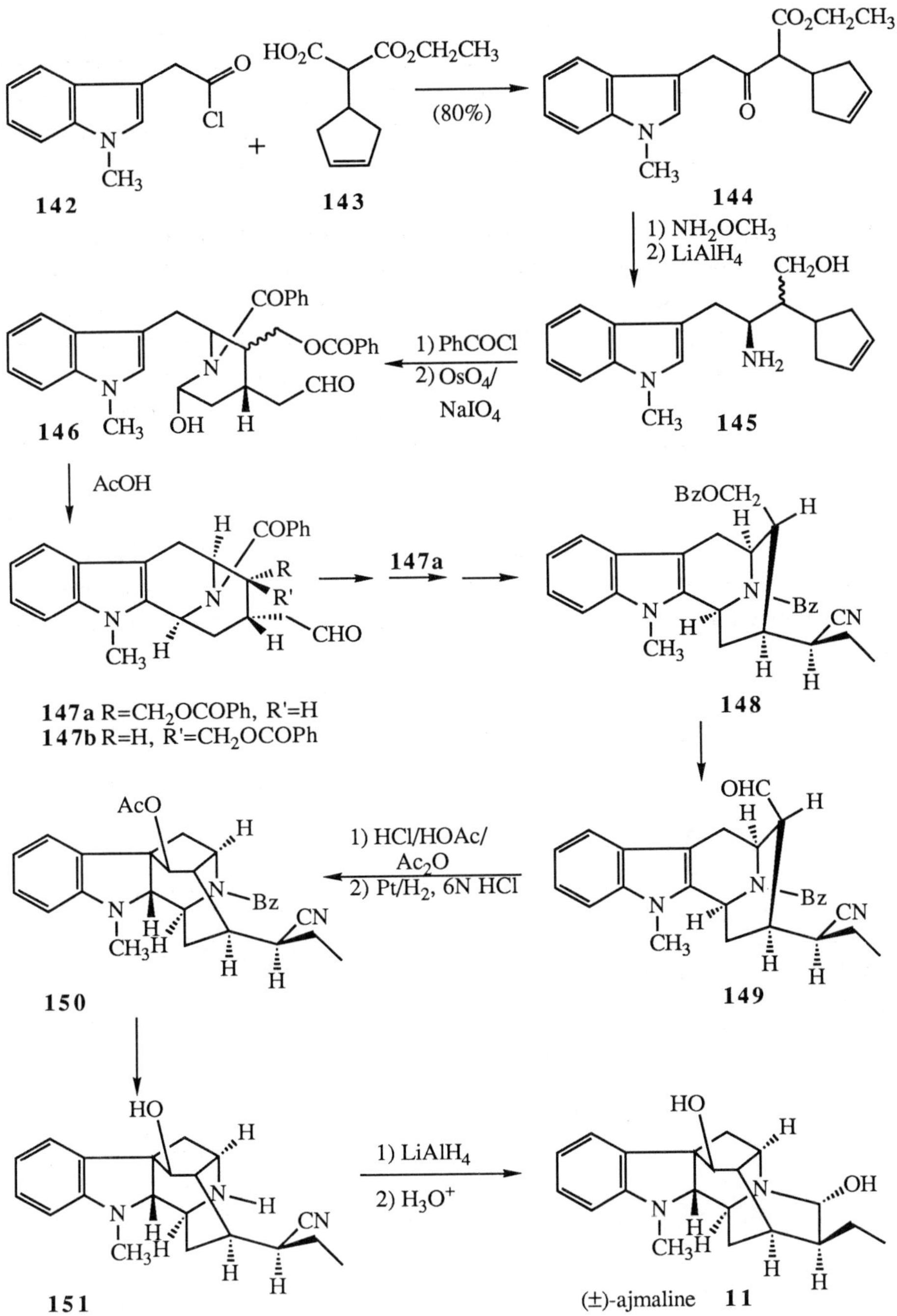

Scheme 23. The Masamune synthesis of (±)-ajmaline.

*N*a-methylindole-3-acetyl chloride **142** provided the ketoester **144**. The ketoester **144** was treated with methoxylamine followed by reduction with lithium aluminum hydride to provide the epimeric mixture of (±)-amino alcohols **145**. Benzoylation of the amino and alcohol moieties in **145** followed by oxidative cleavage of the carbon-carbon double bond of the cyclopentene unit afforded the mixture of epimeric aldehydes **146** which upon stirring with acetic acid, cyclized to the *N*b-seco-sarpagine derivative **147** (**a**/**b**, 4:5). The tetracyclic aldehyde **147a** was then converted into (±)-ajmaline in eleven steps. The steps involved for the latter transformations included conversion of the aldehyde functionality in **147a** into the cyano group of **148**. After deprotection of the benzyloxy function of the nitrile **148**, the alcohol moiety which resulted was oxidized to the aldehyde **149**. Acid-catalyzed cyclization of aldehyde **149** followed by catalytic hydrogenation provided the indoline **150**. Deprotection of both the secondary amine and hydroxyl group afforded the aminonitrile **151** which had previously been converted into ajmaline **11** with lithium aluminum hydride (150); consequently, the synthesis of (±)-ajmaline had been completed.

In 1969 Mashimo and Sato (151, 152) reported the synthesis of (±)-isoajmaline, a naturally occurring compound stereoisomeric with ajmaline at C(20) and C(21). Their synthesis of isoajmaline and also of ajmaline (153) converged with the Masamune route (153) in the latter stages of the pathway. The synthesis of isoajmaline **152**, as outlined in Scheme 24,

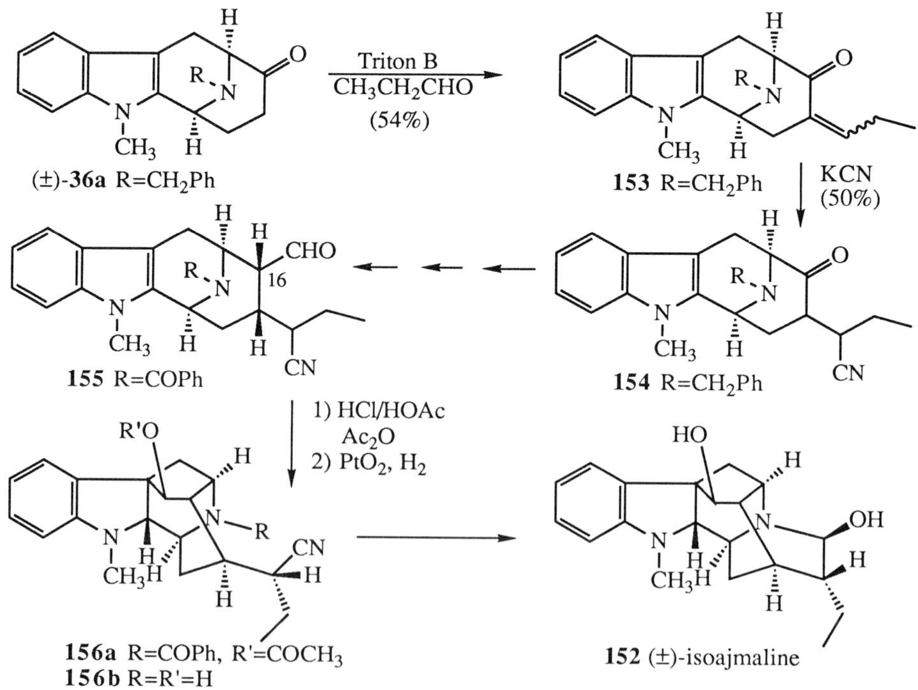

Scheme 24. Synthesis of (±)-isoajmaline.

began with the (±)-tetracyclic ketone prepared *via* the route of Yoneda (79). Condensation of the (±)-tetracyclic ketone **36a** with *n*-propanal in the presence of Triton B afforded the propylidene derivative **153** in 54% yield. Hydrocyanation of the double bond in **153** gave the cyano derivative **154**. Through a series of standard transformations, the cyanoaldehyde **155** was formed in 10% overall yield from **153**. It should be pointed out that the aldehyde function at C(16) of **155** was opposite in configuration to that required for the cyclization to provide the ajmaline skeleton. However under conditions of equilibration, an acid-catalyzed cyclization, which was later followed by catalytic hydrogenation, furnished the C(7)-C(17) functionalized isoajmaline system **156a**. The cyanoacetate **156a** was converted into its *N*b-H analog to afford the aminonitrile **156b** that had been converted into isoajmaline **152** earlier by Robinson *et al.* (150).

As mentioned earlier, Mashimo and Sato extended the above synthetic route to include a formal total synthesis of (±)-ajmaline. Mashimo was able to achieve this synthesis *via* the preparation of the Masamune intermediate **147**. Illustrated in Scheme 25 is the formal synthesis of ajmaline reported by Mashimo *et al.* The route began with (±)-*N*a-methyl,*N*b-benzyltetracyclic ketone **36a** employed previously in the preparation of isoajmaline.

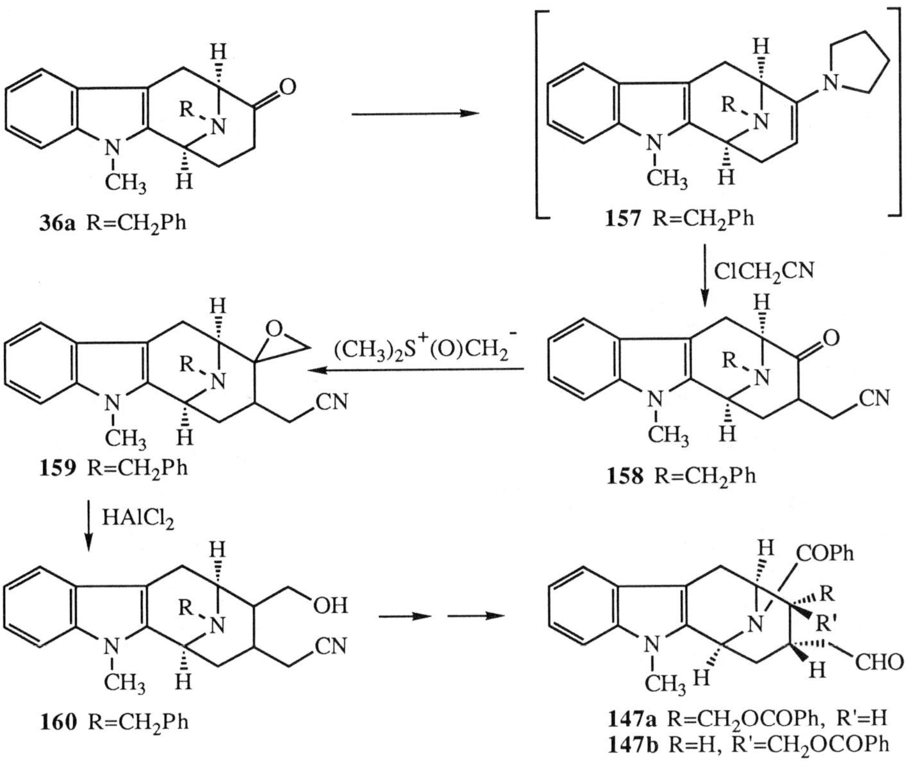

Scheme 25. Formal synthesis of (±)-ajmaline.

Alkylation of the pyrrolidine-enamine **157** with chloroacetonitrile afforded the nitrile **158** in 50% yield. Epoxide formation *via* the Corey sulfur ylid gave oxirane **159**, and this was followed by reductive cleavage to furnish carbinol **160**. Reductive debenzylation and subsequent dibenzoylation then furnished the Masamune intermediate (±)-**147**, which completed a formal total synthesis of (±)-ajmaline.

In 1970 van Tamelen (52, 154) reported the biogenetic-type total synthesis of ajmaline. For specific details on the synthetic strategy presented in Scheme 26, the reader is referred to the full paper of van Tamelen *et al.* (52) in which the methodology employed for each transformation is thoroughly described. Reductive alkylation of (±)-N_a-methyltryptophan **161**

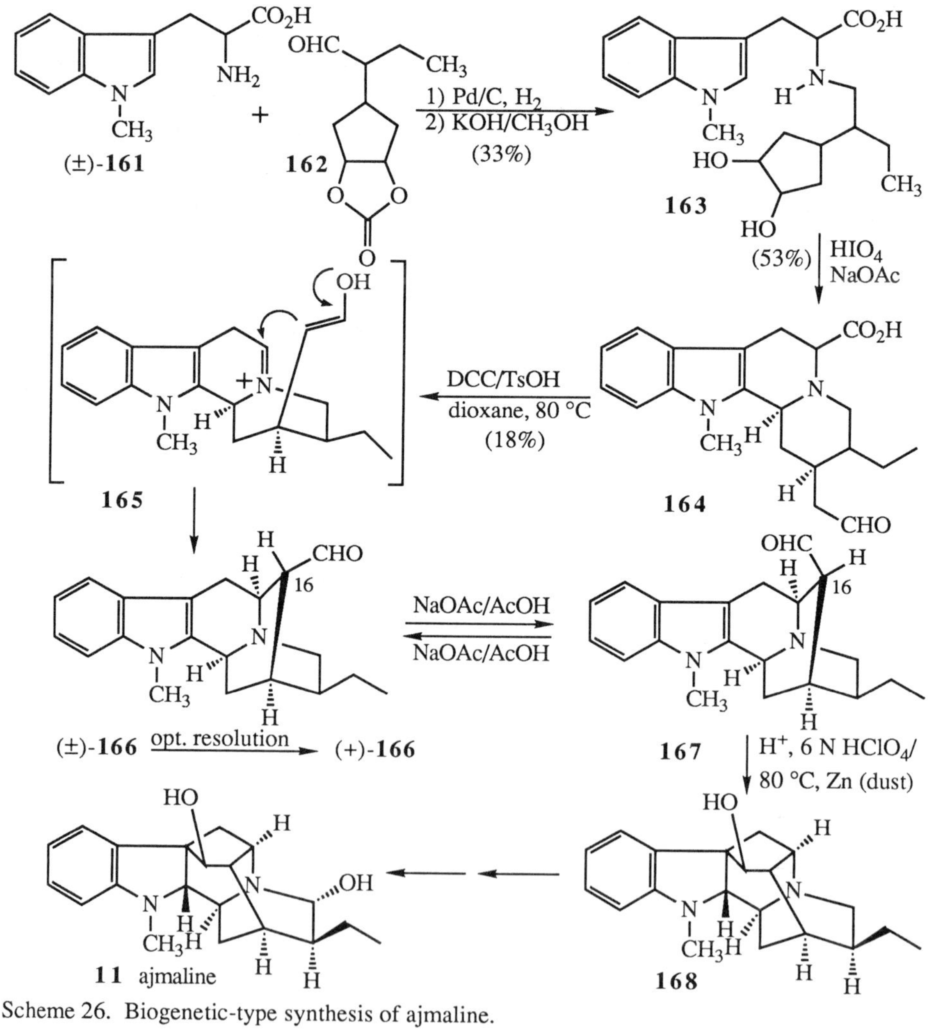

Scheme 26. Biogenetic-type synthesis of ajmaline.

with aldehyde **162** (prepared in four steps from cyclopent-3-enyl tosylate) furnished the N_b-substituted amino acid. The carbonate protecting group was removed with alkali to provide diol **163**. Cleavage of the glycol **163** gave the corresponding dialdehyde, which spontaneously cyclized to provide the sarpagine/ajmaline skeleton **166**. Again, this aldehyde **166** contained the incorrect configuration at C(16) and had to be epimerized in order to promote cyclization. An equilibrium was established at a ratio of 85:15 in favor of the undesired aldehyde **166** when aldehyde **166** was treated with acetic acid and sodium acetate. Deoxyajmalal **166** was also resolved with D-camphor-10-sulfonic acid to provide optically active deoxyajmalal B which was further equilibrated to provide optically active deoxyajmalal A **167**. Since the (±)-aldehyde **167** had previously been converted into (±)-deoxyajmaline **168** by Taylor *et al.* and functionalization of (±)-**168** at C(21) had been completed using the methods of Hobson and McCluskey (155), the biogenetic-type synthesis of natural (+)-ajmaline was considered complete.

Note that all of the total syntheses of ajmaline reported to date are in the racemic series or required the optical resolution of (±)-deoxyajmalal B **166** late in the biogenetic-type route. This fact alone indicates the complex nature of the transformations involved in the preparation of this biologically important antiarrhythmic alkaloid. The interest in the biological properties of natural ajmaline has prompted additional studies toward a better understanding of the biosynthesis of this alkaloid (156). The synthesis of the optically active (-)-tetracyclic ketone **36a** reported in 1988 by Zhang (82) and later in 1993 by Bailey (103) provides a route for an enantiospecific approach towards (+)-ajmaline. This strategy will become clearer later in the section on raumaclines and in the section on the biological aspects of macroline/sarpagine/ajmaline alkaloids.

2.3.2. Raumaclines

Since the complete structure of (+)-ajmaline **11** has been well documented (49, 141, 157) and confirmed by X-ray crystallography (158), this commercially available (159) alkaloid has been widely used as a starting material for the preparation of other alkaloids.

During the structure determination of suaveoline **3**, Potier *et al.* carried out a partial synthesis of N_b-methylsuaveoline **169** from ajmaline **11** (Scheme 27). In brief, ajmaline **11** was treated with benzoyl chloride, and this was followed by oxidation with lead tetraacetate to furnish aldehyde **170**. The aldehyde **170** was then heated with methyl iodide to form a quaternary salt which was subsequently heated with aqueous ammonia to provide N_b-methylsuaveoline **169**, presumably formed from cyclization of the intermediate dialdehyde **171**. Since N_b-methylsuaveoline was also prepared from suaveoline **3** (43), this transformation served to correlate the absolute stereochemistry of suaveoline **3** with that of ajmaline **11** at C(3) and C(5).

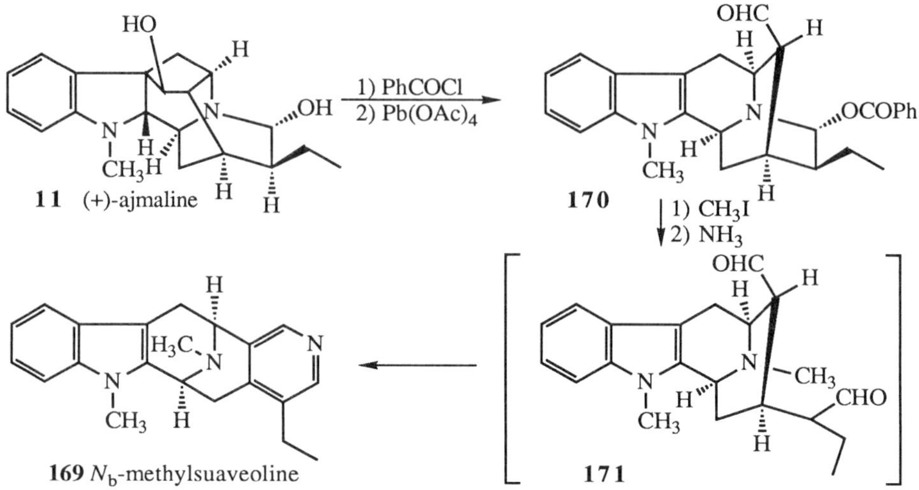

Scheme 27.

Recently, Stöckigt, Sakai, *et al.* (75, 160, 161) isolated six new alkaloids (Figure 6), termed the raumaclines, from cell cultures of *Rauwolfia serpentina* Benth. after feeding experiments with ajmaline **11**. Raumacline **27** and N_b-methylraumacline **28** were first detected as products of these feeding experiments in 1990 (160). The structures of these alkaloids were elucidated by spectroscopic methods and partial synthesis from ajmaline **11** (160). This biomimetic synthesis from ajmaline is represented in Scheme 28. Sodium borohydride reduction of ajmaline **11** afforded dihydroajmaline **172a** in 78% yield. The required carbamate **172b** was formed when dihydroajmaline **172a** was treated with benzyl chloroformate. Oxidative ring opening of **172b** provided an aldehyde which cyclized spontaneously with the C(21)-OH group to furnish N_b-CBZ raumacline **173**. Removal of the CBZ group *via* catalytic hydrogenolysis afforded raumacline **27** identical in all respects with the natural compound. Methylation of raumacline **27** with sodium cyanoborohydride in formaldehyde in the presence of a catalytic amount of acetic acid furnished N_b-methylraumacline **28**. This biomimetic approach of Sakai *et al.* (75, 160) again illustrates the important use of (+)-ajmaline for stereochemical correlations.

Later, Stöckigt *et al.* isolated (75, 161) four more raumacline alkaloids from *Rauwolfia serpentina* Benth. cells cultivated in the presence of ajmaline (+)-**11**. The first of the four new alkaloids, 6α-hydroxyraumacline **29**, was also synthesized from ajmaline *via* raumacline **27**. Depicted in Scheme 29 is the conversion of raumacline **27** into 6α-hydroxyraumacline **29**. The other three raumacline alkaloids represented in Figure 6 are 6α-methoxyraumacline **30**, 19-hydroxy-N_b-methylraumacline **31** and isoraumacline **32**, the latter alkaloid contains the opposite configuration at C(20) to raumacline **27**. Their structures were elucidated by the latest

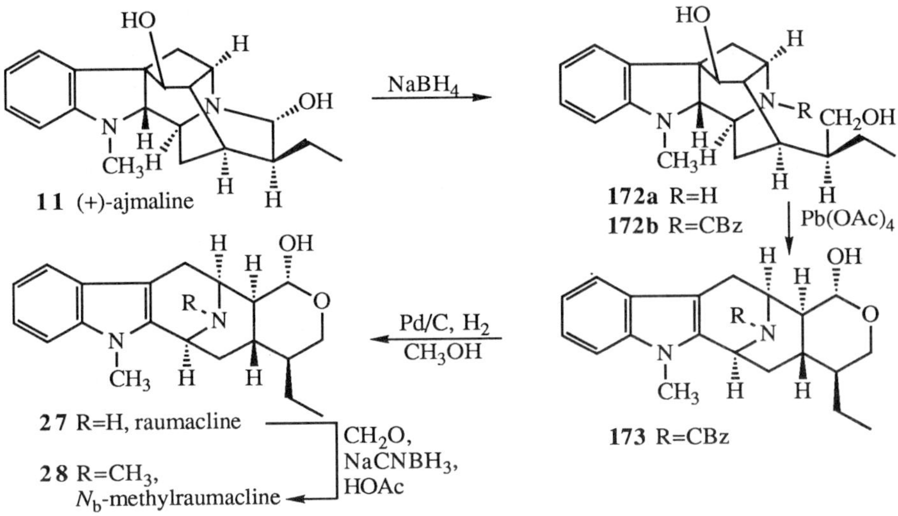

Scheme 28.

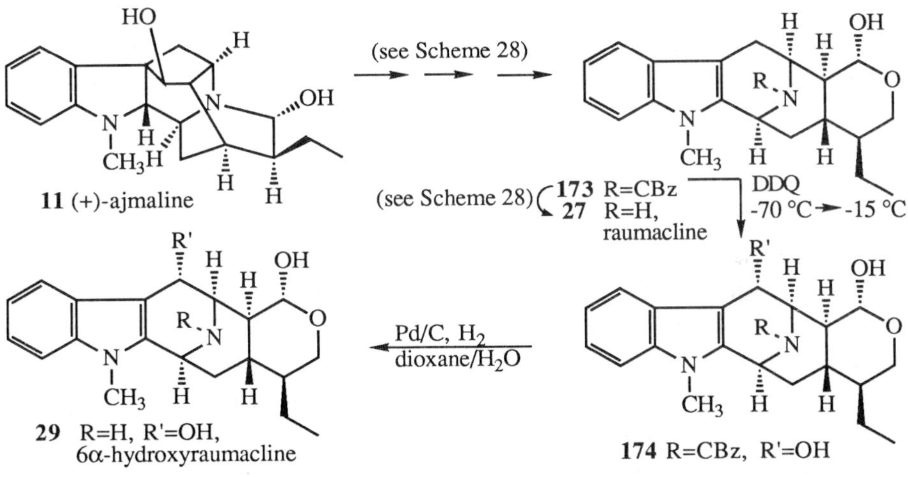

Scheme 29.

spectroscopic methods; however, to these authors' knowledge partial syntheses from ajmaline or raumacline **27** have not yet been reported.

Soon after the isolation of this new class of alkaloids, Fu (114) reported the enantiospecific synthesis of (-)-raumacline **27** and (-)-N_b-methylraumacline **28**. The enantiospecific nature of the total synthesis of **27** and **28** is important for several reasons. The relationship between raumacline **27** and ajmaline **11** has now been established chemically by Sakai *et al.*; therefore, entry into other ajmaline alkaloids including the total synthesis of the unnatural antipode (-)-ajmaline can be envisaged. The synthetic route executed by Fu employed the same oxyanion Cope rearrangement developed for the total synthesis of (-)-suaveoline **3** (114) and *N*b-methylsuaveoline **61**. It is interesting to note that in the *N*b-benzylazabicyclo[3.3.1]nonane system **36a**, the Claisen rearrangement (123) occurred principally from the desired α-face of the enone **68** (>4:1), while the ortho ester Claisen rearrangement took place in a related *N*b-benzyl system with a stereoselectivity of 13:1 from the top face of the allylic system principally *via* boat transition states (113). On the other hand the allylic alcohol **59** underwent the anionic oxy-Cope rearrangement to provide the same C(15) functionalized tetracyclic systems **60a,b** and **60c** in a ratio of 3:2 (Scheme 30) that had been obtained earlier by 1,4-addition to (-)-**57** (suaveoline, Scheme 5). This demonstrated the importance of the pseudosymmetric secondary Grignard reagent **58**, the olefinic bond of which also served as a latent aldehyde function (refer to Scheme 5). The diastereofacial selectivity for the oxyanion Cope rearrangement in the simpler *N*b-benzyl system **175** also favored attack from the desired bottom face of the double bond, again in a ratio of 3:2. Apparently the ethyl substituents which originate from the pseudosymmetric Grignard reagent **58**, do not affect the diastereoselectivity in this pericyclic process. Since the diastereoselectivity in the Claisen rearrangement (*N*b-methyl **36b** *vs* *N*b-benzyl **36a**) was greater from the desired face in the case of the smaller *N*b-alkyl group (123), studies are underway to determine the diastereoselectivity of the oxyanion Cope rearrangement in the *N*b-H series (162).

The formyl group of the mixture of aldehydes **60a,b** was protected as the ethylene acetal and this was followed by oxidative cleavage of the double bond to provide two epimeric aldehydes **179a,b** in excellent yield (Scheme 31). Aldehyde **179a** possesses the desired chirality (*S*) at C(20) for the synthesis of (-)-raumacline **27** and *N*b-methylraumacline **28**. For this reason, **179b** was treated with base and converted into an equilibrium mixture of **179a** and **179b** (1:1), which was again separated easily by flash chromatography on silica gel. The(-)-(*S*)-aldehyde **179a** was reduced to the alcohol **180**. This was followed by deprotection of the aldehyde function and cyclization under acidic conditions to provide (-)-*N*b-benzylraumacline **181**. It is worth noting in this last sequence that the formation of **181** from aldehyde **179a** was stereospecific. Catalytic debenzylation of the hydrochloride salt of (-)-**181** in ethanol furnished (-)-raumacline **27** in 91% yield. When the base (-)-**181** was subjected to catalytic debenzylation with excess Pd/C and hydrogen in methanol, an 85% yield of (-)-*N*b-methylraumacline **28** was realized. The recent isolation of isoraumacline **32** (Figure 6) which bears the opposite configuration (*R*) at C(20) to that of raumacline **27** is significant for the synthesis of isoraumacline **32** would simply require conversion of the (*S*)-aldehyde **179a** into (*R*)-aldehyde **179b**. This can be accomplished by a similar set of equilibration experiments (115). These two syntheses of (-)-**27** and (-)-**28** and that of (-)-suaveoline **3** represent the first enantiospecific synthesis of members of the ajmaline family of indole alkaloids and demonstrate that the strategy employed for the preparation of the macroline related sarpagine alkaloids can be extended to other families of indole alkaloids.

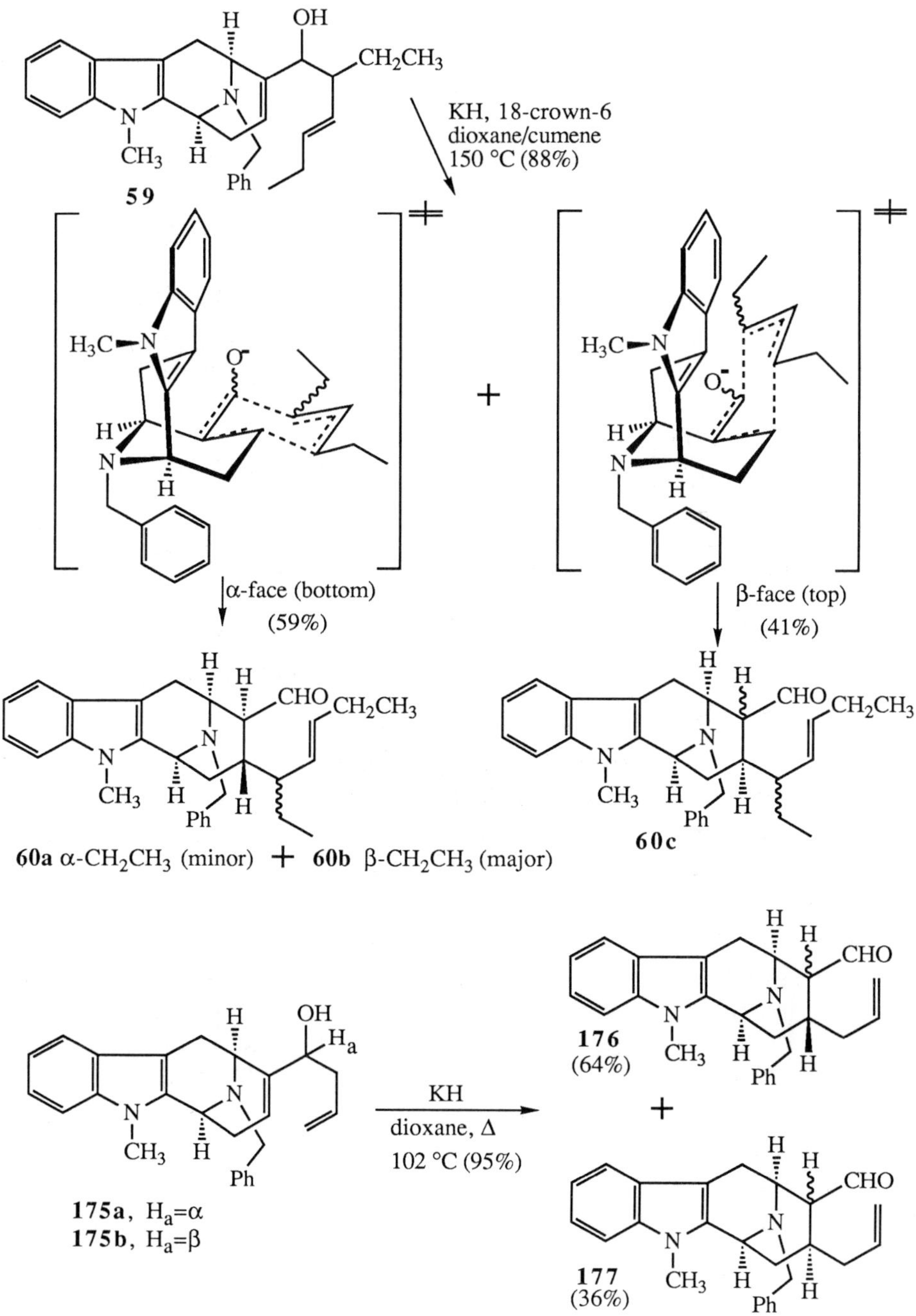
59
KH, 18-crown-6
dioxane/cumene
150 °C (88%)
α-face (bottom)
(59%)
β-face (top)
(41%)
60a α-CH2CH3 (minor) + 60b β-CH2CH3 (major)
60c
KH
dioxane, Δ
102 °C (95%)
175a, Ha=α
175b, Ha=β
176
(64%)
177
(36%)

Scheme 30.

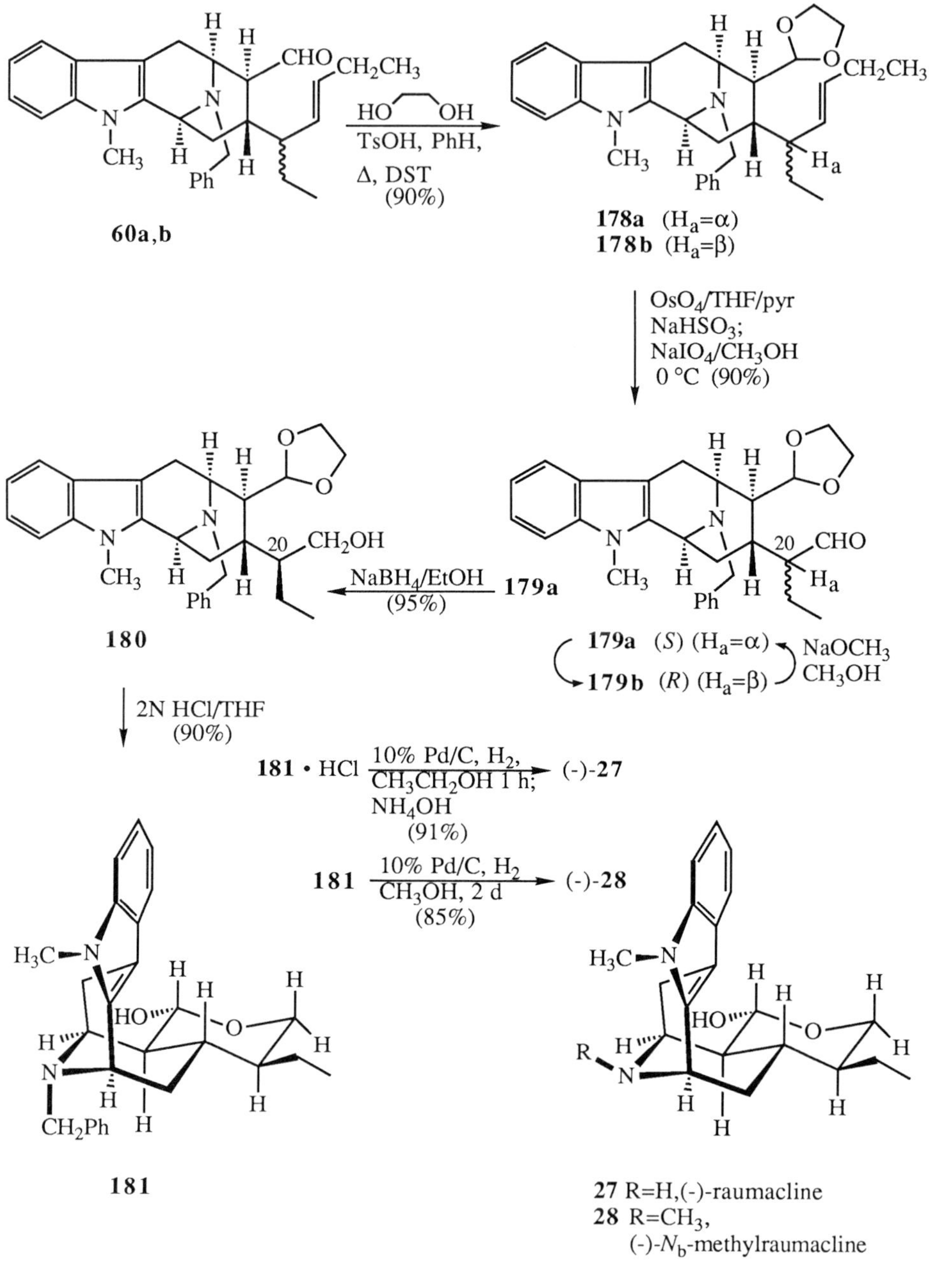

Scheme 31. Synthesis of (-)-raumacline **27** and (-)-N_b-methylraumacline **28**.

3. Biological Aspects of Alkaloids

3.1. Macroline/Sarpagine Alkaloids

Since Cinchona bark became known to the Europeans three centuries ago, its preparations and its principle alkaloid, quinine, have been used in the treatment of malaria. During an epidemic in Manila, dita bark, an extract of *Alstonia scholaris*, was said to have surpassed quinine as a drug to treat malaria (163). Two species, *A. scholaris* and *A. constricta*, were recognized in the British Pharmacopoeia in 1914, moreover *A. scholaris* as well as other representatives of the genus have been used in West Africa, Malaya and the Philippine Islands. A number of species of *Alstonia* (Apocynaceae), including *Alstonia angustifolia* Wall, are employed in South East Asia for the treatment of malaria and dysentery (164). As a result of the medicinal properties of these plants, much attention has been devoted towards the isolation of the alkaloidal components responsible for the biological activity (13, 14, 165).

In the case of the *Catharanthus* alkaloids vinblastine **90** and vincristine **91**, the bisindole structure appears to be essential for full cytotoxic potency. Of the known monomeric alkaloids in this series, only lochnericine and lochnerinine exhibit even the slightest activity (166). Although this pattern in activity may not be true in all cases, it is not surprising to see additional reports of the biological activity of bisindole alkaloids in the recent literature. With respect to the macroline/sarpagine alkaloids, Wright *et al.* (165) reported their findings on the antiprotozoal activity of nine alkaloids from *Alstonia angustifolia* against *Entamoeba histolytica* and *Plasmodium falciparum in vitro*. There is a need for new therapeutic agents in the tropics where diseases caused by protozoa and/or resistant strains of parasites are responsible for many deaths. Three bisindole alkaloids, macralstonine acetate **182**, macrocarpamine **183**, and villalstonine **129** (Figure 14) were found to have significant activity against both protozoa mentioned above. However, the monomeric alkaloids alstonerine **65**, alstophylline **78**, 11-methoxyakuammicine **184**, norfluorocurarine **185**, pleiocarpamine **130**, and vincamajine **186** were all considerably less active than the above described dimers. The bisindole macralstonine **79** was found to be inactive against the two protozoa. Each test employed emetine as a standard amoebicidal drug and the results are presented in Table 4. Macrocarpamine **183** was found to be the most active antiamoebic alkaloid with an antiamoebic activity one fourth of the standard drug emetine **187**. Villalstonine **129** was found to be the most potent alkaloid against *P. falciparum* and was about fifteen times less potent than the antimalarial drug chloroquine. The acetate of macralstonine **182** was much more active against both types of protozoa when compared with the parent macralstonine **79**. It was suggested that the ester function of **182** facilitated the transport of the acetate **182** across cell membranes as a result of increased lipophilicity (165). These results, therefore, explain the use of *Alstonia angustifolia* in traditional medicine for the treatment of amoebic dysentery and for the treatment of malaria, although the potencies of even the most active alkaloids are less than

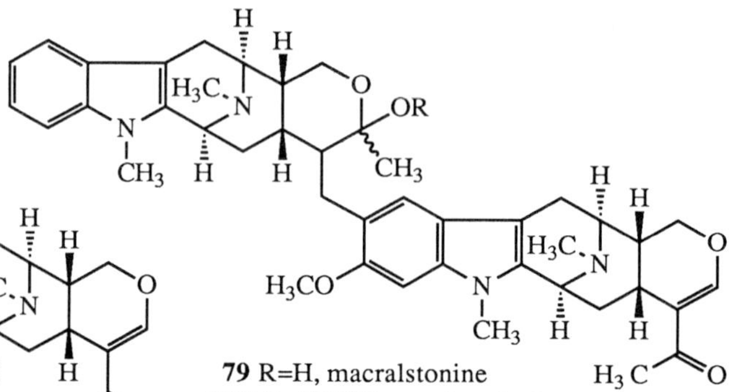

79 R=H, macralstonine
182 R=$COCH_3$, semisynthetic

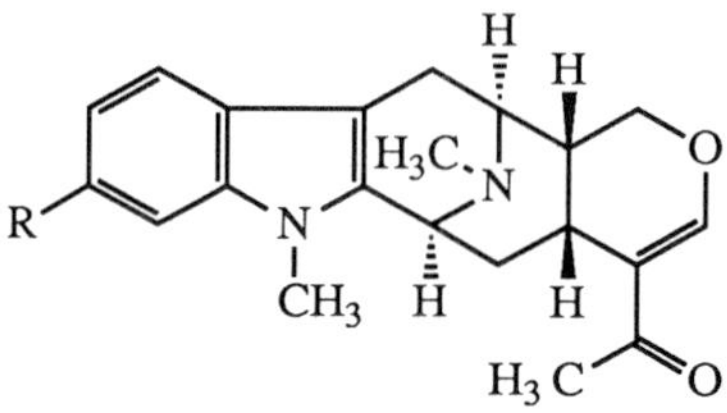

65 R=H, alstonerine
78 R=OCH_3, alstophylline

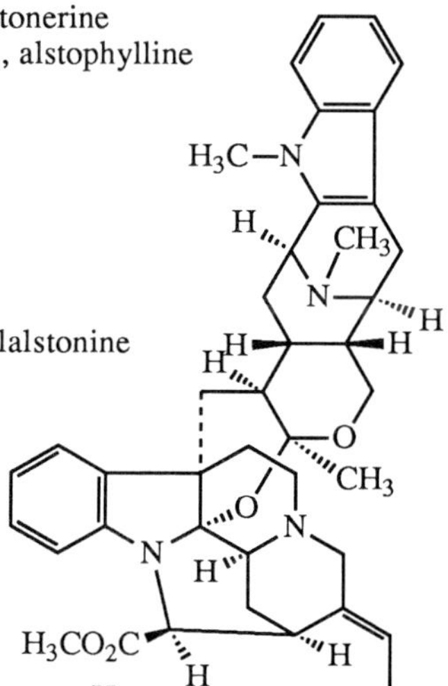

129 villalstonine

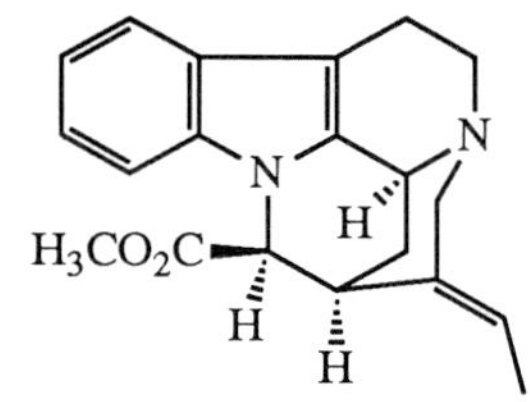

130 pleiocarpamine

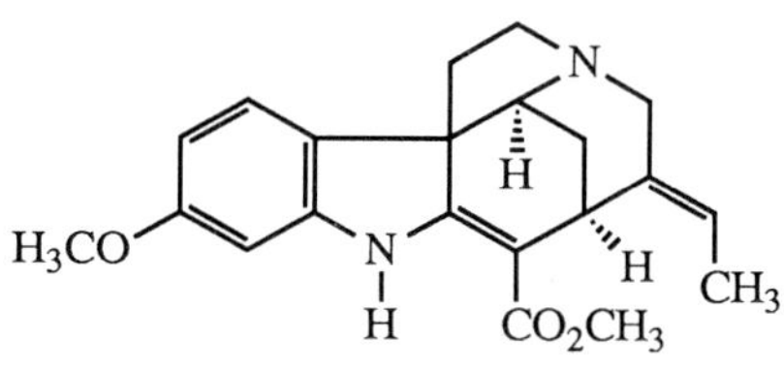

184 11-methoxyakuammicine

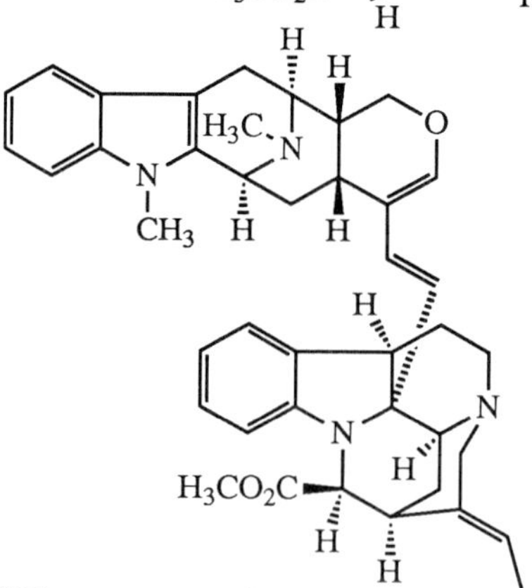

183 macrocarpamine

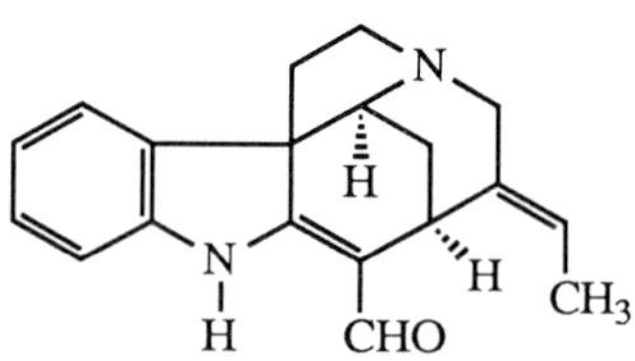

185 norfluorocurarine

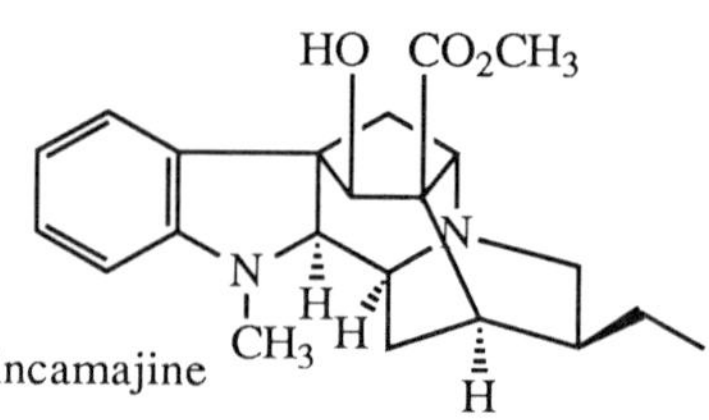

186 vincamajine

Figure 14.

the standard drugs tested. Wright *et al.* also assessed the toxicity of villalstonine **129** against KB cells (human epidermoid cancer of the mouth) using a microdilution method similar to the antiamoebic test used above. The cytotoxic activity of villalstonine **129** against KB cells [ED_{50}(95% C.I.)=11.6(10.2-13.0) μM] was found to be similar to its antiamoebic activity. This similarity suggests there is no selective toxicity for amoebae in this series. However, the standard antiamoebic drug emetine **187** is highly toxic to KB cells [ED_{50}=0.673μM(SEM=0.20)] but is three times less toxic to amoebae than to KB cells. Therefore, villalstonine **129** appears to have a more favorable antiamoebic/cytotoxic ratio as compared with emetine **187** (165). When the structures of the alkaloids are compared to their activity some interesting hypotheses arise. Macralstonine **79** is comprised of a unit of macroline **2** and a unit of alstophylline **78**, yet macralstonine **79** is more active than alstophylline **78** or alstonerine **65**. Villalstonine **129** and pleiocarpamine **130** also exhibit a similar relationship. Since the monomeric alstonerine **65** and pleiocarpamine **130** elicit practically no antiprotozoal activity, at least part of both ring systems of the bisindoles appears to be necessary for activity (165). With the recent completion of the synthesis (123) of macroline **2** and a stable equivalent of macroline **77**, additional tests for antiprotozoal activity can be performed. Studies on the alkaloids of *Alstonia angustifolia* may lead to more selective antiprotozoal agents in the future.

Table 4.

Alkaloid	*E. histolytica* ED_{50}(95% C.I.) μM	*P. falciparum* ED_{50}(95% C.I.) μM
alstonerine **65**	75.3(65.0-85.6)	46.3(27.7-77.3)
alstophylline **78**	67.7(57.2-78.2)	82.5(65.9-102)
macralstonine **79**	inactive at 70	inactive at 178
macralstonine acetate **182**	15.51(14.78-16.24)	3.43(1.86-6.34)
villalstonine **129**	11.8(11.7-12.0)	2.92(1.11-3.14)
pleiocarpamine **130**	47.4(46.8-52.9)	20.5(12.6-33.17)
macrocarpamine **183**	8.12(7.76-8.48)	9.36(7.20-12.1)
11-methoxyakuammicine **184**	70.5(65.3-75.6)	41.3(26.5-64.3)
norfluorocurarine **185**	84.1(82.6-89.6)	129(70.2-239)
vincamajine **186**	inactive at 70	138(79.3-238)

Manalo *et al.* (167) reported that the hypotensive activity of *Alstonia muelleriana* Domin. crude extract affected a gradual fall in the mean arterial pressure reaching a maximum of 32% in one hour with no significant changes in respiration, intestinal mobility, or the electrocardiogram. The activity of this extract of *A. muelleriana* promoted interest in the activity of its constituents (163). Macralstonine **79**, isolated from *Alstonia macrophylla* and *A. muelleriana*, was shown to lower blood pressure in dogs (126, 167). Manalo *et al.* (167) reported that macralstonine **79** caused a progressively increasing fall of blood pressure when administered to dogs beginning with higher doses up to 5 mg/kg. In all cases respiration was unaffected. In studies with 5 mg/kg, the fall of blood pressure was prompt and consistent throughout with a drop to 50-60% of control levels. No instance of tachypylaxis was observed. However, until recently (165), the paucity of isolable material from these *Alstonia* species has retarded biological studies on the macroline related alkaloids.

In the sarpagine series normacusine B exhibits sedative and ganglion blocking activity (168). Lochnerine has been reported to exert hypoglycemic activity (169), while affinisine demonstrated weak antibacterial activity and moderate analgesic and CNS depressant properties (170). More recently, normacusine B, pericyclivine, 16-epi-affinine and pleiocarpamine were evaluated in studies on antimicrobial activity against gram-negative bacteria (171). Macusine B, a quaternary sarpagine alkaloid, was also one of 46 alkaloids isolated from various *Strychnos* species that were tested for cytotoxic activity (172). Unfortunately macusine B was not active against the four tumor cell lines tested. Kinghorn *et al.* has recently reported their findings on the evaluation of natural products as inhibitors of human immunodeficiency virus type 1 (HIV-1) reverse transcriptase (173). Among the many alkaloids tested for activity ajmaline, lochnerine, *N*-acetylperivine, perakine, pericyclivine and sarpagine were chosen as representatives of the macroline/sarpagine/ajmaline family. Studies were also begun to evaluate these alkaloids for activity against HIV-2 (174). At present, none of these alkaloids has exhibited significant activity against HIV-1 or -2 (174).

3.2. Ajmaline and Related Compounds

In the introduction of this chapter the most prominent effect of (+)-ajmaline was mentioned. The well known antiarrhythmic activity of (+)-ajmaline was reviewed by Creasey (29), and the following section was taken from that review. "The most prominent action of ajmaline is an antiarrhythmic effect on the heart that is less pronounced than that of propranolol (30), but is superior in terms of the ratio of the refractory phase over reduced conduction to that of procaine amide and quinidine (31). For this reason the drug has been extensively used in Europe to treat arrhythmias. In a study that involved 900 patients with acute or subacute myocardial infarction, ajmaline was found to be useful in the management of both ventricular and supraventricular arrhythmias (32). It should be noted, however, that successful treatment with return to normal sinus rhythm in 85% or more of the subjects required two drugs, electrolyte replacement therapy, and the administration of thiamine. Care is needed when the drug is used in those patients with liver disease. The action of ajmaline involves a dose-dependent reduction in the maximum rate of rise of the muscle action potential, without affecting the resting potential." Studies still continue on (+)-ajmaline in order to develop new treatments for a variety of cardiovascular diseases. However, the antipode (-)-ajmaline has never been isolated or synthesized, the use of which might provide a new adjunct to antiarrhythmic therapy. An important review by Creasey has detailed the ganglionic blocking activity of other ajmaline related alkaloids (29).

4. Summary

During the last several years over seventy indole alkaloids have been isolated from various species of *Alstonia* (36, 118). Many of these bases are related to the sarpagine/ajmaline alkaloids and contain a unit of macroline as a common structural feature. The macroline bases consist of both monomeric and bisindole alkaloids, the majority of which have not fallen to total synthesis. Recently, the formation of 1,3-disubstituted β-carbolines (>98% ee, HPLC) *via* the stereocontrolled Pictet-Spengler reaction with complete transfer of chirality has been developed (82, 86). Furthermore, execution of a Dieckmann reaction gave the required indolo-substituted azabicyclo[3.3.1]nonane intermediate **36a** in optically active form. Synthesis of (-)-**36a** has led to the total synthesis of a variety of macroline/sarpagine alkaloids. The alkaloid (-)-suaveoline **3** was recently synthesized using the above methodology including an oxyanion Cope rearrangement (114, 115). Upon completion of the synthesis of (-)-suaveoline **3** and (-)-N_b-methylsuaveoline **61**, Fu reported the optical rotation of (-)-suaveoline **3**, which had been previously misassigned in the literature (43, 44). The indole alkaloid (-)-alstonerine **65** was synthesized in an enantiospecific fashion from the same tetracyclic ketone (-)-**36a**, a Michael addition and a diastereoselective Claisen rearrangement were important steps in this route. The chemistry developed in this sequence by Zhang can be employed for the enantiospecific synthesis of alstophylline **78**, the nonmacroline portion of macralstonine **79**. With the recent completion of the synthesis of (+)-macroline **2** and the stable macroline equivalent **77**, a route for the total synthesis of the macroline related bisindole alkaloids can be envisaged. The biomimetic coupling reaction of macroline with other indole alkaloid bases to provide bisindoles was reported by LeQuesne, consequently the enantiospecific synthesis of these bisindoles reverts to the synthesis of macroline and the complementary monomeric alkaloid. The multigram scale synthesis of 6-methoxytryptophan reported by Allen *et al.* (139) may provide entry into 6-methoxy substituted indole alkaloids many of which are known to possess biological activity, as well as a route to alstophylline **78** and related *Alstonia* alkaloids. The interest in the total synthesis of bisindoles arises from their potential biological activity and the macroline/sarpagine class of alkaloids is no exception.

The macroline/sarpagine alkaloids are related, structurally, to the ajmaline class of bases. From a synthetic perspective this relationship opens up many routes toward the preparation of potential Class I antiarrhythmic agents (23, 29) in the ajmaline series, including (+)-ajmaline **11**. The pharmacological data reported on many of the macroline/sarpagine/ajmaline alkaloids in this chapter will hopefully emphasize the importance of a general, enantiospecific synthetic route to this large class of indole alkaloids. In Figure 15 the realization of this strategy is illustrated from the optically pure tetracyclic ketone **36a**. The synthesis of the macroline related alkaloids (-)-alstonerine **65** and macroline **2** have been realized from (-)-**36a**, moreover the ajmaline related alkaloids (-)-suaveoline **3**, (-)-raumacline **27** and N_b-methylraumacline **28** have also been prepared in greater than 98% ee. In addition, a partial synthesis of the bisindole villalstonine **129** has recently been completed from synthetic macroline (123) and natural pleiocarpamine. The synthesis of other members of these bisindoles are underway and will be reported in due course.

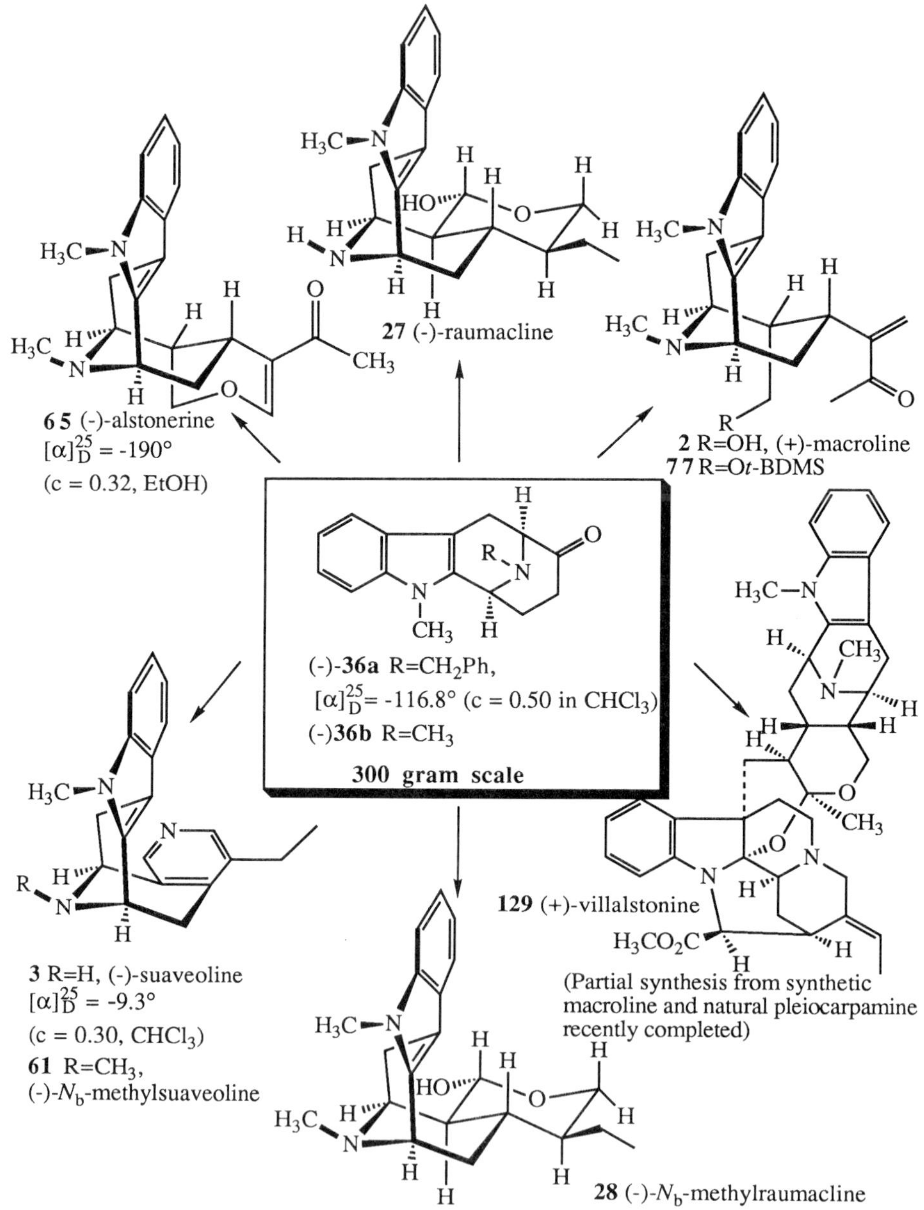

Figure 15.

Acknowledgement. We wish to express our sincere gratitude to all the coworkers whose names appear in the references and in this chapter. Their diligence and enthusiasm are responsible for the successes detailed in this endeavor.

References

1. The Alkaloids. New York: Academic Press, 1950-1968(RHF Manske, ed. vol I-XI).
2. KW Bentley. The Alkaloids. In: The Chemistry of Natural Products. New York: John Wiley and Sons, 1957.
3. SW Pelletier. Chemistry of the Alkaloids. New York: Van Nostrand Reinhold Book Corporation, 1970(SW Pelletier, ed.).
4. RF Raffauf. A Handbook of Alkaloids and Alkaloid-Containing Plants. New York: Wiley-Interscience, 1970.
5. Specialist Periodical Reports. The Alkaloids. London: The Royal Society of Chemistry, 1971-1982.
6. JS Glasby. Encyclopedia of the Alkaloids. New York: Plenum Press, 1975.
7. KC Rice, JM Farah, NG Grayson. U.S. Patent Application filed August 24 1989.
8. KC Rice, JM Farah, NG Grayson. U.S. Patent Application filed August 20 1990.
9. AE Jacobson. In: LS Harris, ed. Problems of Drug Dependence 1989. Washington D.C.: NIDA Research Monograph 95, Published 1990: 556-577.
10. CB Pert, JA Danks, MA Channing, WC Eckelman, SM Larson, JM Bennett Jr., TR Burke, KC Rice. FEBS Lett. 1984;177:281-286.
11. AH Newman, N Ostrowski, MA Channing, KC Rice. Abs. Soc. Neurosci. 1989;15:703.
12. R Kawai, Y Sawada, M Channing, AH Newman, B Dunn, KC Rice, RG Blasberg. J. Pharm. Exp. Ther. 1990;255:826-835.
13. JM Cook, PW LeQuesne. Phytochemistry 1971;10:437.
14. A Chatterjee, J Banerji, A Banerji. J. Indian Chem. Soc. 1949;51:156.
15. ML Chatterjee, MS De. Bull. Cal. S.T.M. 1962;10(1):15.
16. C Mela. Fitoterapia 1967;38(1):2-15.
17. S Ozima, T Watanabe, M Kato, Y Ishii, T Narusawa. Sogo Rinsho 1971;20(3):537.
18. GN Novikova, IA Gotsiridzh, GV Abuladze. Akademii nauk Gruzinsko Seriia Biologichesksia 1984;10:54.
19. L Szatmáry, E Czakó, F Rényi-Vámos, E Bodor, G Rablóczky. Acta Medica Hungarica 1986;43(3):321-331.
20. K Okumura, Y Hashimoto, M Yasuhara, R Hori. Br. J. Pharmacol. 1988;93:827-832.
21. C Köppel, J Tenczer, I Arndt. Eur. J. Drug Metab. Pharmacokinet. 1989;14(4):309-316.
22. C Köppel, U Oberdisse, G Heinemeyer. Clin. Toxicol. 1990;28(4):433-444.
23. M Ito, S Onodera, H Noshiro, H Odakura, S Yasuda, M Nagashima, S Shinoda, H Suzuki. J. Electrocardiol. 1990;23(4):323-329.

24. J Thorman, J Hüting, P Kremer, J Wissemann, H Bahawar, M Schlepper. J. Cadiovasc. Pharmacol. 1990;16:182-190.

25. J Thormann, J Hüting, P Kremer, V Mitrovic, J Wissemann, H Bahawar, M Schlepper. Z. Kardiol. 1990;79:706-716.

26. X Chen, M Borggrefe, C Hief, W Haverkamp, A Martinez-Rubio, G Breithardt. Eur. Heart J. 1991;12:177.

27. R Mletzko, W Jung, M Manz, B Lüderitz. Z. Kardiol. 1991;80:459-462.

28. R Padrini, L Compostella, D Piovan, A Javarnaro, F Cucchini, M Ferrari. Clin. Pharmacokinet. 1991;21:150-154.

29. WA Creasey. The Monoterpenoid Indole Alkaloids. New York: John Wiley and Sons, 1983:783-829. (JE Saxton, ed. Heterocyclic Compounds; Indole Series; vol 25).

30. HJ Mest, J Winkler, W Foerster. Acta Biol. Med. Ger. 1977;36(7-8):1193-1196.

31. HF Benthe. Naunyn Schmiedegergs Arch. Exptl. Pathol. Pharmakol. 1956;229:82-91.

32. VS Graslin, AI Romanov, II Bykov, VI Palii. Kardiologiya 1977;17:5.

33. A Pfitzner, J Stöckigt. Tetrahedron Lett. 1983;24(47):5197-5200.

34. WI Taylor. The Alkaloids. New York: Academic Press, 1968:Chapter 2. (RHF Manske, ed. vol 11).

35. JE Saxton. The Alkaloids. New York: Academic Press, 1970:Chapter 3. (RHF Manske, ed. vol 12).

36. Y Bi, LK Hamaker, JM Cook. Studies in Natural Products Chemistry. Amsterdam: Elsevier Science, 1993:383-432. (FZ Basha, A Rahman, ed. Bioactive Natural Products, Part A; vol 13).

37. RL Garnick, PW LeQuesne. J. Am. Chem. Soc. 1978;100:4213.

38. M Hesse, H Hürzeler, CW Gemenden, BS Joshi, WI Taylor, H Schmid. Helv. Chim. Acta 1965; 48:689.

39. M Hesse, F Bodmer, CW Gemenden, BS Joshi, WI Taylor, H Schmid. Helv. Chim. Acta 1966;49:1173.

40. DE Burke, JM Cook, PW LeQuesne. J. Am. Chem. Soc. 1973;95:546.

41. DE Burke, CA DeMarkey, PW LeQuesne, JM Cook. J. Chem. Soc., Chem. Commun. 1972:1346.

42. RL Garnick. [Ph.D.]. Northeastern University, 1977.

43. SP Majumdar, P Potier, J Poisson. Tetrahedron Lett. 1972:1563.

44. SP Majumdar, J Poisson, P Potier. Phytochemistry 1973;12:1167.

45. BA Akinloye, WE Court. J. Ethnopharmacol. 1981;4:99.

46. MA Amer, WE Court. Phytochemistry 1981;20:2569.

47. AMAG Nasser, WE Court. Phytochemistry 1983;22:2297.

48. AMAG Nasser, WE Court. J. Ethnopharmacol. 1984;11:99.

49. RB Woodward. Angew. Chem. 1956;68:13.

50. MF Bartlett, BF Lambert, HM Werblood, WI Taylor. J. Am. Chem. Soc. 1963;85:475-477.

51. S Masamune, SK Ang, C Egli, N Nakatsuka, SK Sarkar, Y Yasunari. J. Am. Chem. Soc. 1967;89(10):2506-2507.

52. EE van Tamelen, LK Oliver. Bioorg. Chem. 1976;5:309-326.

53. SA Minina, YV Kostin, EI Gendenshtein, EI Molokhova. Khim. Farm. Zh. 1987;21(5):559-561.

54. SP Hollinshead, DS Grubisha, DW Bennett, JM Cook. Heterocycles 1989;29:529.

55. N Aimi, E Yamanaka, J Endo, S Sakai, J Haginiwa. Tetrahedron 1973;29:2015-2021.

56. H Takayama, M Kitajima, K Ogata, S Sakai. J. Org. Chem. 1992;57:4583-4584.

57. CE Nordman, K Nakatsu. J. Am Chem. Soc. 1963;85:353.

58. A Rahman, WSJ Silva, KA Alvi, KTD DeSilva. Phytochemistry 1987;26:865.

59. A Rahman, MM Qureshi, A Muzaffar, KTD DeSilva. Heterocycles 1988;27(3):725-732.

60. H Takayama, C Phisalaphong, M Kitajima, N Aimi, S Sakai. Tetrahedron 1991;47(8):1383.

61. J Naranjo, M Pinar, M Hesse, H Schmid. Helv. Chim. Acta 1972;55:752.

62. ZM Khan, M Hesse, H Schmid. Helv. Chim. Acta 1967;50:1002.

63. T Kishi, M Hesse, CW Gemenden, WI Taylor, H Schmid. Helv. Chim. Acta 1965;48:1349.

64. JA Martinez, H Velez, T Santana. Phytochemistry 1989;28(3):961-962.

65. J Garnier, J Mahuteau. Planta Med. 1986:66-67.

66. JA Martinez, C Gomez, T Santana, H Velez. Planta Med. 1989;55:283-285.

67. TM Pinchon, JM Nuzillard, B Richard, G Massiot, L Le Men-Oliver, T Sevenet. Phytochemistry 1990;29:3341-3344.

68. P Clivio, B Richard, JR Deverre, T Sevenet, M Zeches, L Le Men Oliver. Phytochemistry 1991;30(11):3785-3792.

69. M Quaisuddin. Bangladesh J. Sci. Ind. Res. 1980;15(1-4):35-44.

70. A Rahman, SA Abbas, F Nighat, G Ahmed, MI Choudhary, KA Alvi, H Rehman, KTD DeSilva, LSR Arambewela. J. Nat. Prod. 1991;54(3):750-754.

71. M Lin, D Yu, X Liu, F Fu, Q Zheng, C He, G Bao, C Xu. Yaoxue Xuebao 1985;20(3):198-202.

72. MM Iwu, WE Court. Planta Med. 1982;45:105-111.

73. R Mukherjee, MdFF Melo, CAdM Santos, E Guittet, BC Das. Heterocycles 1990;31:1819.

74. A Morfaux, D Guillaume, G Massiot, L Le Men-Oliver. C.R Acad. Sci. Paris Series II 1989;309:33-36.

75. S Endress, S Suda, H Takayama, M Kitajima, N Aimi, S Sakai, J Stöckigt. Planta Med. 1992;58(1):A618.

76. H Takayama, M Kitajima, S Suda, N Aimi, S Sakai, S Endress, J Stöckigt. Tetrahedron 1992;48(13):2627-2634.

77. A Rahman, F Nighat, A Nelofer, K Zaman, MI Choudhary, KTD DeSilva. Tetrahedron 1991;47(18/19):3129-3136.

78. R Mukherjee, BS Da Silva, BC Das, PA Keifer, JN Shoolery. Heterocycles 1991;32(5):985-990.

79. N Yoneda. Chem. Pharm. Bull. 1965;13(10):1231.

80. D Soerens, J Sandrin, F Ungemach, P Mokry, GS Wu, E Yamanaka, L Hutchins, M DiPierro, JM Cook. J. Org. Chem. 1979;44:535.

81. ML Trudell. [Ph.D.]. University of Wisconsin-Milwaukee, 1989.

82. LH Zhang, JM Cook. Heterocycles 1988;27:1357, 2795.

83. LH Zhang. [Ph.D.]. University of Wisconsin-Milwaukee, 1990.

84. D Soerens. [Ph.D.]. University of Wisconsin-Milwaukee, 1978.

85. J Campbell. Aldrichimica Acta 1972;5(2):29.

86. LH Zhang, Y Bi, F Yu, G Menzia, JM Cook. Heterocycles 1992;34(3):517.

87. A Pictet, T Spengler. Ber. 1911;44:2030.

88. W Whaley, T Govindachari. Organic Reactions. New York: John Wiley and Sons, 1951:Chapt. 3. (R Adams, ed. vol VI).
89. R Abramovitch, I Spenser. Advances in Heterocyclic Chemistry. New York: Academic Press, 1964:79. (vol 3).
90. R Sundberg. The Chemistry of Indoles. New York: Academic, 1970:236.
91. F Ungemach, JM Cook. Heterocycles 1978;9:1089.
92. M Jawdosiuk, JM Cook. J. Org. Chem. 1984;49:2699.
93. J Sandrin, SP Hollinshead, JM Cook. J. Org. Chem. 1989;54:5636-5640.
94. K Narayanan, JM Cook. J. Org. Chem. 1991;56:5733-5736.
95. M Shimizu, M Ishikawa, Y Komoda, T Nakajima, K Yamaguchi, N Yoneda. Chem. Pharm. Bull. 1984;32(2):463.
96. M Shimizu, M Ishikawa, Y Komoda, T Nakajima, K Yamaguchi, S Sakai. Chem. Pharm. Bull. 1984;32(4):1313-1325.
97. F Ungemach, M DiPierro, R Weber, JM Cook. J. Org. Chem. 1981;46:164.
98. R Plate, RHM van Hout, H Behm, HCJ Ottenheijm. J. Org. Chem. 1987;52:555.
99. KM Czerwinski, L Deng, JM Cook. Tetrahedron Lett. 1992;33(33):4721-4724.
100. P Magnus, B Mugrage, M DeLuca, GA Cain. J. Am. Chem. Soc. 1989;111:786.
101. P Magnus, B Mugrage, MR DeLuca, GA Cain. J. Am. Chem. Soc. 1990;112:5220.
102. P Magnus. private communication.
103. PD Bailey, NR McLay. J. Chem. Soc., Perkin Trans. 1 1993:441.
104. LH Zhang, JM Cook. J. Am. Chem. Soc. 1990;112:4088.
105. SP Hollinshead. [Ph.D.]. University of York, 1987.
106. PD Bailey, SP Hollinshead, NR McLay, K Morgan, SJ Palmer, SN Prince, CD Reynolds, SD Wood. J. Chem. Soc., Perkin Trans. 1 1993:431.
107. G Tóth, C Szántay Jr., G Kalaus, G Thaler, G Snatzke. J. Chem. Soc., Perkin Trans. II 1989:1849.
108. J Sandrin, D Soerens, JM Cook. Heterocycles 1976;4(7):1249.
109. F Ungemach, D Soerens, R Weber, M DiPierro, O Campos, P Mokry, JM Cook. J. Am. Chem. Soc. 1980;102:6976.
110. PD Bailey, NR McLay. Tetrahedron Lett. 1991;32(31):3895-3898.
111. IS Cloudsale, AF Kluge, NL McClure. J. Org. Chem. 1982;47:919.
112. JD Hobson, J Raines, RJ Whiteoak. J. Chem. Soc. 1963:3495.
113. ML Trudell, JM Cook. J. Am. Chem. Soc. 1989;111:7504.
114. X Fu, JM Cook. J. Am. Chem. Soc. 1992;114:6910-6912.
115. X Fu, JM Cook. J. Org. Chem. 1993;58:661-672.
116. ML Trudell, D Soerens, RW Weber, L Hutchins, D Grubisha, D Bennett, JM Cook. Tetrahedron 1992;48:1805.
117. RW Weber. [Ph.D.]. University of Wisconsin-Milwaukee, 1984.
118. JL Ingham, A Koskinen, M Lounasmaa. Progress in the Chemistry of Organic Natural Products. New York: Springer-Verlag, 1983: 268-346.

119. RE Gilman. [Ph.D.]. University of Michigan, 1959.

120. RC Elderfield, RE Gilman. Phytochemistry 1972;11:339.

121. JM Cook, PW LeQuesne, RC Elderfield. J. Chem. Soc., Chem. Commun. 1969:1306.

122. LH Zhang, ML Trudell, SP Hollinshead, JM Cook. J. Am. Chem. Soc. 1989;111:8263.

123. Y Bi, JM Cook. Tetrahedron Lett. 1993;34(28):4501-4504.

124. A Stoll, A Hofmann. Helv. Chim. Acta 1953;36:1143.

125. CK Ratnayake, LSR Arambewela, KTD DeSilva, A Rahman, KA Alvi. Phytochemistry 1987;26(3):868.

126. SK Talapatra, N Adityachaudhury. Science and Culture 1958;24:243.

127. RC Elderfield. Scientist 1960;48:193.

128. HA Lazar. [Ph.D.]. University of Michigan, 1972.

129. DE Burke, GA Cook, JM Cook, K.G. Haller, HA Lazar, PW LeQuesne. Phytochemistry 1973;12:1467.

130. LK Hamaker, JM Cook. unpublished results.

131. A Brossi, M Suffness. The Alkaloids. New York: Academic Press, 1990 (vol 37).

132. RB Woodward, FE Bader, H Bickel, AJ Frey, RW Kierstead. J. Am. Chem. Soc. 1956;78:2023, 2657.

133. PA Wender, JM Schaus, AW White. J. Am. Chem. Soc. 1980;102:6157.

134. SF Martin, H Rüeger, SA Williamson, S Grzejszczak. J. Am. Chem. Soc. 1987;109:6124.

135. G Stork. Pure Appl. Chem. 1989;61:439.

136. G Mukhopadhyay, B Mukherjee, A Patra, A Chatterjee, R Ghosh, P Roychowdhury, H Kawazura. Phytochemistry 1991;30(7):2447.

137. TM Sharp. J. Chem. Soc. 1934:1227.

138. T Kishi, M Hesse, W Vetter, CW Gemenden, WI Taylor, H Schmid. Helv. Chim. Acta 1966;49:946.

139. MS Allen, LK Hamaker, AJ LaLoggia, JM Cook. Syn. Comm. 1992;22(14):2077.

140. U Schöllkopf, R Lonsky, P Lehr. Liebigs Ann. Chem. 1985:413.

141. MF Bartlett, R Sklar, WI Taylor, E Schlittler, RLS Amai, P Beak, NV Bringi, E Wenkert. J. Am. Chem. Soc. 1962;84:622.

142. M Lounasmaa, A Nemes. Tetrahedron 1982;38(2):223-243.

143. GA Cordell. The Monoterpenoid Indole Alkaloids. New York: John Wiley and Sons, 1983:539-728. (JE Saxton, ed. Heterocyclic Compounds; Indole Series; vol 25).

144. JP Kutney. Nat. Prod. Rep. 1990:85-103.

145. Y Bi, JM Cook. unpublished results .

146. EE Waldner, M Hesse, WI Taylor, H Schmid. Helv. Chim. Acta 1967;50:1926.

147. JM Nuzillard, TM Pinchon, C Caron, G Massiot, L Le Men-Oliver. C.R. Acad. Sci. Paris, Serie II 1989;309:195.

148. G Massiot, JM Nuzillard, B Richard, L Le Men-Oliver. Tetrahedron Lett. 1990;31(20):2883.

149. JA Joule. The Monoterpenoid Indole Alkaloids. New York: John Wiley & Sons, 1983: 201-264. (JE Saxton, ed. Heterocyclic Compounds; Indole Series; vol 25).

150. FAL Anet, D Chakravarti, R Robinson, E Schlittler. J. Chem. Soc. 1954:1242.

151. K Mashimo, Y Sato. Tetrahedron Lett. 1969:901.

152. K Mashimo, Y Sato. Tetrahedron 1970;26:803.

153. K Mashimo, Y Sato. Tetrahedron Lett. 1969:905.

154. EE van Tamelen, LK Oliver. J. Am. Chem. Soc. 1970;92:2136.

155. JD Hobson, JG McCluskey. J. Chem. Soc. 1967:2015.

156. L Polz, W Fahn, H Schübel, J Stöckigt. The Enzymatic Formation of *Rauwolfia* Alkaloids in Plant Cell Cultures. 5th DECHEMA Annual Meeting of Biotechnologists. Frankfurt, Germany: VCH publishers, 1987: 321-326.

157. R Robinson. Angew. Chem. 1957;69:40.

158. R Prewo, JJ Stezowski. Acta Cryst. 1978;B34:454.

159. Available from Sigma Chemical Co. (1993) .

160. L Polz, J Stöckigt, H Takayama, N Uchida, N Aimi, S Sakai. Tetrahedron Lett. 1990;31(46):6693.

161. S Endress, S Suda, H Takayama, N Aimi, S Sakai, J Stöckigt. Planta Med. 1992;58:410.

162. AC Peterson. private communication.

163. TA Henry. The Plant Alkaloids. (4th ed.) London: J. and A. Churchill LTD., 1949.

164. IH Burkhill. A Dictionary of Economic Products of the Malay Peninsular. London: Crown Agents for the Colonies, 1935:113.

165. CW Wright, D Allen, Y Cai, JD Phillipson, IM Said, GC Kirby, DC Warhurst. Phytother. Res. 1992;6:121-124.

166. NR Farnsworth, RN Blomster, JP Buckley. J. Pharm. Sci. 1967;56:23.

167. N Isidro, GD Manalo. J. Phillipine Pharm. Assoc. 1967; 53: 8.

168. M Hava. The *Vinca* Alkaloids. New York: Marcel Dekker, Inc., 1973:305-338. (WI Taylor, NR Farnsworth, ed.) The Pharmacology of *Vinca* species and Their Alkaloids.

169. H Svoboda, DA Blake. New York: Marcel Dekker, Inc., 1974:45-83. (WI Taylor, NR Farnsworth, ed.) The Phytochemistry and Pharmacology of *Catharanthus roseus*, L.G. Don.

170. IW Southon, J Buckingham. Dictionary of Alkaloids. New York: Chapman and Hall, 1989(IW Southon, J Buckingham, ed.).

171. TA van Beek, R Verpoorte, A Baerheim Svendsen. J. Nat. Prod. 1985;48(3):400-423.

172. J Leclercq, M-C de Pauw-Gillet, R Bassleer, L Angenot. J. Ethnopharm. 1986;15:305-316.

173. GT Tan, JM Pezzuto, AD Kinghorn, LSH Hughes. J. Nat. Prod. 1991;54(1):143-154.

174. GT Tan, JF Miller, AD Kinghorn, SH Hughes, JM Pezzuto. Biochem. Biophys. Res. Comm. 1992;185(1):370-378.

Chapter Three

Erythrina Alkaloids

Amrik Singh Chawla and Vijay K. Kapoor
Department of Pharmaceutical Sciences,
Panjab University,
Chandigarh 160 014, India

CONTENTS

This chapter is dedicated to the memory of the late Professor A. H. Jackson.

1. INTRODUCTION

The *Erythrina* genus comprises one hundred and eight species of orange or red-flowered trees, shrubs and herbaceous plants [1]. They are found throughout the tropical and semi-tropical regions of the world [1,2]. The species are closely related to one another, and can be intercrossed to produce hybrids that have more than 50 per cent fertile pollen. Several of the more common species are used for decorative purposes in gardens and city streets. The genus *Erythrina* (Fabaceae) is a part of the Papilionaceae sub-family of the Leguminosae, and the botanical relationships between the various species within the genus have been classified by the late B.A. Krukoff [3] for whom they were a life-long study. A collection of over seventy species of *Erythrina* has been established at the Pacific Tropical Botanical Garden and the Waimea Arboretum and Botanical Garden in Hawaii.

The genus has been divided into five sub-genera and twenty-six sections, largely on the basis of morphology, especially the colour and shapes of the flowers and fruits, and of fluorescence characteristics. There are twenty-eight species and three sub-species in Mexico, twenty-six species in Central America, twenty-two species in South America, five species and two varieties in the West Indies, thirty species and sub-species in tropical Africa, six species in South Africa, seven species in continental Asia and seven species in Malaysia and the Pacific [3]. The external and internal topographic features of about 3500 seeds from 573 samples, representing 101 species have been studied [4]. Chromosomal examinations have revealed a remarkable uniformity in number, size and ploidy level. Based on $x = 21$, most of the species are diploid with two tetraploid and one hexaploid species; the chromosomes are very small ranging from 1-2 μ [5].

The chemical structure of the alkaloids found in *Erythrina* species is based on a tetracyclic spiroamine system, erythrinane (**1**), which is almost unique to the

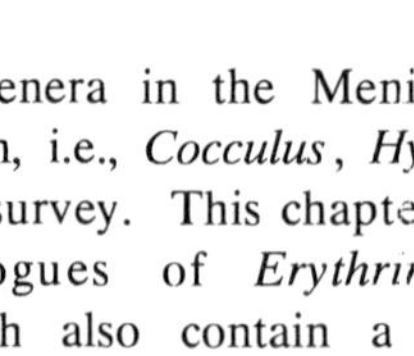

1

genus. However, three other genera in the Menispermaceae family also contain alkaloids with the same skeleton, i.e., *Cocculus*, *Hyperbaena* and *Pachygone* ; these have also been included in this survey. This chapter covers the literature published till February, 1992. Homologues of *Erythrina* alkaloids, the so-called "homoerythrina" alkaloids, which also contain a spiroamine skeleton, occur in *Cephalotaxus*, *Dysoxylum* , *Phelline* and *Schelhammera* species; these are not included in this review.

Early work on the chemistry of *Erythrina* alkaloids in the 1930's and 1940's was stimulated by the discovery of the curare-like action of extracts of the seeds of various species [6-9]. In these early studies, the seeds of about fifty species were found to contain alkaloids with muscle paralysing activity [10-12]. Moreover, the *Erythrina* alkaloids were tertiary bases, whereas other alkaloids with curarising activity were all quaternary salts. In recent years, the field has expanded dramatically and around one hundred alkaloids have been identified (Tables 1-5); the structures of most of these have been established by a combination of spectroscopic methods and chemical correlations with alkaloids of known structures.

Like other alkaloids containing the isoquinoline ring system, the *Erythrina* alkaloids are biosynthesized from tyrosine, and it has been established that (+) - *S*-norprotosinomenine is a specific precursor. The chemistry of the alkaloids has been extensively reviewed [13-18] several times since the first survey in 1952. A series of annual or biennial reviews have also been published by the Chemical Society and Royal Society of Chemistry [19-23]. In the present chapter, we shall briefly review the earlier structural work leading to the elucidation of the basic skeleton of the *Erythrina* alkaloids and extensive coverage of the more recent, largely spectroscopic studies, as well as their syntheses, biosynthesis and pharmacology.

2. OCCURRENCE AND ISOLATION

Nearly 100 alkaloids have now been characterised from *Erythrina* species, the seeds of the plants usually being the most prolific source (ranging from 0.1-1% dry weight); the alkaloids are also present in the root, stem, bark, leaves and flowers of many species. They occur both in the free base form and also in 'combined' form as glycosides or as esters of thioacetic acid. In the early work on these alkaloids, the prefix 'erythr-' was used in the naming of free bases, e.g. erythroidine, erythraline, etc., and prefix 'erysothio-' to name those containing the thioacetic acid residue, e.g. erysothiovine and erysothiopine. Mild acid hydrolysis of the combined alkaloids followed by basification affords the so-called 'liberated' alkaloids which again were originally designated by the specific prefix 'eryso-', e.g. erysodine, erysovine etc.; the linkage to the thioacetic acid residue, or the glucose unit is *via* a phenolic hydroxyl group.

The *Erythrina* alkaloids are usually classified into three main groups: dienoid (Table 1), alkenoid (Table 2) and lactonic (Table 3).

There are also several structural variants of the dienoid and alkenoid types, e.g. quaternary salts, *N*- oxides and dehydrogenated derivatives (Table 4). Recently, a new type of *Erythrina* alkaloid has been characterised which is structurally similar to the dienoid series but contains a pyridine ring instead of a benzene ring [24]. In addition, a number of other alkaloids, mainly tetrahydroisoquinolines, which do not contain the erythrinan skeleton,have also been isolated from various *Erythrina* species (Table 5). The simple indole alkaloid, hypaphorine, is also present in almost every species studied.

Table 1. Dienoid Alkaloids

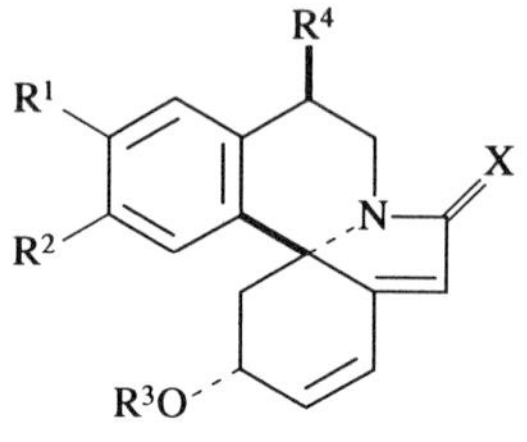

	R^1	R^2	R^3	R^4	X
2. Erysodine	OH	OCH_3	CH_3	H	H_2
3. Erysovine	OCH_3	OH	CH_3	H	H_2
4. Erysopine	OH	OH	CH_3	H	H_2
5. Erysoline	OCH_3	OH	H	H	H_2
6. Erysonine	OH	OCH_3	H	H	H_2
7. Erysotrine	OCH_3	OCH_3	CH_3	H	H_2
8. Erysotramidine	OCH_3	OCH_3	CH_3	H	O
9. Erythravine	OCH_3	OCH_3	H	H	H_2
10. Erythraline	$-OCH_2O-$		CH_3	H	H_2
11. Erythrinine	$-OCH_2O-$		CH_3	OH	H_2
12. Erythrocarine	$-OCH_2O-$		H	H	H_2
13. Erythrascine	OCH_3	OCH_3	CH_3	OAc	H_2
14. Erythrartine	OCH_3	OCH_3	CH_3	OH	H_2
15. Erythristemine	OCH_3	OCH_3	CH_3	OCH_3	H_2
16. 11-Hydroxyerysodine	OH	OCH_3	CH_3	OH	H_2
17. 11-Hydroxyerysovine	OCH_3	OH	CH_3	OH	H_2
18. 11-Methoxyerysodine	OH	OCH_3	CH_3	OCH_3	H_2
19. 11-Methoxyerysovine	OCH_3	OH	CH_3	OCH_3	H_2
20. 11-Oxoerysodine	OH	OCH_3	CH_3	O	H_2
21. 11-Oxoerysovine	OCH_3	OH	CH_3	O	H_2
22. 11-Oxoerysopine	OH	OH	CH_3	O	H_2
23. 11-Methoxyerysopine	OH	OH	CH_3	OCH_3	H_2
24. 11-Oxoerythraline	$-OCH_2O-$		CH_3	O	H_2
25. 11-Methoxyerythraline	$-OCH_2O-$		CH_3	OCH_3	H_2
26. 8-Oxoerythraline	$-OCH_2O-$		CH_3	H	O
27. 8-Oxoerythrinine	$-OCH_2O-$		CH_3	OH	O
28. Glucoerysodine	a	OCH_3	CH_3	H	H_2
29. 11-Methoxygluco-erysodine	a	OCH_3	CH_3	OCH_3	H_2
30. 11-Methoxygluco-erysovine	OCH_3	a	CH_3	OCH_3	H_2

Table 1. Continued

	R^1	R^2	R^3	R^4	X
31.Rhamnoerysodine	b	OCH_3	CH_3	H	H_2
32.Erysothiovine	OCH_3	c	CH_3	H	H_2
33.Erysothiopine	OH	c	CH_3	H	H_2
34.Erysophorine	OCH_3	d	CH_3	H	H_2
35.Erysodinophorine	d	OCH_3	CH_3	H	H_2
36.Erysopinophorine	d	OH	CH_3	H	H_2
37.Iso-erysopino-phorine	OH	d	CH_3	H	H_2
38.Coccuvinine	H	OCH_3	CH_3	H	H_2
39.Coccuvine	H	OH	CH_3	H	H_2
40.Coccolinine	H	OH	CH_3	H	O
41.Coccoline	H	OCH_3	CH_3	H	O

a: 1-β-glucosyl; b: 1-β-rhamnosyl; c: $HOOC-CH_2-SO_3$-; d: hypaphorine ester

Table 2. Alkenoid Alkaloids

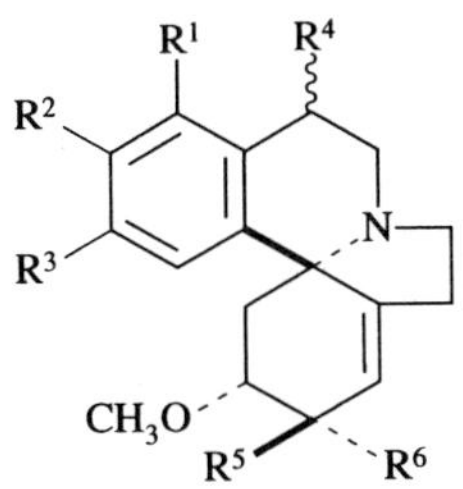

	R^1	R^2	R^3	R^4	R^5	R^6
42.Erythratidine	H	OCH_3	OCH_3	H	H	OH
43.Epi-erythratidine	H	OCH_3	OCH_3	H	OH	H
44.Erythratidinone	H	OCH_3	OCH_3	H	O	
45.Erythramine	H	$-OCH_2O-$		H	H	H
46.Erythratine	H	$-OCH_2O-$		H	OH	H
47.11-Hydroxyerythra-tine	H	$-OCH_2O-$		OH	OH	H

Table 2. Continued

	R^1	R^2	R^3	R^4	R^5	R^6
48. Erythratinone	H	$-OCH_2O-$		H	O	
49 11-Hydroxyerythra-tidine	H	OCH_3	OCH_3	OH	H	OH
50 11-Hydroxy-epi-erythratidine	H	OCH_3	OCH_3	OH	OH	H
51 11-Methoxyerythra-tidine	H	OCH_3	OCH_3	OCH_3	H	OH
52. Erysotine	H	OH	OCH_3	H	OH	H
53 Erysotinone	H	OH	OCH_3	H	O	
54. 11-Hydroxyeryso-tinone	H	OH	OCH_3	OH	O	
55 Erysosalvine	H	OCH_3	OH	H	OH(H)	
56 Erysosalvinone	H	OCH_3	OH	H	O	
57 Erysopitine	H	OH	OH	H	OH(H)	
58 Erysoflorinone	H	OH	OH	H	O	
59 Dihydroerysodine	H	OH	OCH_3	H	H	H
60 Dihydroerysovine	H	OCH_3	OH	H	H	H
61 Dihydroerysotrine	H	OCH_3	OCH_3	H	H	H
62 Coccutrine	OCH_3	H	OH	H	H	H
63 Cocculidine	H	H	OCH_3	H	H	H
64 Cocculine	H	H	OH	H	H	H
66. Cocculitinine	H	H	OH	H	OH	H
66. Cocculitine	H	H	OCH_3	H	OH	H
67. Cocculidinone	H	H	OCH_3	H	O	
68. Erythroculine	H	OCH_3	$COOCH_3$	H	H	
69. Erythlaurine	OH	OCH_3	$COOCH_3$	H	H	H
70. Erythramide	H	OCH_3	$CONH_2$	H	H	H

The early structural studies on *Erythrina* alkaloids were mainly carried out by Folkers and his group during the late 1930's and early 1940's. Subsequently, several other laboratories became involved and, following a series of extensive degradative studies, the structures of the aromatic (dienoid) alkaloids were deduced by Prelog's group in Zürich [25] and of the erythroidines by Boekelheide's group in Rochester, U.S.A.[26,27]. Both the novelty of the chemical structures, and the remarkable curare-like activity of the alkaloids provided the stimulus for the post-war resurgence in chemical studies.

In the early work, the alkaloids were isolated from the plant material by methanol extraction, after most of the lipid material had been removed by extraction

Table 3. Lactonic Alkaloids

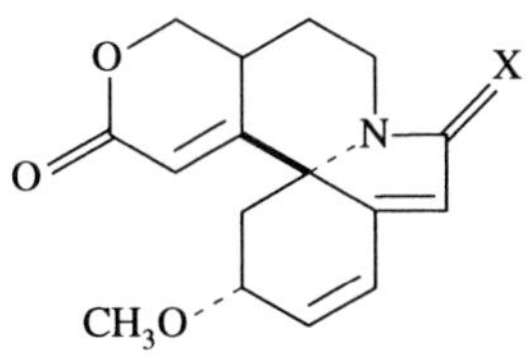

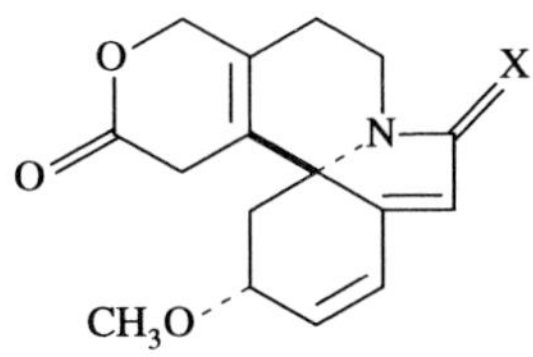

71 X = H_2 α-Erythroidine
72 X = O 8-Oxo-α-erythroidine

73 X = H_2 β-Erythroidine
74 X = O 8-Oxo-β-erythroidine

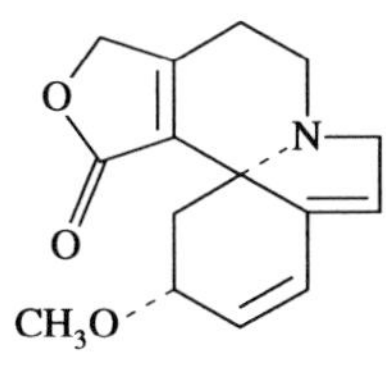

75 Cocculolidine

with light petroleum. The crude product obtained by evaporation of the methanol was then dissolved in dilute hydrochloric acid. Basification of the hydrochloric acid solution with sodium bicarbonate and extraction with chloroform then afforded the crude 'free' alkaloid fraction. The aqueous layer from this extraction was then acidified with hydrochloric acid and heated (60-70°) to hydrolyse the combined alkaloids; basifying and extraction with chloroform yielded the corresponding 'liberated' alkaloids. Fractional crystallisation of the salts was often originally employed to separate the mixtures of the alkaloids obtained, but in more recent years column chromatography and lately semipreparative high performance liquid chromatography (HPLC) have been the predominant methods of isolation and purification.

During the last twenty years or so, several groups have carried out major investigations of the alkaloid content of a variety of *Erythrina* species, notably those of Barton (in England) [28-31], Ito (in Japan) [32-43] and of Singh and Chawla [44-50] and Ghosal (in India) [51- 55]. Systematic screening of the seeds of a wide range of species has been pioneered by Rinehart's group in Illinois [56,57] followed by Jackson's group in Cardiff [58-65] using gas chromatography-mass spectrometry (GC-MS) as the main technique to facilitate chemotaxonomic studies. The extraction procedures utilised in all these studies were essentially the same as those developed by Folkers, and both the 'free' and 'liberated' fractions were investigated in each case; any free hydroxyl groups in the alkaloids were derivatised as their trimethylsilyl ethers prior to GC-MS. In addition Rinehart *et al.*. [56,57] and Singh and Chawla [44] found that a significant amount of the alkaloids were also present in the hexane fraction having been co-extracted with the lipids. In more recent studies in Cardiff the hexane fraction has been regularly extracted with acid to obtain the alkaloids present; these usually correspond largely with those in the 'free'alkaloid fraction although the proportions are rather different; so that although the hexane fraction was not examined in much of the earlier work this is only likely to have affected the quantitative aspect of the work.

Table 4. Alkaloids with Other Structural Modifications

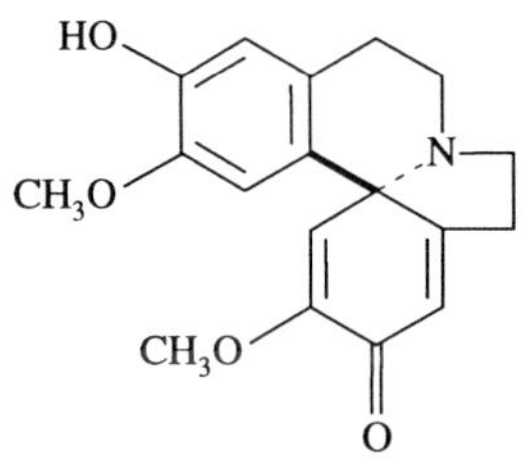

76 Erysodienone

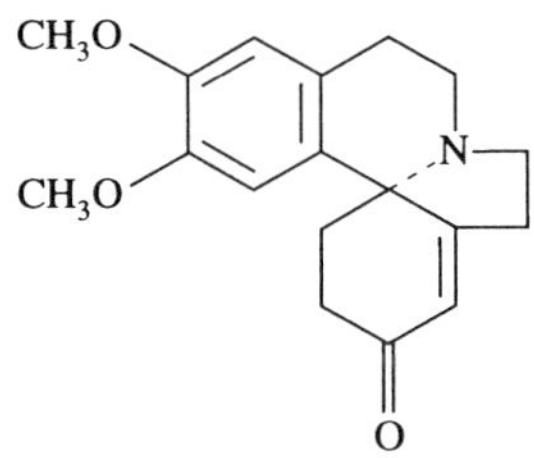

77 3-Demethoxyerythratidinone

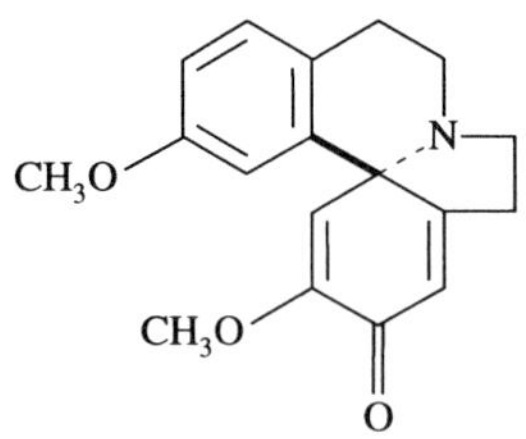

78 Coccudienone

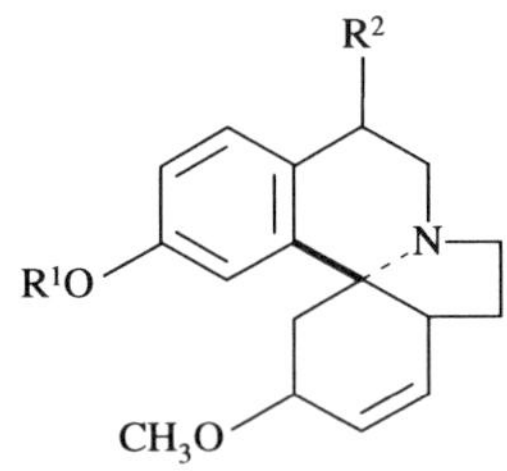

79 R^1=H, R^2=OCH_3 Cocculimine
80 R^1=CH_3, R^2=H Isococculidine
81 R^1=R^2=H Isococculine

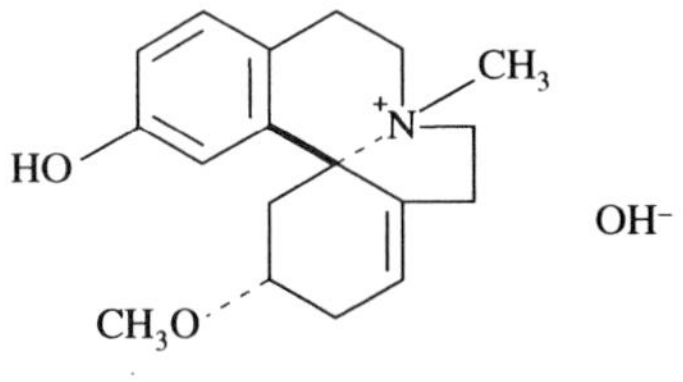

82 Pachygonine

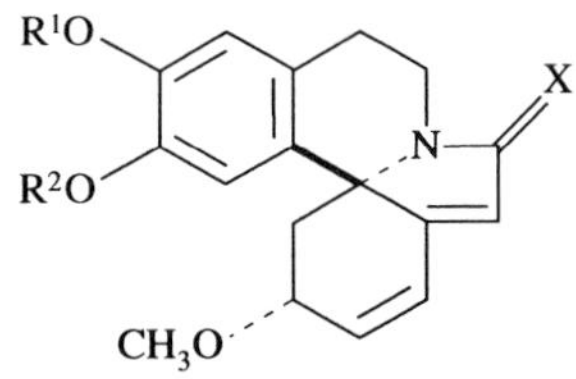

83 $R^1=R^2=CH_3$, X=O
Erytharbine

84 R^1 R^2= —CH_2—, X=O
Crystamidine

85 R^1=H, $R^2=CH_3$, $X=H_2$
10, 11-Dehydroerysodine

86 $R^1=CH_3$, R^2=H, $X=H_2$
10, 11-Dehydroerysovine

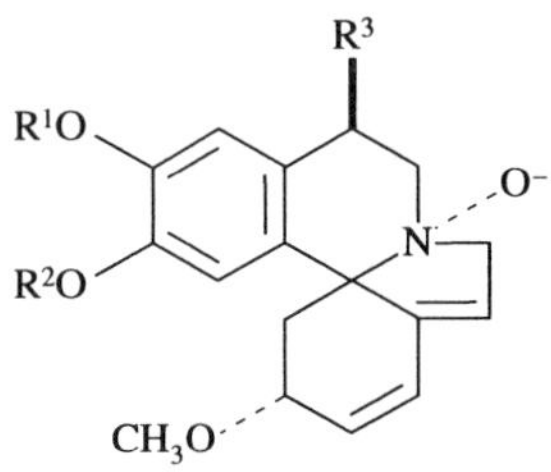

87 $R^1=R^2=CH_3$, R^3=H
Erysotrine *N*-oxide

88 R^1 $R^2=CH_3$, R^3=OH
Erythrartine *N*-oxide

89 R^1 R^2= —CH_2—, $R^3=OCH_3$
11 β-Methoxyerythraline *N*-oxide

90 $R^1=R^2=R^3=OCH_3$
Erythristemine *N*-oxide

The results of these various studies especially those involving systematic screening (e.g. by GC-MS) have shown that individual species are often distinctive in their alkaloid content, although the sub-genera and sections are not clearly marked. Erysodine (**2**), erysovine (**3**) and erysopine (**4**) are the most widely distributed alkaloids, erysodine and erysovine usually also being the most abundant alkaloids in almost every species. A significant difference between most of the American species of *Erythrina* and those occurring in other parts of the world is that the American species generally do not contain any alkaloids oxygenated at the 11-position of the erythrinan nucleus [58]. An exception however, is the occurrence of erythrinine (**11**) in *E.cristagalli* [40]. The lactonic alkaloids α- and β−erythroidines occur in relatively few species, but when they are found they are usually the major components of the alkaloid fraction. The alkenoid alkaloids occur fairly

Table 5. Miscellaneous ***Erythrina*** **Alkaloids**

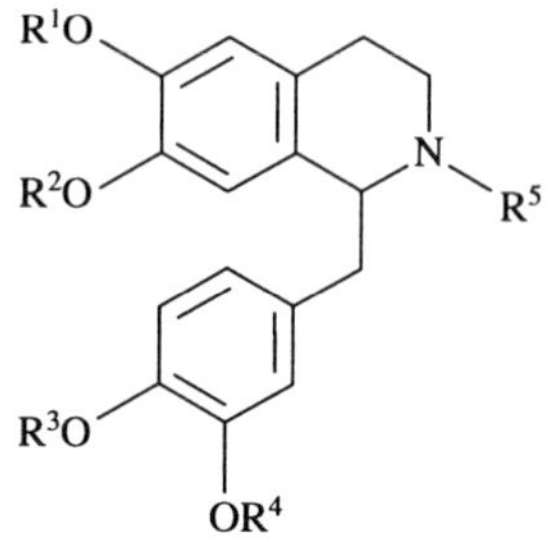

	R^1	R^2	R^3	R^4	R^5
91. *N*-Norprotosino-menine	H	CH_3	CH_3	H	H
92. Protosinomenine	H	CH_3	CH_3	H	CH_3
93. *N*-Nororientaline	CH_3	H	H	CH_3	H
94. Orientaline	CH_3	H	H	CH_3	CH_3

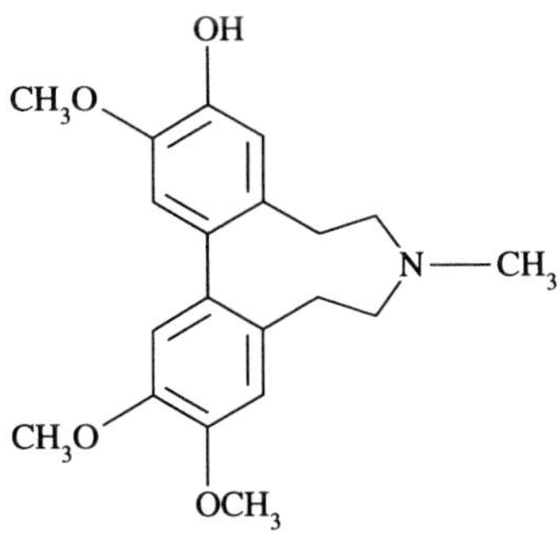

95 Erybidine

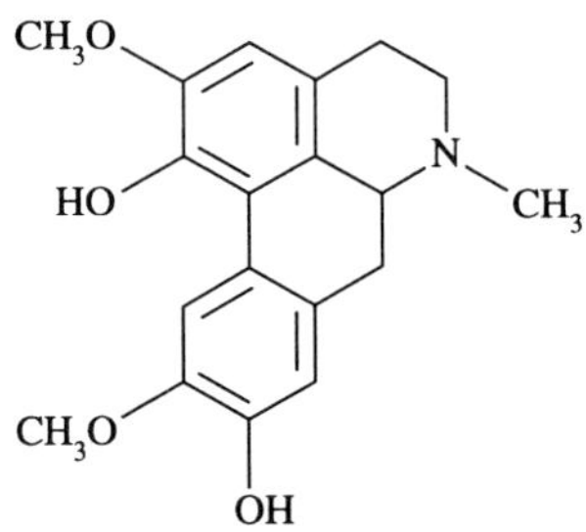

96 Isoboldine

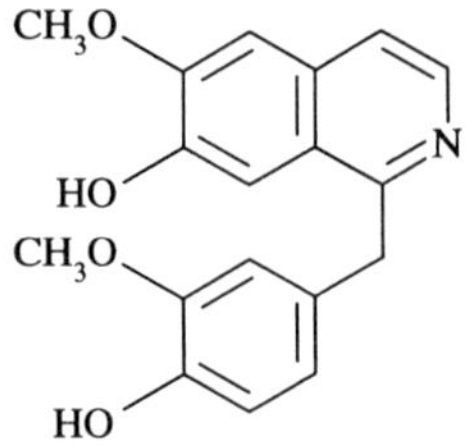

97 Cristadine

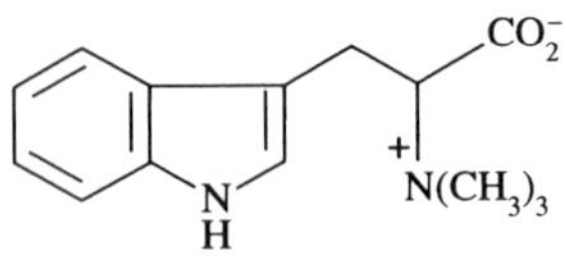

98 Hypaphorine

widely throughout the genus but they are usually much less abundant than the dienoid or lactonic alkaloids. There is also an increasing number of reports of the presence of 8-oxo alkaloids of the dienoid and lactonic types [61,63,66-69].

New quaternary alkaloids (*N*-methylated) have been isolated from the pod walls of *E. arborescens* [70-72] and from the roots of *Pachygone ovata* (Menispermaceae) [73]; the latter is the first report of occurrence of an *Erythrina* alkaloid outside the *Erythrina* and *Cocculus* genera. Another alkaloid, characterised as 3-demethoxy-2α,3α-methylenedioxyerythroculine, has been isolated from *Hyperbaena columbica* (Menispermaceae) [74]. The isolation of *N*-oxides **(87,88)** of erysotrine **(7)** and erythrartine **(14)** from *E. mulungu* [75]; 11β−methoxyerythraline *N*-oxide **(89)** from *E. cristagalli* [66] and *E. suberosa* [67]; and erythristemine *N*-oxide **(90)** and erysotrine *N*-oxide from *E. xbidwillii* [76] have been reported. The occurrence of erythrartine from *E. herbacea* [77] and *E. xbidwillii* [76] has been described. Recently, three new glycodienoid alkaloids,11β-methoxyglucoerysodine **(29)**, 11β-methoxyglucoerysovine **(30)** and rhamnoerysodine **(31)**, have been isolated from *E. lysistemon* [78]. The spectral data for more than 90 *Erythrina* alkaloids originating mainly from *Erythrina* and *Cocculus* plant species have been reported [79].

Several other isoquinoline alkaloids have been found in *Erythrina* species, e.g. isolation of dibenzo[*d,f*]azonine base erybidine (a biosynthetic precursor of *Erythrina* alkaloids) in *E. arborescens* [37], *E. cristagalli* [42,60], *E. herbacea* [80] and *E. xbidwillii* [33] as well as in the leaves of *E. poeppigiana* [61] has been reported. Isoboldine, an aporphine alkaloid has also been detected in *E. abyssinica* [31] and *E. poeppigiana* [61], and cristadine **(97)**, a papaverine type alkaloid from *E. cristagalli* [81].

Whilst the alkaloid profile of various *Erythrina* species is generally characteristic, significant variations have been found in the alkaloid content of samples collected from different places, or at different times from the same place [60]. Differences between the alkaloid content of the seed, bark, leaves etc. of the plant have also been observed [51,53,82,83]. Some striking variations have been reported, e.g. the GC-MS investigation [57] of *E. folkersii* did not reveal the presence of erythraline **(10)**, whereas Folkers and Koniuszy [84] had isolated it from the same plant. Similarly, Ghosal *et al.* [51] reported that erysotrine was the major alkaloid in *E. variegata* bark together with minor amounts of erysodine and erysovine, whereas Singh *et al.* [48] isolated only erysovine. Recently, Chawla et al. [85] characterised erysotine **(52)**, erythratidine **(42)**, epi-erythratidine **(43)** and 11-hydroxy-epi-erythratidine **(50)** in *E. variegata* bark in addition to those previously reported. Two samples of *E. suberosa* seeds collected from different places in India yielded different alkaloids [45].

There is also evidence of chemical variants within species. Letcher [86] inferred that there were two varieties of *E. lysistemon* which yielded either erysotrine or 11-methoxyerythraline, but not both.

3. STRUCTURE DETERMINATION

The structure determination of *Erythrina* alkaloids has been carried out by chemical as well as spectroscopic methods.

3.1 Chemical Methods

The key reactions in the early structural studies were hydrogenation and acid-catalysed degradation. Thus, hydrogenation of dienoid alkaloids **(99)** led to a marked diminution in the UV absorption at 235 nm due to saturation of double bonds, whilst mild acid treatment afforded a strong UV absorption at 313 nm due

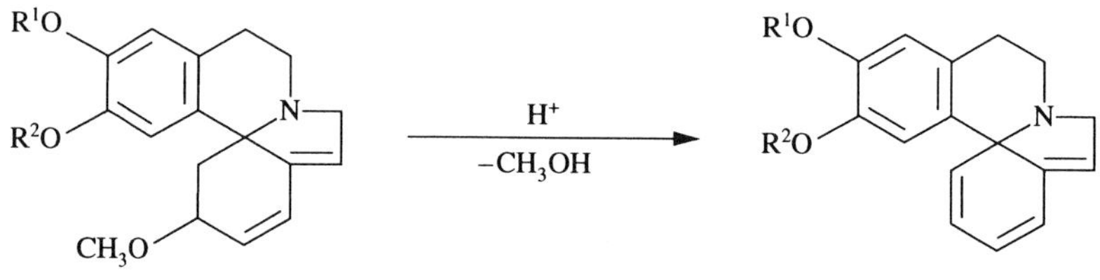

99 **100**

to elimination of methanol and formation of a conjugated triene system **(100)** [27,87].

3.1.1 The Apo Rearrangement

The triene **(100)** obtained above on heating with strong acid undergoes rearrangement resulting in aromatisation of ring A(**101**). This arrangement is known

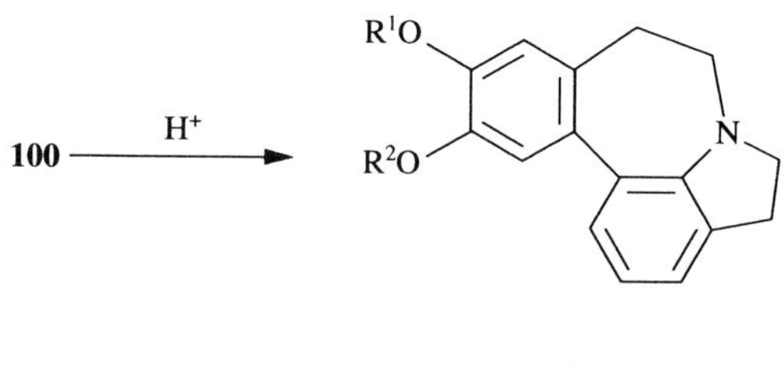

101

as the apo rearrangement. A similar rearrangement in β-erythroidine **(73)** results in apo-β-erythroidine **(102)**. This rearrangement results in the destruction of chirality at position 3 and 5 and the apo alkaloids are optically inactive. The structures of apo erysopine and apo-β-erythroidine have been confirmed by synthesis [88,89]. These results led to the postulation of the spirocyclic erythrinan skeleton and this was confirmed by synthesis of the parent nucleus by Belleau [90].

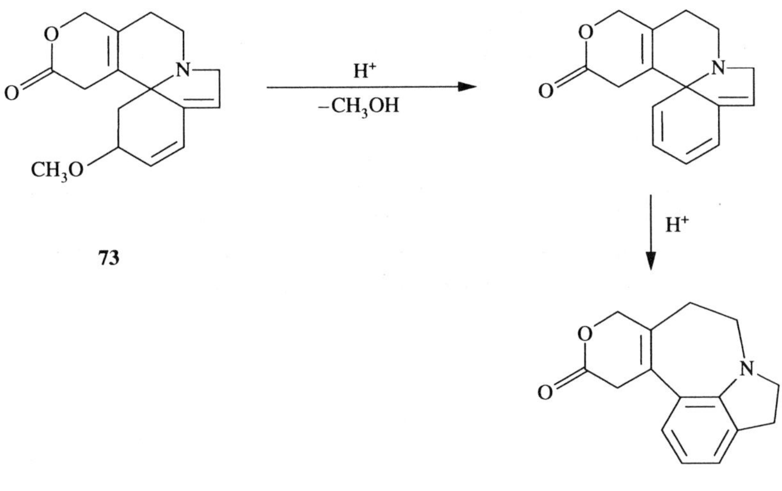

73

102

The double bond in ring D of **73** is stable and does not isomerize into conjugation with the lactone carbonyl as long as the spiroamine system is intact. However,

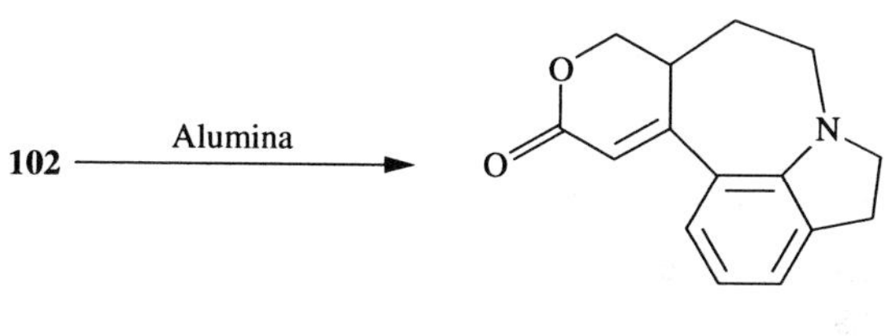

103

the double bond in **102** readily isomerizes into conjugation with the lactone carbonyl giving isoapo-β-erythroidine **(103)** under mild conditions as chromatography over alumina [91].

3.1.2 Hofmann Degradation

An important technique extensively used in the structure elucidation of *Erythrina* alkaloids is the Hofmann exhaustive methylation followed by degradation. This technique has been of value in giving primary evidence for the structure of erythroidines.

β-Erythroidine **(73)** responds poorly to Hofmann degradation but the reaction proceeds with ease in the case of dihydro-β-erythroidine **(104)**. Hofmann degradation of **104** through conversion to the quaternary betaine **(105)** followed by thermal decompostion, gave the methine base **(106)** and subsequently decarboxylation and dehydration afforded an oxygen-free base **(107)**. Oxidation of **107** results in the formation of *N*-methylphthalimide. Ozonolysis of **107** gave formaldehyde. To indicate the points of attachment of the lactone ring to ring C, **104** was reduced with lithium aluminium hydride to give the diol **(108)** which on Hofmann degradation gave**109** with ring A being aromatized. The structure of **109** has also been confirmed by Hofmann degradations.

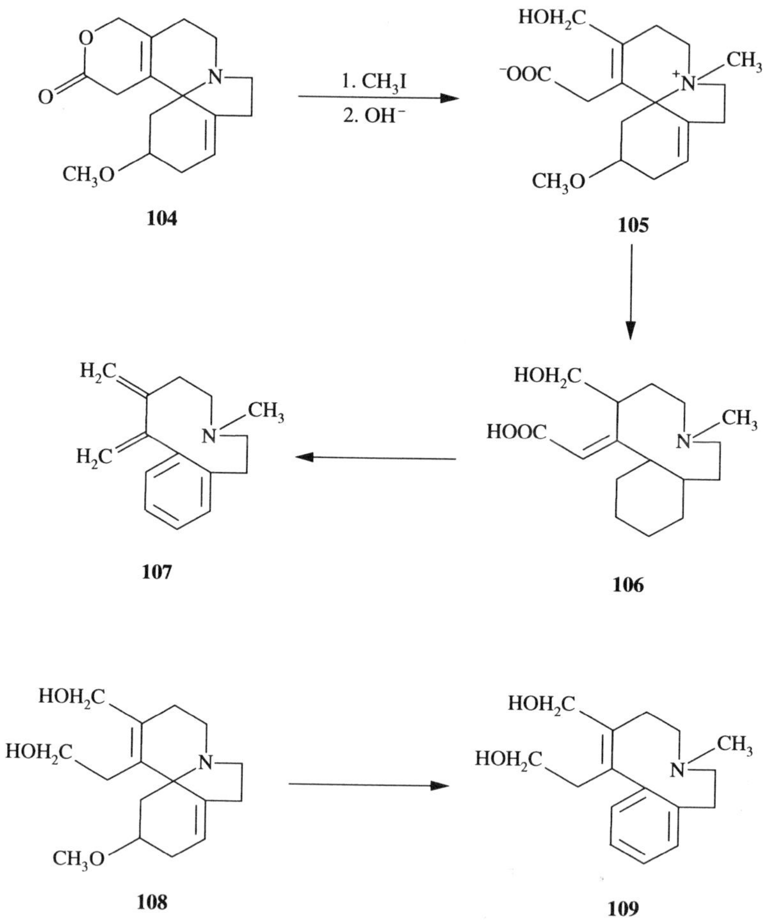

With regard to the structure of α-erythroidine **(71)**, again the primary evidence has come from Hofmann degradation of its reduction product **(110)**. The Hofmann reaction is accompanied by loss of methanol to generate an aromatic ring to give **111**. Further Hofmann degradation gave the optically active tetrahydrofuran derivatives **(112)** and **(113)**. These two derivatives have subsequently been synthesized [92].

The evidence for the position of the aliphatic methoxyl group at C-3 was provided only by **73**. Hofmann degradation of β-erythroidinol **(114)** gave

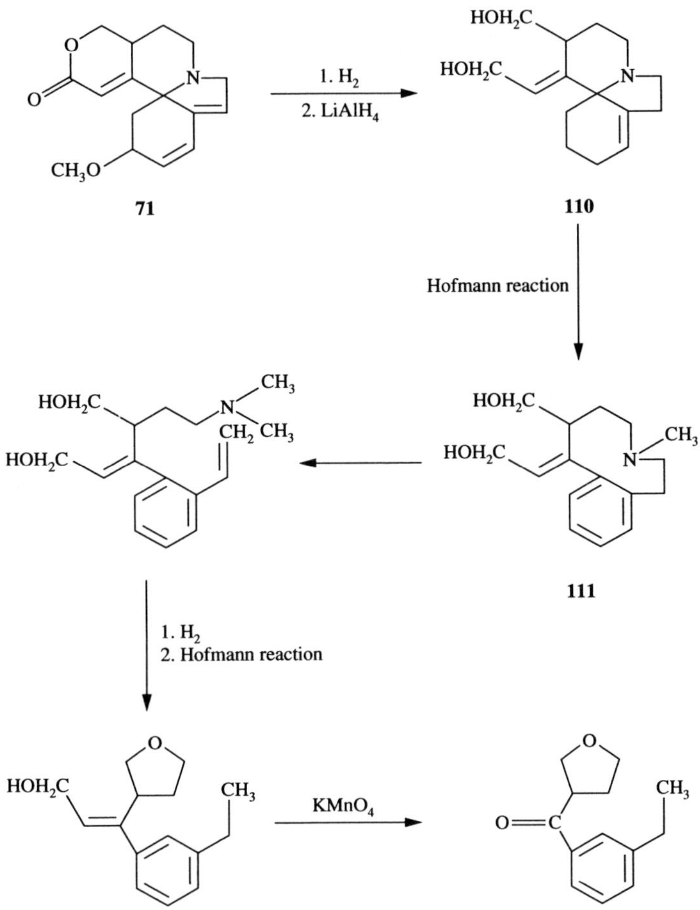

71 110 111 112 113

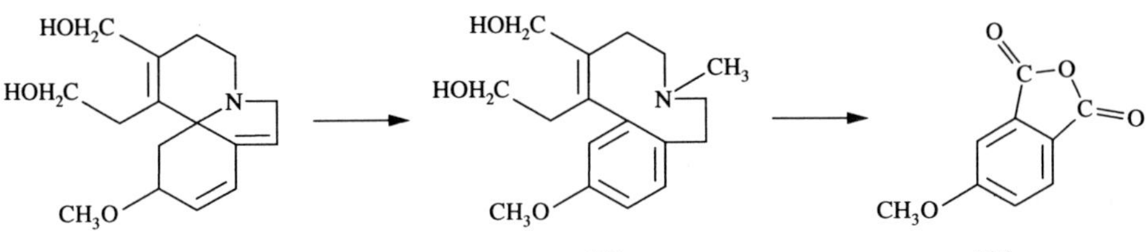

114 115 116

115 in which aromatization of ring A has taken place without the loss of the methoxyl group. Oxidation of **115** with potassium permanganate gave

3-methoxyphthalic anhydride **(116)**. As **73** does not behave as an enol ether, the possibility of the methoxyl group being present at C-2 was ruled out. The position of the methoxyl group at C-3 was confirmed by X-ray analysis of erythraline hydrobromide [93].

3.1.3 The Spiro System

The confirmation of the spirocyclic erythrinan skeleton of the *Erythrina* alkaloids came through the cyanogen bromide degradation of dihydroerysotrine (61) [94]. After reduction of the crude product with lithium aluminium hydride the optically inactive base (117) was obtained, which on oxidative degradation gave the dicarboxylic acid (118). The apo rearrangement of the alkaloids giving dihydroindole derivative such as 101 has already been discussed. The origin of these products can only be explained if there is a spiroamine structure present in the alkaloids. However, further proof was provided by the synthesis of the parent nucleus [90].

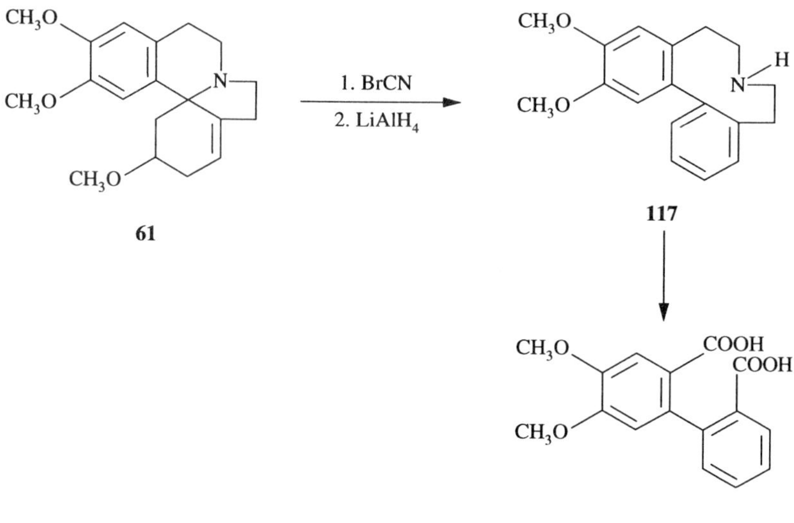

61 **117** **118**

3.2 Spectroscopic Methods

3..2.1 Infrared and Ultraviolet Spectra

The aromatic dienoid alkaloids show IR absorbances at 1610 cm^{-1} and UV absorbances around 285-290 nm (dioxygenated aromatic ring) and 235-240 nm (diene component). The alkenoids absorb in the UV around 225 nm, whereas the enone group usually shows absorbance around 230 nm and IR absorbance in the region 1675-1698 cm^{-1}. The lactonic alkaloids, α- and β-erythroidines exhibit absorption at 1720 cm^{-1} (lactonic carbonyl group) in the IR spectra and at 224 and 238 nm, respectively, in the UV spectra. The 8-oxo erythroidines show a lactam absorbance at 1745 cm^{-1} (five-membered lactam carbonyl group) and an additional UV absorbance at 253 nm arising from the dienone chromophore [63].

3.2.2 NMR Spectra

The ^1H-NMR spectra can give detailed information on structure and stereochemistry, especially with the availability of very high field instruments, even with small quantities of the sample. The dienoid alkaloids are readily distinguished by the presence of an ABX system corresponding to three olefinic protons, whereas the alkenoid alkaloids only have a single olefinic resonance and the lactonic alkaloids lack the two aromatic singlet resonances. The use of ^1H-NMR spectroscopy has enabled one to distinguish between erysodine **(2)** and erysovine **(3)** [29] and to establish the structures of erythratine **(46)** [29] and 11-methoxyerythraline **(25)** [86]. INDOR and NOE methods promise to be very useful techniques in strucutural studies, the positions of aromatic hydroxyl and methoxyl groups were shown to be C-15 and C-16, respectively, for dihydroerysovine **(60)** [95]. The usefulness of the technique [96] has also been demonstrated with cocculidine **(63)** whose structure is already established [97].

The stereochemistry at C-2 and C-3 in erythratidine **(42)** and erythratine **(46)** were established by ^1H-NMR [29,31]. With the aid of INDOR technique the value of $J_{3,4}$ was found to be 5.5 and 12 Hz for both alkaloids suggesting that 3-H was axial in both cases. The values $J_{1,2}$ (4.25 Hz) and $J_{2,3}$ (4.25 Hz) in **42** suggested that 2-H is pseudo-equatorial and the stereochemistry is that of **A** (as shown Fig. 1). In **46** and its C-2 epimer (epi-erythratine) the values of $J_{2,3}$ were found to be 7.5 and 3-4 Hz, respectively, thereby suggesting the stereochemistry

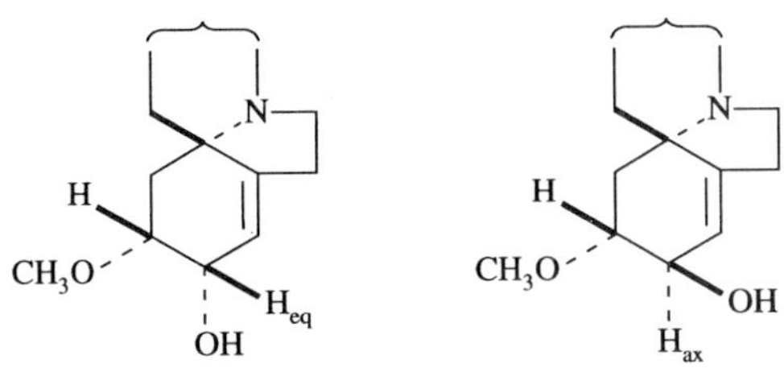

A **B**

Fig. 1. Stereochemistry in ring A

B and **A** at C-2 for erythratine and epi-erythratine, respectively. Thus **42** and **46** have opposite stereochemistry at C-2.

In 11-oxygenated alkaloids the stereochemistry at C-11 has also been determined with this technique and has been further confirmed by X-ray determination [42,82].

The ^1H-NMR characteristics of α- and β-erythroidines have been reported [63,97].

The ^{13}C-NMR chemical shifts of a number of dienoid, alkenoid and lactonic alkaloids have been reported [64]; these have been assigned by internal comparisons within the series of related compounds and also by the use of model compounds. In addition, a few selective decoupling and deuteration experiments were carried out.

3.2.3 Mass Spectra

Mass spectrometry has played a vital role in the identification of *Erythrina* alkaloids when present in very small quantities. Mass spectrometry, especially in conjunction with gas chromatography, has proved particularly useful in preliminary screening of *Erythrina* species for their alkaloid content [56-58]. A detailed analysis of the mass spectra of a range of *Erythrina* alkaloids showed [98] that the predominant fragmentation of the dienoid and lactonic alkaloids corresponded to cleavage of methyl or methoxyl from the C-3 methoxyl unit (Fig.2).

The alkenoid alkaloids also showed similar fragmentations; another prominent peak at M^{+}-58 corresponding to a retro-Diels-Alder fragmentation with loss of C-3, C-4 and C-3 methoxyl group also occurred (Fig. 3). In some instances mass spectrometry allows one to distinguish between epimers, e.g. erythratidine and epi- erythratidine [99].

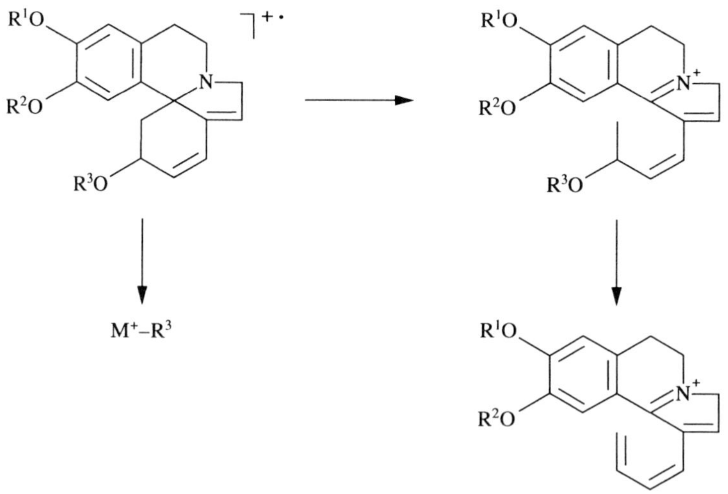

Fig.2 Mass spectral fragmentation pattern of dienoid *Erythrina* alkaloids

The fragmentation patterns of enone alkaloids [57,98], 8-oxo derivatives [100] and 11-oxygenated alkaloids [30,32,58,86] have been described.

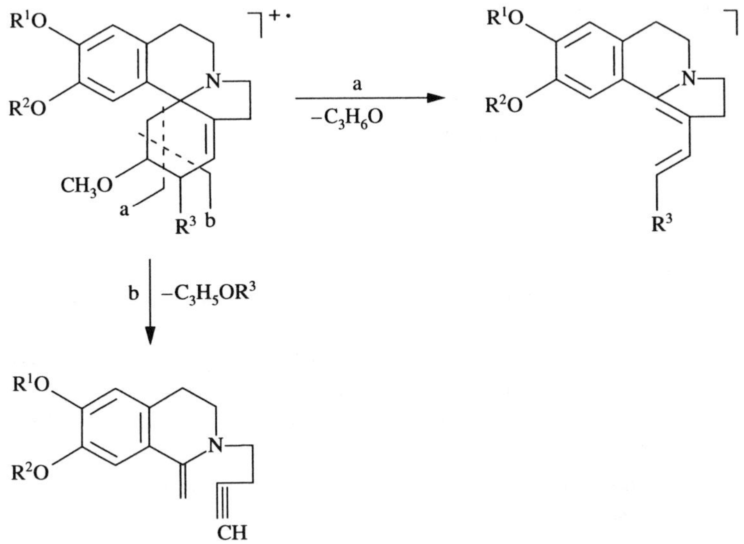

Fig. 3. Mass spectral fragmentation of alkenoid *Erythrina* alkaloids

3.2.4 Chiroptical Analysis

The dienoid alkaloids show a positive Cotton effect. The application of the transoid diene rule to the ORD data led to the assignment of absolute configuration of these alkaloids opposite to that found by X-ray analysis. It was explained that the allylic methoxyl system (at C-3) has helical chirality opposite to that of the diene chromophore, and it appears that the sign of the Cotton effect is determined by the former group [101].

The alkenoid alkaloids also show a positive Cotton effect and this pattern was utilized to assign the absolute configuration of cocculine **(64)** and cocculidine **(63)**, later supported by X-ray analysis [102,103].

3.2.5 X-Ray Crystallography and Absolute Stereochemistry

The absolute configuration of lactonic and aromatic alkaloids has been established to be (3*R*,5*S*) [104-106]. An X-ray analysis of the 2-bromo-4,6-dinitrophenolate salt of erythristemine **(15)** showed the absolute configuration at C-11 as *S*. The preparation of this derivative constituted a new method and may be applicable in other cases [30,107].

The absolute configurations of the alkenoid alkaloids **64** and coccutrine **(62)** have also been determined by X-ray analysis [108]. These studies showed that ring A exists preferentially in the half chair conformation in the free base, but this was changed to an envelope conformation on protonation of the nitrogen atom.

The crystal structures of alkaloids containing a hydroxyl group at C-2 have not been determined. Barton *et al.* [31] established the absolute configuration of **42** and **46** as 2*S*, 3*R*, 5*S* and 2*R*, 3*R*, 5*S*, respectively ,on the basis of optical rotation and ^{1}H-NMR data. The configuration at C-2 for erysosalvine **(55)**, **52** and erysopitine **(57)** has not been defined.

The absolute configuration of other alkaloids can be determined by comparison of their CD and ^{1}H-NMR characteristics with alkaloids of known stereochemistry as well as on chemical transformations.

3.3 Some Significant Examples of Structure Determination

3.3.1 Dienoid Alkaloids

The dienoid alkaloids may be discussed under simple bases, 8-oxo derivatives and 11-oxygenated derivatives.

Simple Bases. The ambiguities in the structures of most abundantly present alkaloids **2** and **3** have been resolved by Barton and co-workers through ^{1}H-NMR studies [29]. Erysotrine **(7)** earlier known as a synthetic product was first isolated from *E. suberosa* [44,45] in 1969 and subsequently it has been identified in a number of *Erythrina* species [41,51,56,58,59,74,83,109].

GC-MS studies of *E. folkersii* indicated the presence of two new dienoid alkaloids erythravine **(9)** and erysoline **(5)** [57]. The former was characterised as 3-demethylerysotrine and the latter as 3-demethylerysovine on the basis of their mass spectral fragmentations. A novel *Erythrina* alkaloid, erymelanthine **(119)** has been isolated by two different research groups [24,99] from the seeds of *E. melanacantha;* the same alkaloid has also been found in the seeds of *E. merilliana.* The mass spectrum of the alkaloid was typically dienoid, but the molecular ion at m/z 312 ($C_{18}H_{20}N_2O_3$) revealed the unexpected presence of a second nitrogen atom, which was assigned to a pyridine ring on the basis of its ^{1}H-NMR spectrum. Evidence for the methoxycarbonyl group was obtained from IR and ^{13}C-NMR spectra, and its position was deduced by decoupling and

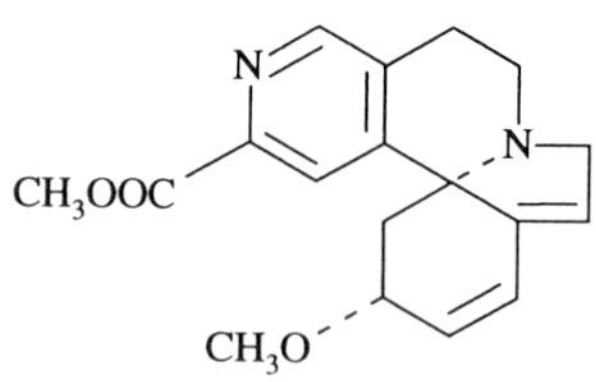

119

NOE experiments. A new alkaloid, erythrocarine **(12)** has been isolated from *E. caribea* [65].

The quaternary alkaloids erysophorine **(34)** [55], erysodinophorine **(35)** [70], erysopinophorine **(36)** [71], and isoerysopinophorine **(37)** [72] have been isolated from *E. arborescens*. The mass spectra of these alkaloids gave no molecular ion but exhibited significant fragments corresponding to the aromatic *Erythrina* skeleton with a conjugated 1,6-diene system and a carboxylated indole-3-alkylamine moiety. The presence of three quaternary *N*-methyl groups and the methoxyl group in each of these alkaloids was evident from the ^{1}H-NMR spectra. Hydrolysis of these alkaloids with 6*M* HC1 readily afforded hypaphorine **(98)** and the corresponding *Erythrina* alkaloid. This type of combined alkaloids have not been isolated from *Erythrina* species.

Recently, three novel glycodienoid alkaloids, 11β-methoxyglucoerysodine **(29)**, 11β-methoxyglucoerysovine **(30)**, and rhamnoerysodine **(31)** have been isolated from *E. lysistemon* and their characterisation was established by the ^{1}H and ^{13}C-NMR data [78]. These studies led to the reversal of the ^{1}H chemical shift assignments for H-1 and H-2 of **2** in particular, and consequently for all the commonly accepted H-1 and H-2 values of *Erythrina* dienoids in general. Additionally, the ^{13}C chemical shifts of C-2 and C-13 in the known glucoerysodine **(28)** have been reversed based upon heteronuclear correlation experiments [64]. The ^{1}H and ^{13}C assignments for the glucose moiety of this alkaloid have also been reported for the first time.

From the flowers of *E. mulungu* [74] there has been isolated erysotrine *N*-oxide [87] and the structure determined by spectroscopic analysis and confirmed by comparing the spectral data with the synthetic *N*- oxide prepared by treating **7** with *m*-chloroperbenzoic acid.

8-Oxo Derivatives. Ito *et al.* [37,38] isolated two new alkaloids erytharbine **(83)** and erysotramidine **(8)** from *E. arborescens*. These are the first examples of natural *Erythrina* alkaloids containing an amide group. The latter

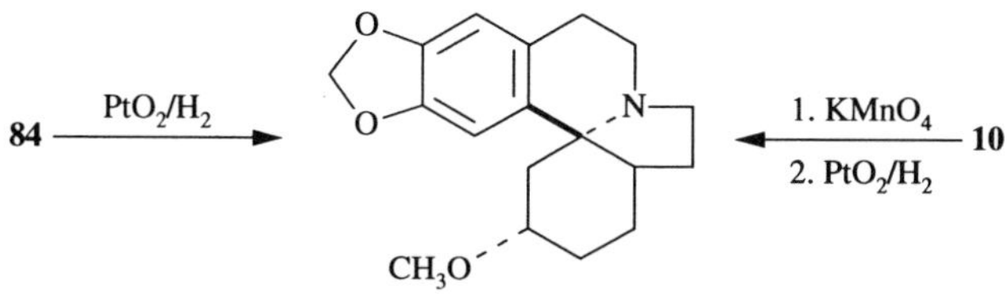

has also been isolated from *E. xbidwillii* [75]. Later on the same group of workers also obtained another alkaloid crystamidine **(84)** with an amide function [42]. The carbonyl group (ν_{max} 1695 cm^{-1}) was placed at C-8 after detailed examination of the ^{1}H-NMR spectrum and the structure finally established by chemical correlation with **10** as shown above. This also established the stereochemistry at C-3 and C-5 in **84**. Other 8-Oxo alkaloids isolated include 8-oxoerythraline from *E. cristagalli* [69] and 8-oxoerythrinine from *E. brucei* [68], *E. cristagalli* [66] and

E. suberosa [67]. The isolation of 8-oxo compounds from *E. berteroana* has been described [61,63]. Their occurrence in *Cocculus laurifolius* [110] has been discussed separately (Section 4). Autoxidation of 1,6-dienes is known to give rise to such functionality and it is probable that these compounds may be produced as artifacts during the drying process.

C-11 Oxygenated Derivatives. Erythristemine **(15)** is the first C-11 oxygenated *Erythrina* alkaloid isolated from *E. lysistemon* from South Africa [30]. The IR spectrum showed no hydroxyl or carbonyl group and the mass fragmentation pattern was consistent with the dienoid alkaloids containing an additional methoxyl group either in ring C or D. The ^{1}H-NMR along with INDOR studies suggested that a methoxyl group was present at C-11. The proposed structure **15** was finally confirmed by X-ray analysis of the 2-bromo-4,6-dinitrophenolate salt [30,107]. The occurrence of this alkaloid has also been reported from *E.arborescens* [54] and *E. xbidwillii [75]*. Reinvestigation of the alkaloid content of *E. lysistemon* has shown the presence of another new alkaloid characterised as 11β−methoxyerysodine **(18)**, largely based on its mass and ^{1}H-NMR spectra and its conversion to the known alkaloid **15** [99]. A little later 11-methoxyerythraline **(25)** was isolated from the same plant occurring in Southern Rhodesia [86]. The spectral data was similar to that of **15** except that a methylenedioxy group was present instead of two aromatic methoxyl groups. It had a similar configuration at C-11 and the same conformation of ring C.

Ito and co-workers reported the isolation of an alkaloid, from *E.* variegata and characterised it as 11-hydroxyerythraline [32]. This was named as erythrinine **(11)**. Catalytic hydrogenation of **11** followed by hydrogenolysis over palladium black in aqueous hydrobromic acid yielded tetrahydroerythraline. This established the basic structure and stereochemistry at C-3 and C-5. The hydroxyl group was placed at C-11 because of the ease of hydrogenolysis and oxidation afforded a ketone (ν_{max} 1680cm^{-1}) in conjugation with the aromatic ring. The isolation of **11** has also been reported from *E. xbidwillii* [34,35], *E. cristagalli* [40,42,66], *E. macrophylla* [61], *Erythrina* species from Singapore [41] and old world species [58]. 11α- and 11β-Hydroxyerysodine **(16)** were isolated from *E. lysistemon* and their structures likewise determined by ^{1}H-NMR spectroscopic comparisons with each other, with **2** and with **18** [91]. From E.*arborescens* 11-acetoxyerysotrine was isolated and named as erythrascine **(13)** [54]. Later from the same plant Ito *et al.* [37] reported the isolation of 11-hydroxyerysotrine (erythrartine, **14**), which subsequently has also been isolated from *E.herbacea* **[76]**, *E. xbidwillii* [75] *and E. variegata* [82].

A large number of 11-oxygenated *Erythrina* alkaloids including 11-oxo derivatives have been detected from different *Erythrina* species [58].

Alkaloid *N*-oxides, erythrartine *N*-oxide **(88)** from*E.mulungu* [74], 11β-methoxyerythraline *N*-oxide **(89)** from *E. cristagalli* [66] and *E.suberosa* [67], and erythristemine *N*- oxide **(90)** from *E. xbidwillii* [75] have been isolated and their structures confirmed by comparing the spectral data with the synthetic compounds obtained by treating the parent alkaloids with *m*-chloroperbenzoic acid.

3.3.2 Alkenoid Alkaloids

The alkenoid alkaloids without an oxygenated function at C-2 and with an oxygenated function at C-2 are discussed below.

Without Oxygenated Function. Dihydroerysodine **(59)** and dihydroerysovine **(60)**, isolated from *Cocculus* species, have been discussed under Section 4. Dihydroerysotrine **(61)** is still known only as a reduction product of **7**.

The structure of erythramine **(45)**, isolated from *E. sandwicensis* (syn. *E. tahitensis*), *E. subumbrans* [111], *E.glauca* (syn. *E. fusca*) [84,112], *E. cristagalli* [29] and also obtained as a reduction product of **10** [113], was finally established by ^{1}H-NMR studies [29]. An alternative synthesis of **45** from **10** has also been described [29]. The conversion of **45** to **10** [106], an alkaloid of known stereochemistry, finally established the position of double bond as well as the configurations at C-3 and C-5 in **45**.

Several other alkenoid alkaloids with abnormal substitution in ring D isolated from *Cocculus* species are discussed under Section 4.

With Oxygenated Function. The position of the double bond in erythratine **(46)** and its stereochemistry was confirmed by ^{1}H-NMR studies and its conversion to **10**. Erythratinone **(48)**, obtained by oxidation of **46**, has been isolated as the major alkaloid from *E. cristagalli* [29]. Further work on the investigation of *Erythrina* species by the same group of workers led to the isolation of two ketonic alkaloids, erythratidinone **(44)** and 3- demethoxyerythratidinone **(77)** from *E. lithosperma* (syn. *E. variegata*) [31]. Sodium borohydride reduction of **44** yielded **42**, and its C-2 epimer. By application of Mills' rule [114] **42** was assigned the 2S configuration which was opposite to that established for **46**[29]. The configuration at C-3 and C-5 were confirmed by dehydration of **42** to **7**. Compound **77** showed spectroscopic properties similar to that of **44**.

Several new alkenoid alkaloids have been reported from *E. salviiflora* during the GC-MS studies of different *Erythrina* species [57]. Erysotinone **(53)** earlier obtained as a racemic synthetic intermediate [115] was isolated from this species and its structure established by conversion of the isolated alkaloid to **59** which was also prepared from **2** [57]. Another gas chromatographic fraction which gave an identical MS to **53** was assigned the isomeric structure and given the name erysosalvinone **(56)**. One more fraction (TMS derivative) exhibiting an enone fragmentation pattern similar to **53** and **56**, but with a molecular ion at 58 amu higher due to the presence of two phenolic hydroxyl groups instead of aromatic methoxyl and hydroxyl groups, was assigned structure **58** and named erysoflorinone [57].

An alkaloid named erysotine **(52)** identical with the reduction product of erysotinone **(53)** has also been isolated; neither **52** nor **53** have earlier been reported from natural sources [116]. The ^{1}H-NMR spectrum of **52** compared favourably with **46** except that the methylenedioxy group of **46** was replaced by aromatic methoxyl and hydroxyl groups. Treatment of **52** with ethereal diazomethane afforded a product that had identical melting point and GC retention time with that of **42** [57,117]. Erysotine like **42** has the 2*S* configuration. The alternative positional isomer related to **56** was also identified and named erysosalvine **(55)**. Recently, **52**, **42**, epi-erythratidine **(43)** and 11-hydroxy-epi-erythratidine **(50)** have been characterised in *E.variegata*[85]. Erysopitine **(57)** was isolated from *E.variegata* [53]. The stereochemistry at C-3 and C-5 was established by converting **57** to **7**, but the configuration at C-2 has not been defined.

3.3.3 Lactonic Alkaloids

Erythroidines **(71,73)** were isolated from *E.americana* and characterised from their spectral properties, formation of quaternary ammonium salts and conversion of **71** to **73** by treatment with base [97]. From *E.berteroana* Chawla *et al.*[63] have reported the isolation of two new alkaloids along with **71** and **73**, the new alkaloids were fully characterised as 8-oxo-α- and 8-oxo-β-erythroidines **(72)** and **(74)** by a combination of spectroscopic methods and decoupling experiments. Both compounds could also be obtained by direct oxidation of the corresponding erythroidines.

3.3.4 Other *Erythrina* Alkaloids

The isolation of *N*-norprotosinomenine **(91)**, protosinomenine **(92)** and erysodienone **(76)** from *E. lithosperma* (syn. *E. variegata*) [52] and erybidine **(95)** from different *Erythrina* species [33,37,42,60,80] is significant from a biosynthetic point of view. Erysodienone had been previously synthesized [98,115,118] but this was the first report of its isolation from plant material. The characterisation of **76** was achieved by its reduction to the known transformation product erysodienol [53]. The structure of **95** has been confirmed by spectral data and conversion to *O*-methylerybidine which in turn was prepared from **76** [33].

Cristadine **(97)**, a new benzylisoquinoline alkaloid was isolated from *E cristagalli* [81]. Methylation of **97** with diazomethane afforded a derivative identical to papaverine. A new quaternary alkaloid, pachygonine **(82)**, was isolated from the roots of *Pachygone ovata* and the structure established on the basis of spectroscopic evidence and its methylation to give cocculidine **(63)** methiodide [73].

4 ABNORMAL *ERYTHRINA* ALKALOIDS

Cocculus (Fam. Menispermaceae) is a genus of about 20 species mostly climbing shrubs distributed in tropical and subtropical regions. Of these only three species namely *C. laurifolius*, *C. trilobus* and *C. carolinus*, have been investigated for alkaloids, and a total of twenty different alkaloids (Table 6) have been reported. The greatest number of alkaloids have been obtained from *C. laurifolius*. The alkaloids isolated from *Cocculus* species are abnormal in the sense that they contain no oxygen function at C-16, the exceptions being dihydroerysodine **(59)**, dihydroerysovine **(60)**, erythroculine **(68)**, erythlaurine **(69)**, and erythramide **(70)**. The alkaloids **68**, **69** and **70** possess a novel feature of having a C_1 unit directly attached to the aromatic ring at C-15. Three alkaloids isococculine **(81)**,

Table 6. ***Erythrina*** **Alkaloids from** ***Cocculus*** **species**

Plant Source and Alkaloid	m.p. (°C)	[α]D (°)	Ref.
Cocculus laurifolius			
Cocculine **(64)**	220-221	+ 252	102, 119
Isococculine **(81)**	182-184	+ 164	110, 120
Cocculidine **(63)**	86-87 (93-95)	+ 260	96, 102 119
Isococculidine **(80)**	95-96	+ 124	119, 121
Coccoline **(41)**	245-246	+ 233	119
Coccolinine **(40)**	174-175		122
Coccuvine **(39)**	137-138		123
Coccuvinine **(38)**	103-104		124
Erythroculine **(68)**	193-196	+ 194	125
Cocculitine **(66)**	142-143	+ 93	126
Dihydroerysodine **(59)**	208-209	+ 224	127
Cocculitinine **(65)**			110
Cocculidinone **(67)**			110
Erthylaurine **(69)**		+ 232	128
Erythramide **(70)**	87-89	+ 262	128
Cocculimine **(79)**			110
Coccudienone **(78)**			110
Cocculus trilobus			
Dihydroerysovine **(60)**		+ 223	95
Coccutrine **(62)**	263-265	+ 232	129
Cocculine **(64)**	205-207	+ 232	129
Cocculolidine **(75)**	144-146	+ 273	130
Cocculus carolinus			
Cocculine **(64)**	207-208	+ 251	131
Cocculolidine **(75)**	140-141	+ 258	131

isococculidine **(80)** and cocculimine **(79)** are of $\Delta^{1,2}$ alkene type. Cocculine **(64)** has been isolated from all the three species of *Cocculus*, while cocculolidine **(75)** has been obtained from *C.trilobus* and *C. carolinus*. Structurally it is a lower homologue of β-erythroidine **(73)**. Coccutrine **(62)** isolated from *C.trilobus* has a unique feature of having an oxygen function at C-17. The structures and stereochemistry of these alkaloids have been established by spectral, and in some instances, by X-ray crystallographic data. Chemical studies done on the alkaloids have supported the structures.

Cocculine (**64**) and cocculidine (**63**) are the first members of abnormal *Erythrina* alkaloids isolated from *C. laurifolius*. The spiro structure of these alkaloids was shown by the formation of an *N,O*-diacetyl derivative (**120**) with acetic anhydride, and ^{1}H-NMR data have shown that these bases are related to the *cis* series of alkenoid *Erythrina* alkaloids [110].

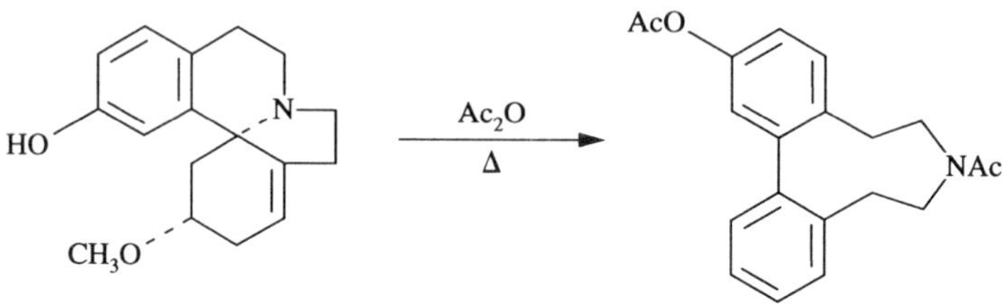

64 **120**

The structure of **68** which has a unique feature of having a methoxycarbonyl group at C-15 was determined by spectroscopic and chemical evidence. The ^{1}H-NMR spectrum of **68** showed the presence of three methoxyl groups and two *para*-oriented aromatic protons. Its reduction with lithium aluminium hydride gave erythroculinol (**121**) which showed only the hydroxyl band but no carbonyl band in the IR spectrum. The disappearance of a methoxyl group in the ^{1}H-NMR spectrum of **121** established the presence of a methoxycarbonyl group. Treatment of **68** with boron trichloride in dichloromethane afforded a phenolic compound (**122**) which showed a remarkable bathochromic shift in the UV spectrum. The large shift of the carbonyl absorption band in the IR spectrum of this compound as compared with **68** suggested the presence of an intramolecular hydrogen bond between the phenolic hydroxy group and ester, thus requiring that these groups be located *ortho* to each other. The environment of nitrogen in **68** was established by Hofmann degradation of **121** to give **123** and the information on the relative positions of the double bond and aliphatic methoxyl group was obtained from its mass spectrum. A classical degradation route of **68** gave the biphenyl derivative (**124**), thus establishing the relative position of methoxycarbonyl group as indicated in **68**. The stereochemistry at C-3 and C-5 was established by conversion of **68** to tetrahydroerysotrine (**126**). Catalytic hydrogenation of **68** followed by lithium aluminium hydride reduction and treatment with silver oxide in 85 per cent phosphoric acid yielded the aldehyde (**125**). The latter was subjected to Baeyer-Villiger oxidation with performic acid. The resulting product was treated with diazomethane to yield **126**. The presence of methoxycarbonyl group in **68** is an interesting feature from a biogenetic point of view.

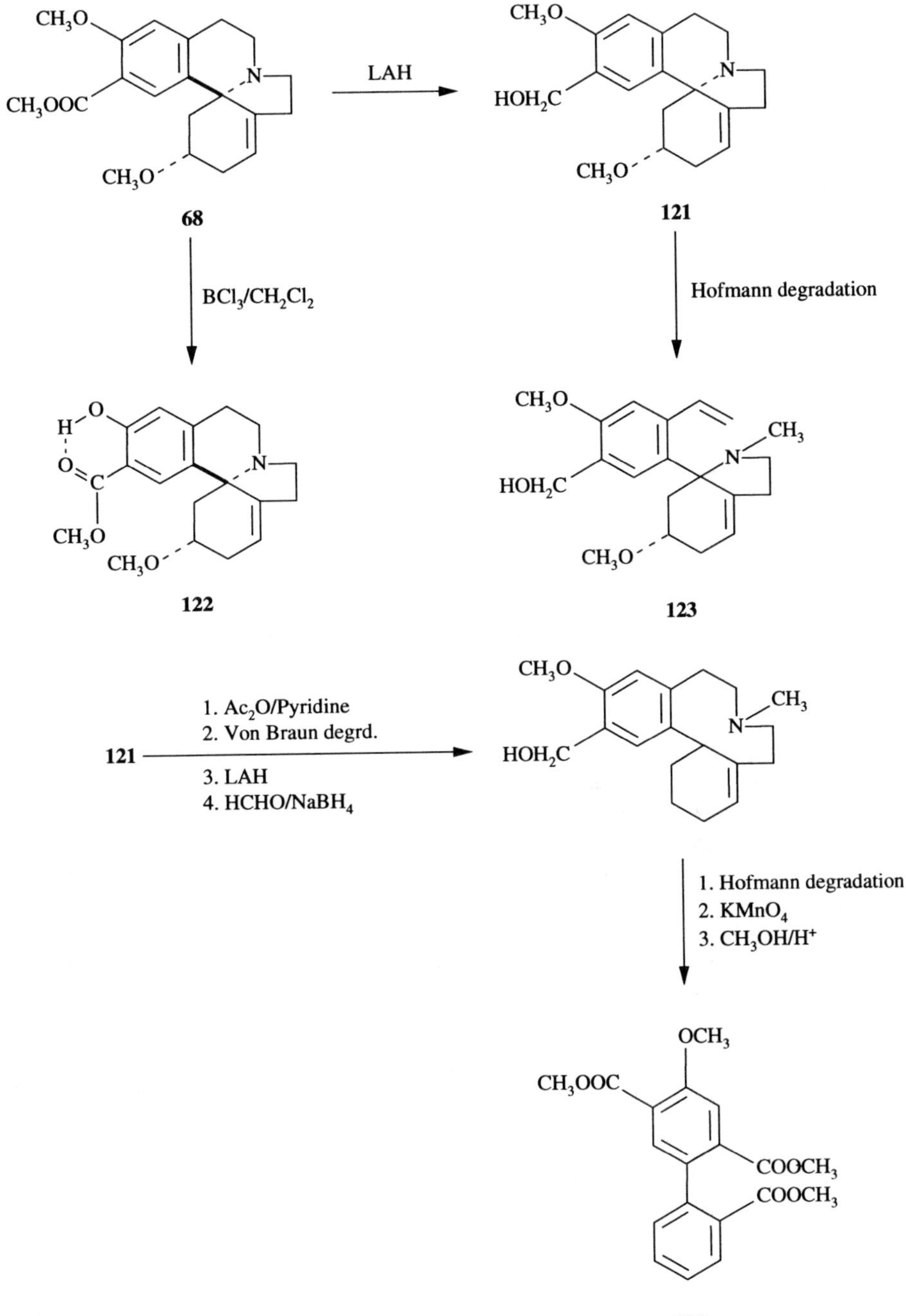
CH3O
CH3OOC
CH3O
68
LAH
CH3O
HOH2C
CH3O
121
BCl3/CH2Cl2
Hofmann degradation
H
O
O
C
CH3O
CH3O
122
CH3O
CH3
HOH2C
CH3O
123
121
1. Ac2O/Pyridine
2. Von Braun degrd.
3. LAH
4. HCHO/NaBH4
CH3O
CH3
HOH2C
1. Hofmann degradation
2. KMnO4
3. CH3OH/H+
OCH3
CH3OOC
COOCH3
COOCH3
124

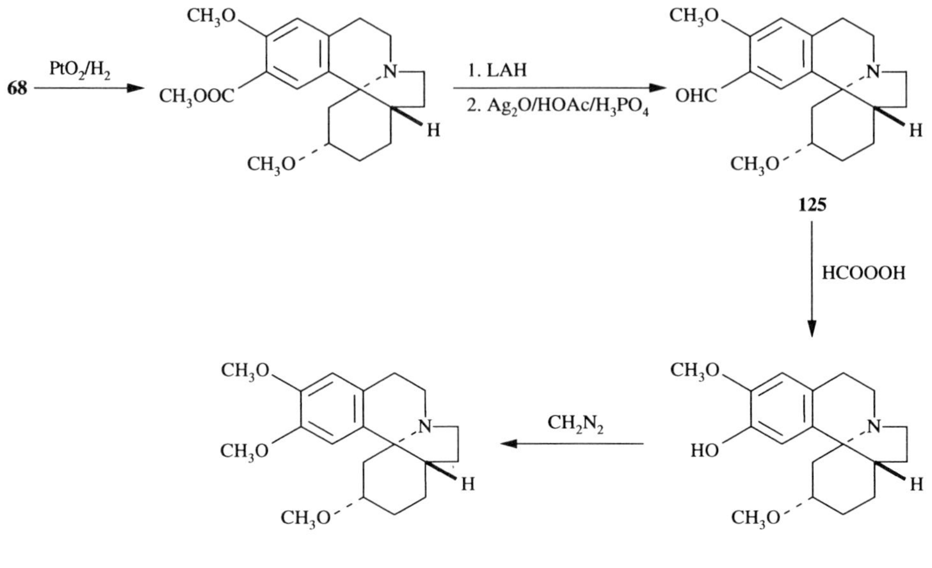

Negative ion mass spectral studies on abnormal *Erythrina* alkaloids have been done [132]. Studies have also been carried out to observe variations in the alkaloidal constituents of the leaves of *C. laurifolius* in different seasons in a year [133].

5. BIOSYNTHESIS

Early experiments on *Erythrina* alkaloid biosynthesis carried out by Leete and Ahmad [134] showed that tyrosine is a precursor of the erythroidines and that a symmetrical intermediate is involved. Several speculative schemes for *Erythrina* alkaloid biosynthesis based on phenolic oxidative coupling were proposed in the late fifties [135], but later a number of experiments carried out by Barton *et al.* [116,136] led to the conclusion that (*S*)-*N*- norprotosinomenine **(91)** was the specific precursor rather than any of its close relatives or an open chain intermediate. The pathway now accepted involves oxidative *para-para* coupling of the phenolic moieties of **91**, ring opening followed by reduction of the imine so generated, to give the dibenzazonine intermediate **(127)**. Further oxidation of **127** to the diphenoquinone **(128)** followed by intramolecular attack of the nitrogen leads to erysodienone **(76)**. Reduction of the $\Delta^{3,4}$ double bond and ketone function produces an alkaloid of the alkenoid type **(52)**, which by 1,4-elimination of water can give a dienoid alkaloid **(2)**. Isolation of the *N*-methyldibenzazonine alkaloid erybidine **(95)** from various *Erythrina* species [33,37,42,80] and the detection of **76** and **91** from *E. lithosperma* (syn. *E. variegata)* [52] provided additional evidence for the pathway as shown in Fig. 4.

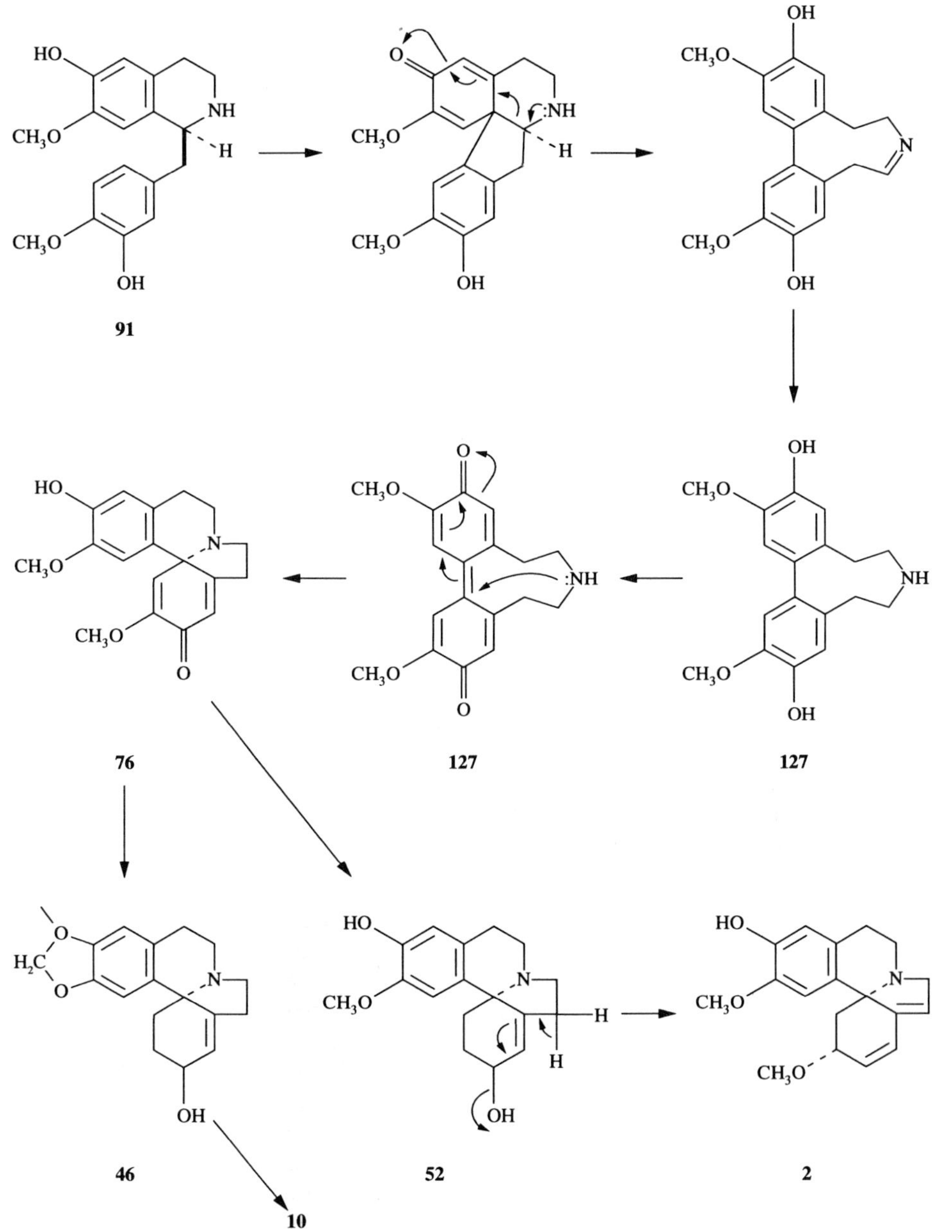

Fig. 4. Biosynthesis of *Erythrina* alkaloids

Barton *et al.* [137] have shown that only (-)-erysodienone, which has the (5-*S*) chirality of the natural alkaloids, is a precursor for erythraline (**10**) and α- and β-erythroidines (**71,73**). The conversion of **91** into **76** involves, formally at least, an inversion of chirality. However, the chirality of **91** may be lost *in vivo*, for the biosynthetic dibenzazonine intermediate **127**, prepared by chemical reduction from chiral **76**, underwent very rapid racemization at room temperature.

Further experiments have established the aromatic *Erythrina* alkaloids as precursors of the lactonic alkaloids and the tracer studies have shown that C-16 was lost in the oxidative cleavage process [137] (Fig. 5).

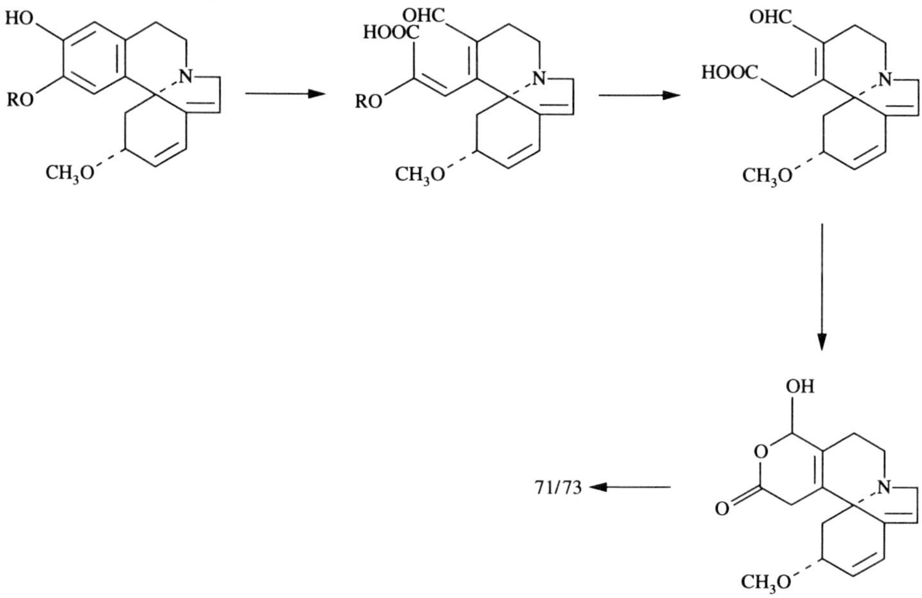

Fig. 5. Biosynthesis of lactonic alkaloids

The feeding experiments with *Cocculus laurifolius* have established that the abnormal *Erythrina* alkaloids isococculidine (**80**), cocculidine (**63**), cocculine (**64**) and coccuvine (**39**) are stereospecifically biosynthesized from **91** [138-140] (Fig. 6).

In the bioconversion of **91** into the abnormal *Erythrina* alkaloids [138] one of the oxygen functions of the precursor can be eliminated by dienone-benzene rearrangement. The occurrence of dibenz[*d,f*]azonine bases in *C. laurifolius* [141] supports indirectly the intermediacy of these bases in the biosynthesis of abnormal *Erythrina* alkaloids. The feeding experiments have established that *O*-demethylation is the terminal step in the biosynthesis of **39** [140]. Feeding of labelled abnormal *Erythrina* alkaloids revealed that **80** is converted into **41** *via* **38** and isococculine (**81**) into **40** *via* **39**.

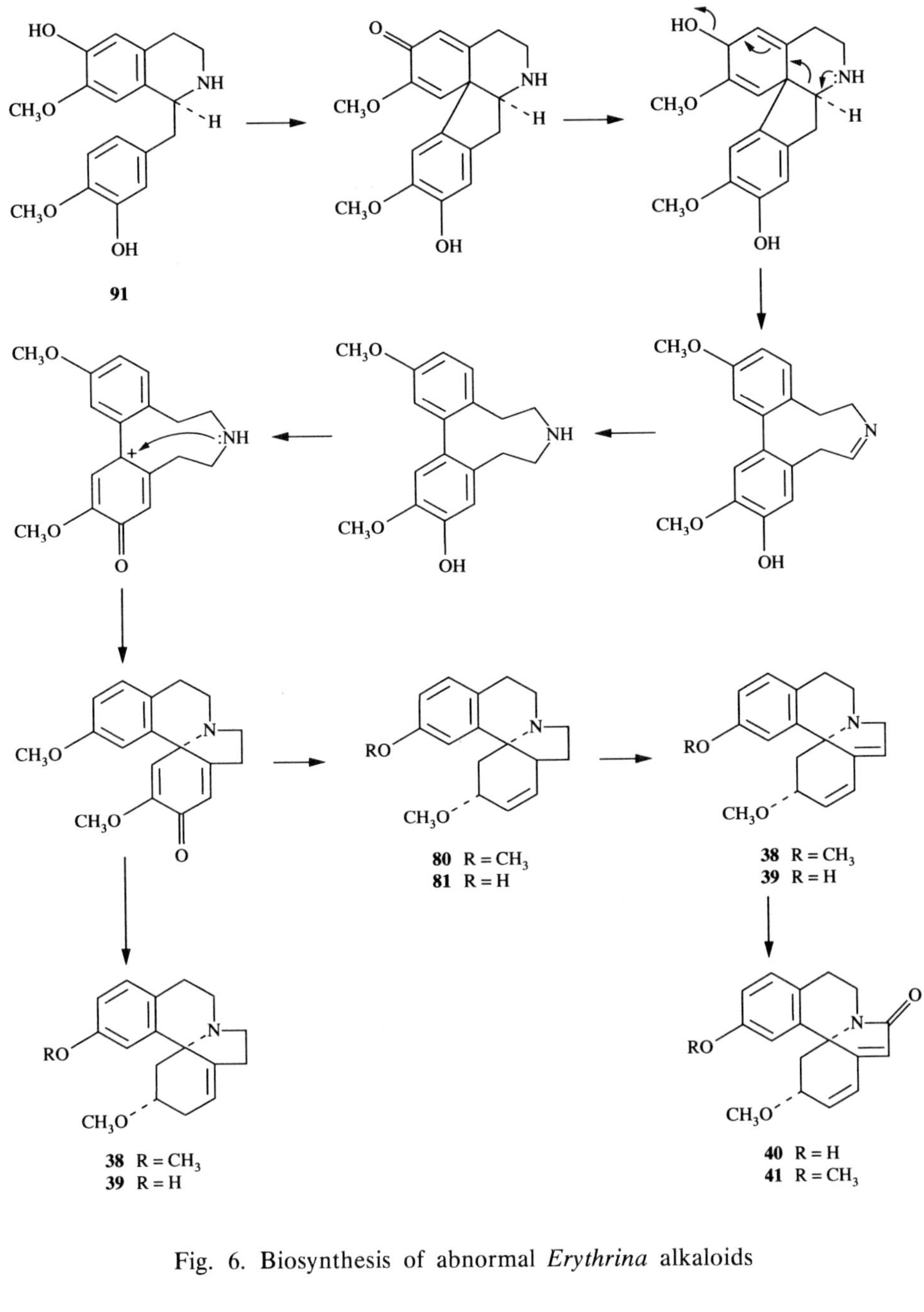

Fig. 6. Biosynthesis of abnormal *Erythrina* alkaloids

6. SYNTHESIS

6.1 *Erythrina* Alkaloids

The synthesis of erythrinan skeleton was reported [90] earlier in 1953, which was later followed by the synthesis of 15,16-dimethoxyerthrinane [142,143]. Similar syntheses involving smooth ring closure were also reported [143-148]. In 1959, Prelog *et al.* [149] employed a different route for the preparation of the erythrinan skeleton and succeeded for the first time in the introduction of a carbonyl group at C-3. Several other approaches to prepare erythrinan derivatives have also been described [150-154]. Tamura *et al.* [155] have reported a synthetic route to erythrinan skeleton from *N*-cyclohexylidene-(3,4-dimethoxyphenyl)ethylamine. Kametani *et al.* [156] have described an alternative synthetic approach to the basic skeleton of *Erythrina* alkaloids (Fig.7). However, although the formation of the basic erythrinan skeleton was achieved, the yield was not satisfactory for this method to be used for its conversion into naturally occurring alkaloids. Ishibashi

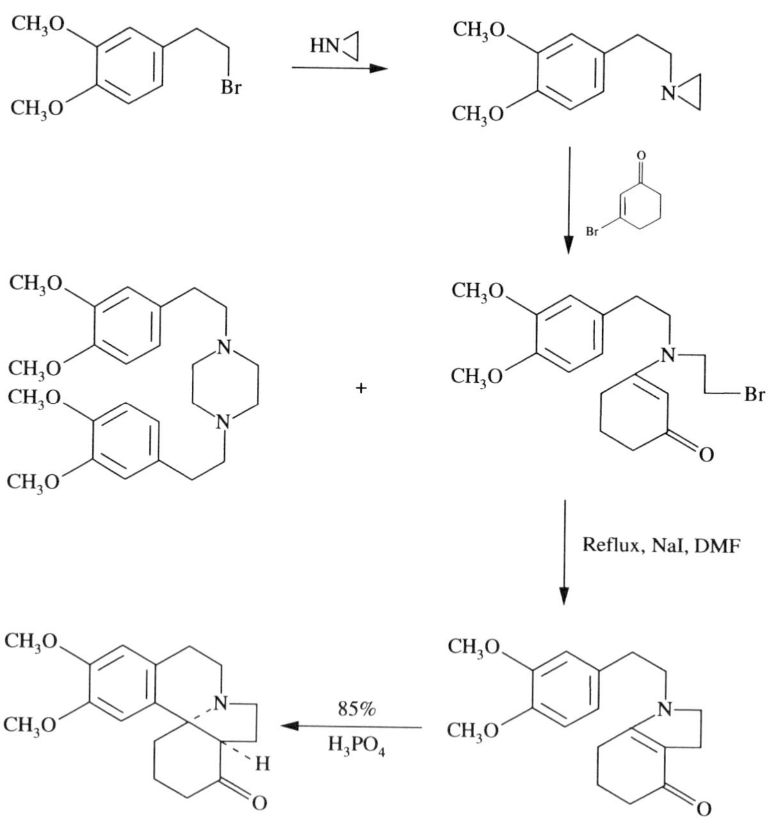

Fig.7.

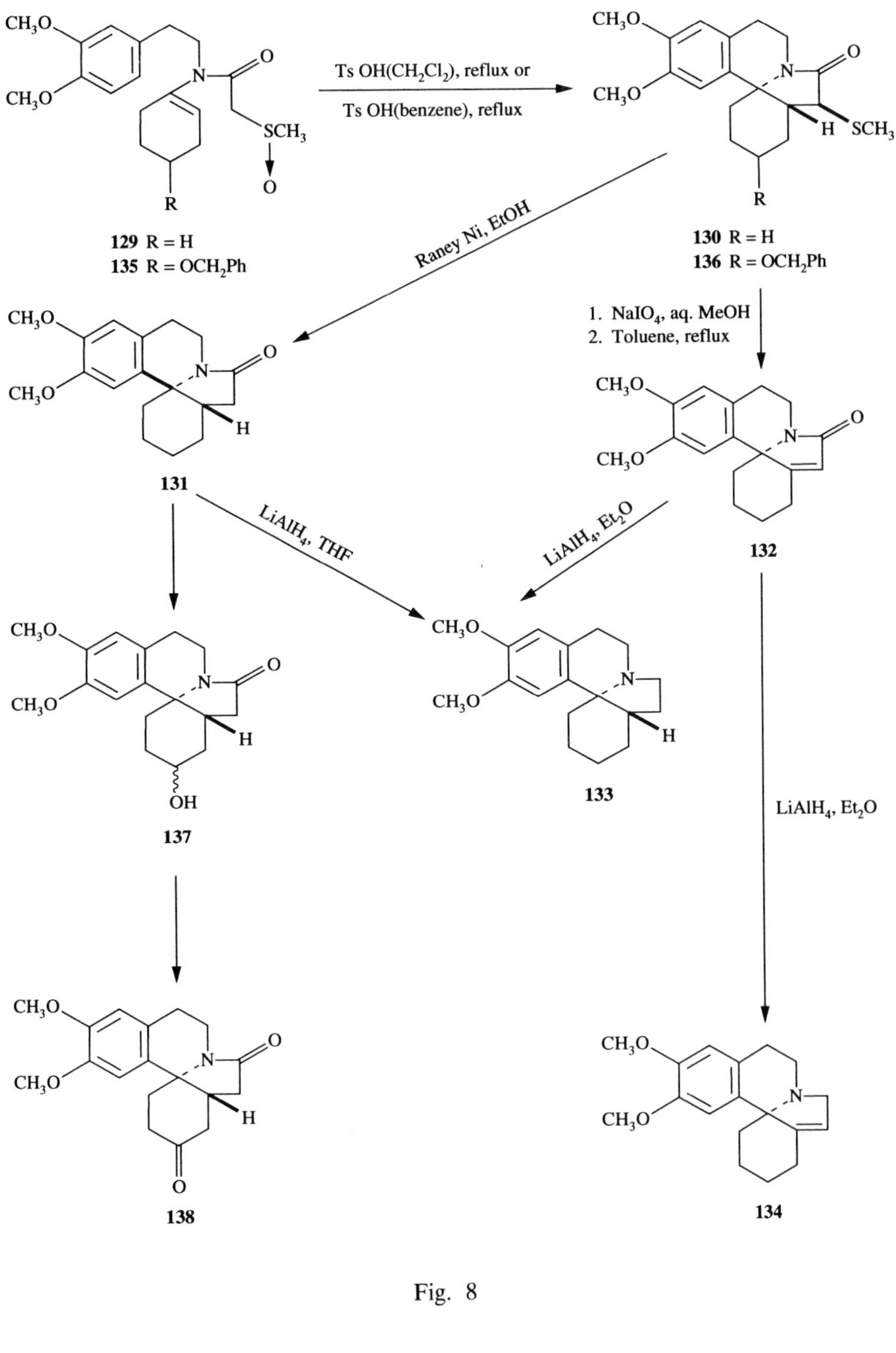

CH3O
CH3O
SCH3
O
R
129 R = H
135 R = OCH2Ph
Ts OH(CH2Cl2), reflux or
Ts OH(benzene), reflux
130 R = H
136 R = OCH2Ph
Raney Ni, EtOH
1. NaIO4, aq. MeOH
2. Toluene, reflux
131
132
LiAlH4, THF
LiAlH4, Et2O
133
137
OH
LiAlH4, Et2O
138
134

Fig. 8

and co-workers [157] have described a novel one-step synthesis of the erythrinan skeleton by utilization of a cyclization that is initiated by thionium ions. If *N*-(cyclohex-1-enyl)- *N*-[2-(3,4-dimethoxyphenyl)ethyl]- α-(methylsulphinyl)acet-mide (**129**) was treated with anhydrous toluene-*p*-sulphonic acid (2 equiv.) in boiling ethylene dichloride it gave **130** in 60 per cent yield; the latter was then transformed into the erythrinan derivatives (**131-134**) as shown in Fig.8. The sulphoxide (**135**) was also converted (*via* **136** and **137**) into the 2-oxoerythrinan derivative (**138**) which has already been converted into the known alkaloid erysotrine (**7**). This synthesis of the erythrinan skeleton is simple, and seems highly promising.

Investigations have been conducted to probe the application of diradical cyclization methods in the synthesis of *Erythrina* alkaloids [158,159]. A novel strategy, based upon an electron-transfer-induced spirocyclization has been reported for constructing the tetracyclic skeleton that is common to the members of the *Erythrina* alkaloids [160]. Another method for the synthesis of erythrinanediones starting from *N*-[2-(3,4-dimethoxyphenyl)ethyl]-*N*-(3-oxo-cyclohex-1-enyl)-α-(methylsulphinyl)acetamide has been described [161]. A synthesis of 15,16-dimethoxyerthrin-6-en-8-one has been accomplished by using an intramolecular Wadsworth- Emmon reaction [162,163]. Synthesis of 15,16-dimethoxy- *cis*-erythrinan-8-one has also been achieved [164] making use of the aminoketone prepared earlier [162].

Dihydroerysodine (**59**) has been synthesized by a biogenetic-like scheme [115]. The synthesis of erysotrine earlier reported in preliminary form has now been published in detail [165-169] and is summarized in Fig. 9. Condensation of homoveratrylamine (**139**) with the glyoxalate derivative of 4-methoxycyclohexa-none gave the enamide (**140**) which, with phosphoric acid, was cyclized to the tetracyclic derivative (**141**). Reduction of **141** with Raney nickel followed by treatment with sulphuric acid gave the oxide (**142**) in which the rings A/B must be *cis*-fused. When **142** was subjected, after *O*-acetylation, to acid treatment, a mixture of two alkenes (**143**) was formed. These two were separated and the one with a double bond at the 1,2 position was epoxidized to **144**. Ring opening of **144** with dimethylamine yielded **145** which, on Cope elimination from the derived *N*-oxide, gave the alkene (**146**). When **146** was treated with acidified methanol allylic rearrangement occurred to give **147** as a mixture of epimers. These were separated by chromatography and each was carried through the remainder of the synthesis. Reduction of the carbonyl group of **147** gave **148**, and this was followed by dehydration to erysotrine.. Finally, resolution of **7** was effected with dibenzoyl-tartaric acid to provide the (+)-isomer, which was identical with erysotrine (**7**) obtained from natural sources.

A new synthetic route to the preparation of *Erythrina* alkaloids involves Birch reduction of the amide (**149**) to the enol methyl derivative (**150**). This when hydrolysed with dilute sulphuric acid in dimethylformamide afforded the β-enamidoketone (**151**), which on cyclization by heating with formic acid gave the tetracyclic compound (**152**) [170]. The isomeric amide (**153**) when similarly treated [171] gave (**154**, 90%), which has a *cis* A/B ring fusion. Ketalization of **154** with ethylene glycol and boron trifluoride, *O*-methylation followed by hydroxylation of the lithium enolate of the lactam afforded the 7β- hydroxyacetallactam. This on epimerization by successive oxidation and reduction gave the corresponding α-hydroxy compound, which was readily converted into the 7α-acetoxy-2- oxo-lactam (**155**). Treatment of **155** with toluene-α-thiol and boron trifluoride etherate in acetic

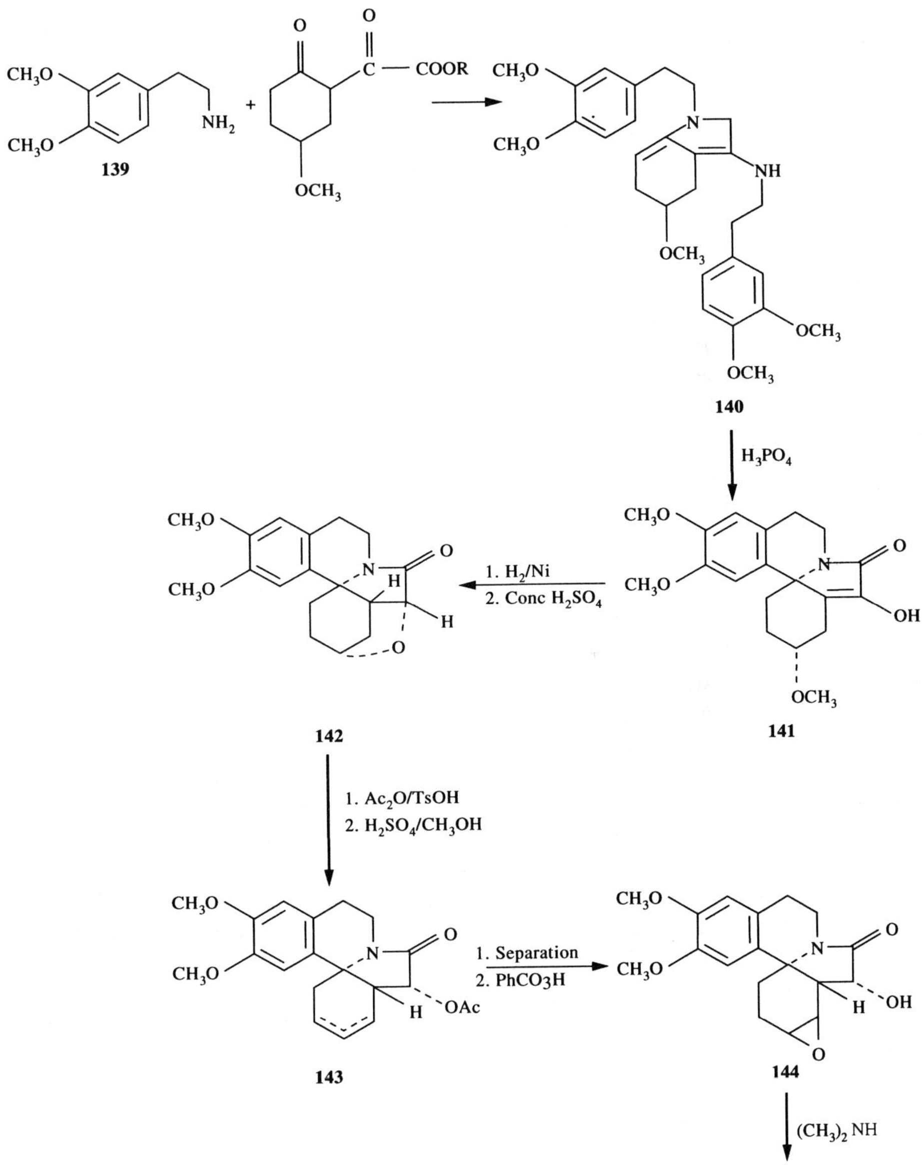

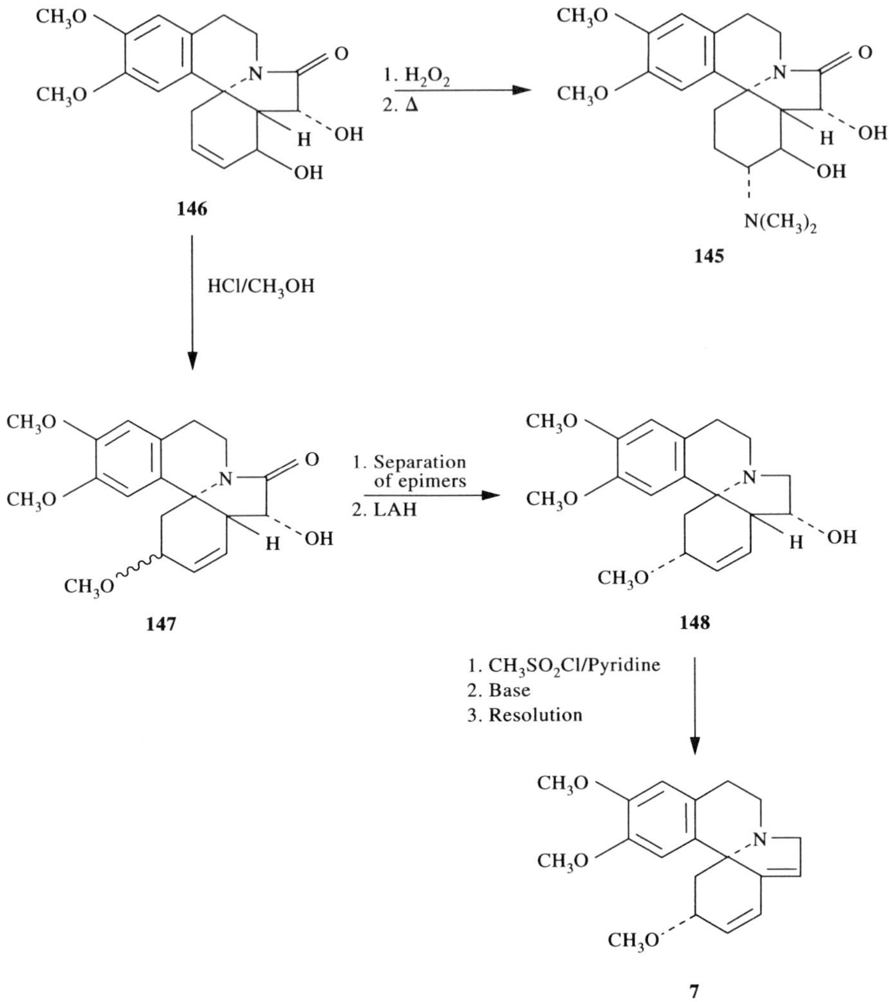

Fig. 9

acid followed by desulphurization with nickel boride [172] gave the required erythrinenone **(156)** in 35 per cent yield, along with the isomeric by-product **(157)** in 55 per cent yield. The conversion of **156** into **7** has already been reported [173].

The partial synthesis of **7** starting from **156** has also been reported [174]. Mondon and Nestler [174] have also described ring-closure reactions leading to the formation of the new cyclic ethers **(158a)**, **(158b)** and **(159)** having the *cis*-erythrinan skeleton. Furthermore, Mondon *et al.* [175] have reported that acetylation of the diol **(160)** yielded the rearranged product **(161a)** which reverted to **160** on hydrolysis. The corresponding *cis*- diol formed a monoacetate without any rearrangement. Assessment of the scope of the reversible rearrangement showed that the rearrangement step consisted of a spontaneous 1,2-shift involving four reaction centers [176]. Further studies [177] revealed that fragmentation of the acetate-mesylate **(161b)** with sodium methoxide led to the β-lactam aldehyde **(162a)**, which was degraded to the β-lactam **(162b)** *via* the carboxylic acid. The constitution of **162a** was determined with the aid of deuteriation experiments, and confirmed by X-ray analysis. Reduction of **162b**, followed by cyclodehydration, yielded the B-nor-*cis*-erythrinan base **(162c)**.

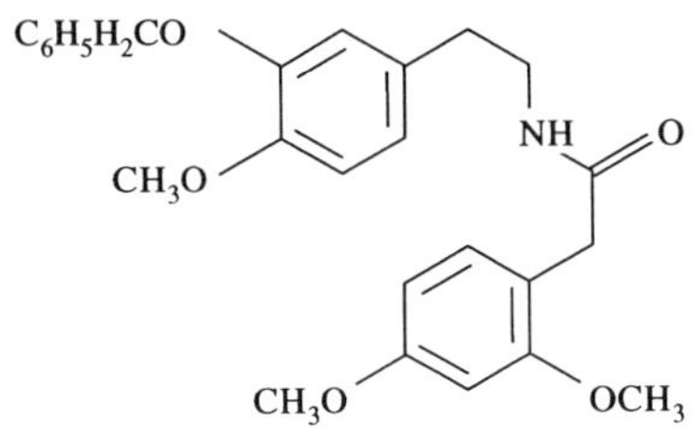

149

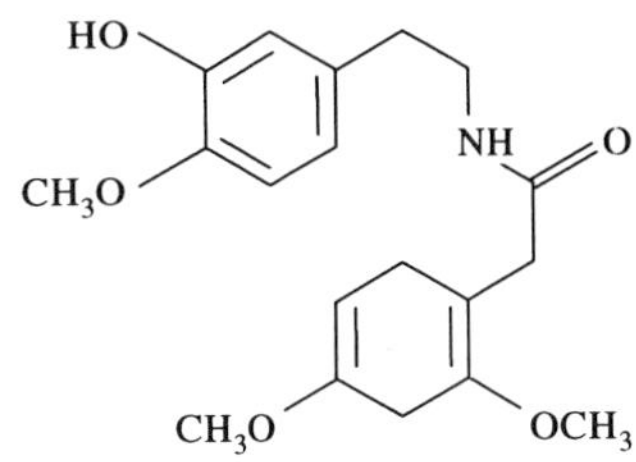

150

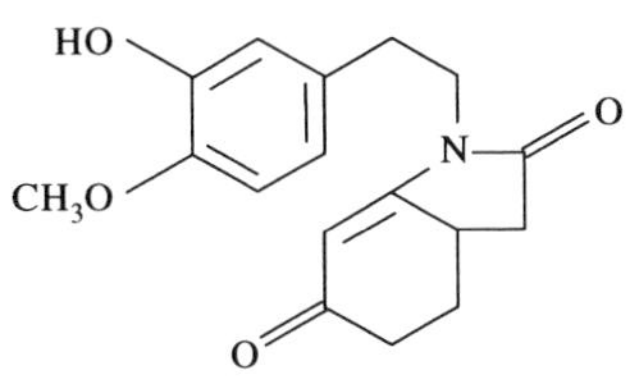

151

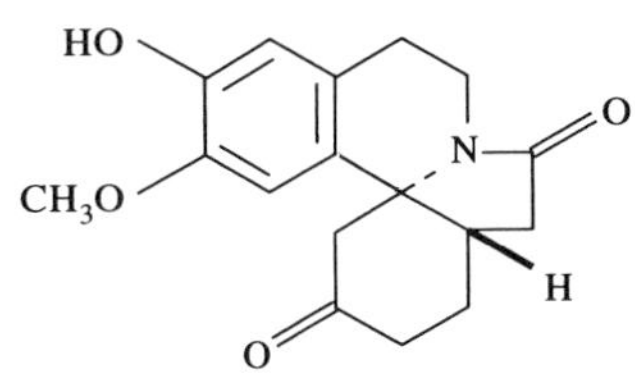

152

Tsuda and co-workers [178] have developed new methods for the synthesis of spiro-type compounds that are related to *Erythrina* alkaloids, either by intramolecular nucleophilic cyclization of dioxopyrrolines or by cycloaddition of activated butadienes to dioxopyrrolines. The conversion of 2-(ethoxycarbonyl)-cycloalkanones **(163)** into the spiro-type compounds **(165)**, in high yield, *via* 2,3-disubstituted dioxopyrrolines **(164)** is shown in Fig. 10 [179]. They also reported that the thermal cycloaddition of butadiene to isoquinolinopyrrolinedione **(166)** proceeded regio- and stereo-selectively to give 1,4- cycloadducts which, on hydrogenation over palladium-carbon, gave a tetracyclic product **(167)**, which was identical with the material obtained by heating **163** (n = 4) with homoveratrylamine followed by treatment with oxalyl chloride and anhydrous phosphoric acid. They have also shown the wide applicability of this method by synthesizing A-nor- and A- homo-erythrinans and other variants in which ring D is heteroaromatic rather than benzenoid. Phenolic erythrinans can be

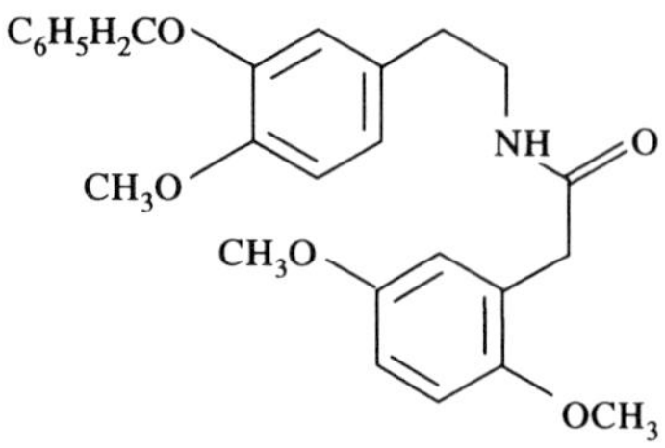

153

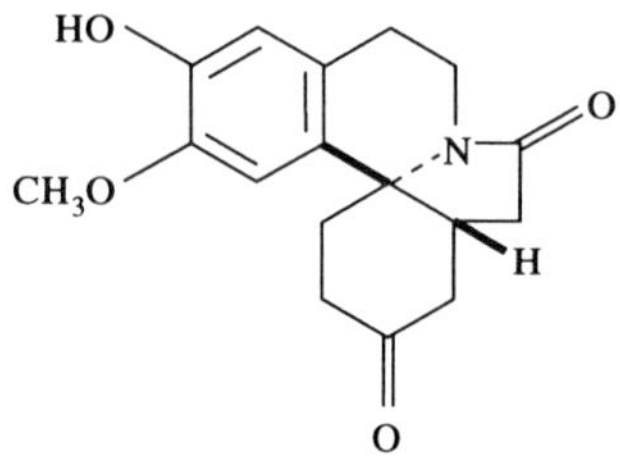

154

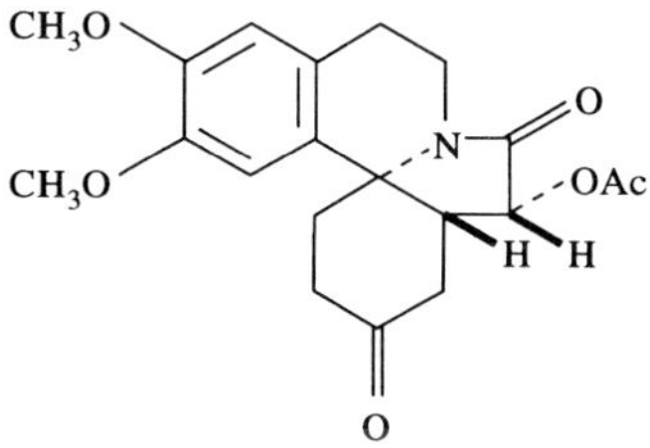

155

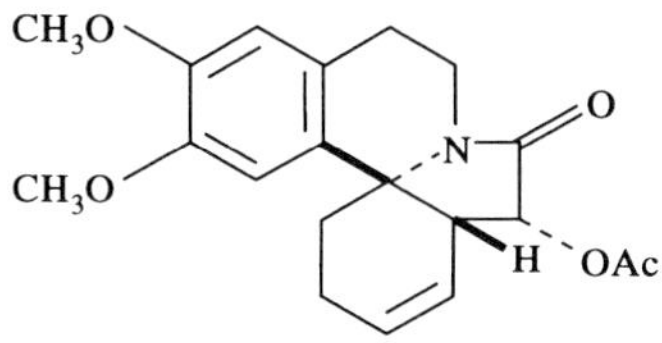

156

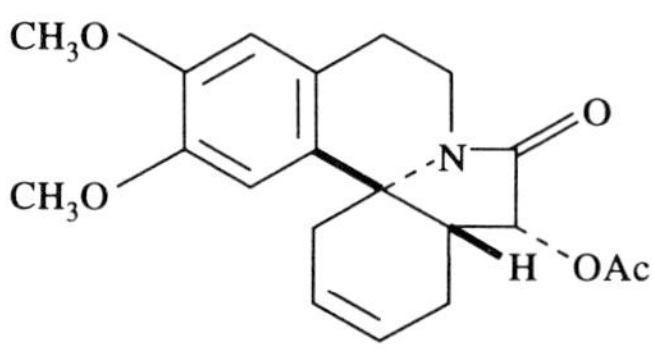

157

synthesized without protection of hydroxyl groups. The dioxopyrroline derivative (**168**) has been cyclized [180] to the corresponding erythrinan (**169a**) and the latter converted into **169b** by decarboxylation with $MgCl_2$ in DMSO [181]. 15,16-Dimethoxy-2,8-dioxo-1, 7-cycloerythrinan earlier prepared by a concerted intermolecular alkylation of the 17β-mesylate [43], has been synthesized from homoveratrylamine in 37 per cent yield [182]. This is useful as an intermediate in the preparation of alkaloids of the dienoid type [43,182]. The Diels-Alder

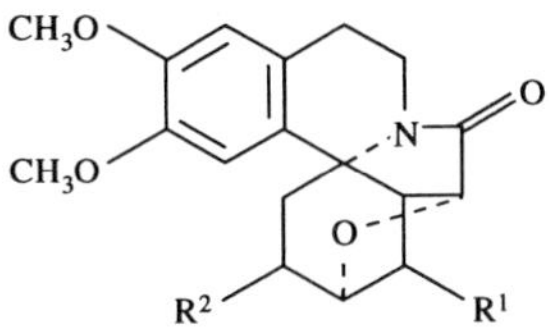

158 a R^1=OH, R^2=H
b R^1=H, R^2=OH

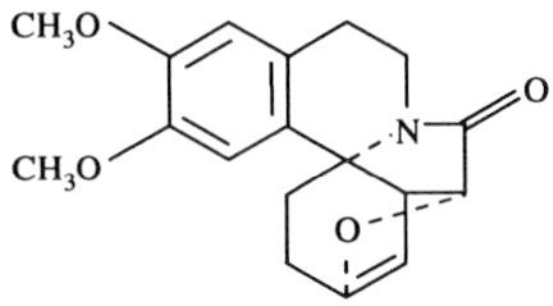

159

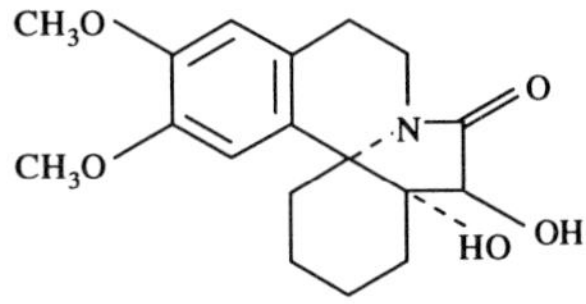

160

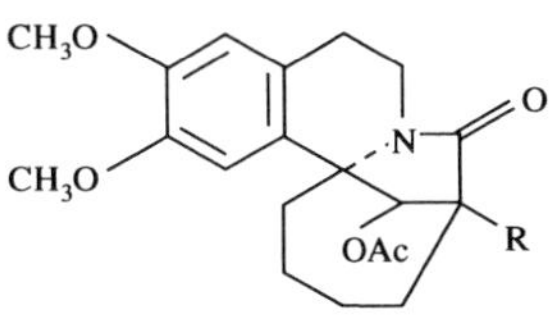

161 a R=OAc
b R=OMs

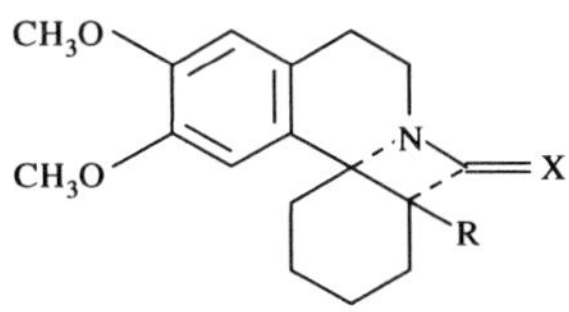

162 a R=CHO, X=O
b R=H, X=O
c R=H, X=H_2

reaction of Δ^2-pyrroline-4,5-diones with activated butadienes which proceeds in a regio- and stereo-selective manner [183] has been applied to the synthesis of ring-D-functionalized erythrinan derivatives in acceptable yields [184-185]. Erysotrine **(7)** and erythraline **(10)** have been synthesized in a stereocontrolled manner using this method [185]. Diels-Alder cycloaddition of 1,3-bis-(trimethylsilyloxy)- butadiene to the dioxopyrroline **(166b)** yielded two adducts; these are the *cis-endo* (with respect to the dioxopyrroline ring) adduct **(170a)**, and the *cis - exo* adduct **(170b)**; the ratio of the products varied, depending on the solvent and the temperature of the reaction [186].

Sano *et al.* [187] have reported the photocycloaddition of trimethylsilyloxy-butadiene to **166a**, followed by treatment of the (2+2) adduct **(171)** with tetra-*n*-butylammonium fluoride (TBAF). To furnish the erythrinan derivative **(169a)** in high yield as a result of a 1,3-shift. Oxidation of 2,8-dioxo-1,7-cycloerythrinan **(172a)** and its 6-ethoxycarbonyl derivative **(172b)** with excess thallium (III) nitrate (TTN) gave the rearranged products **(173a)** and **(173b)**, respectively; their structures were elucidated mainly on the basis of spectral evidence [188].

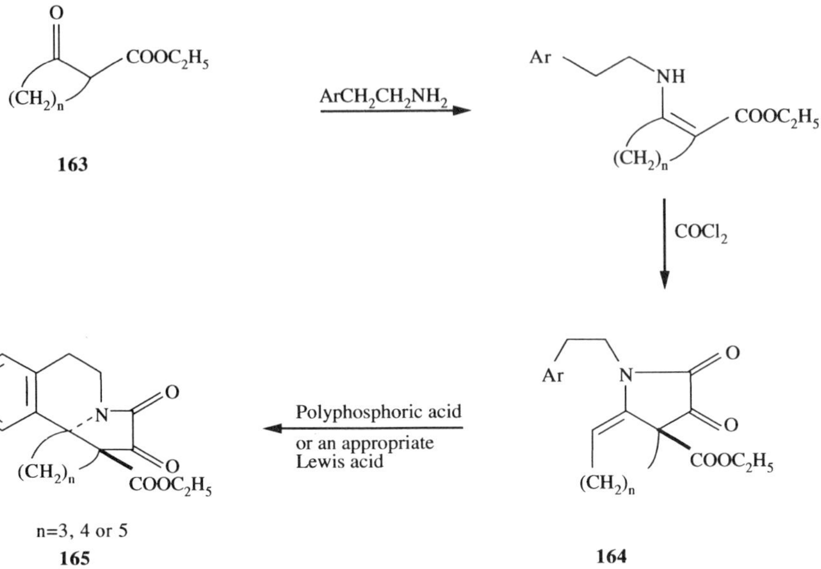

Fig. 10

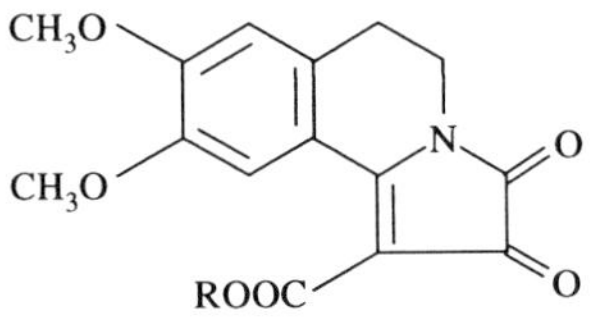

166 a R=C$_2$H$_5$
b R=CH$_3$

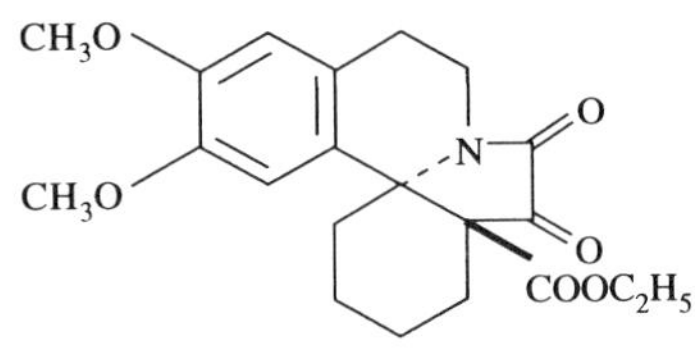

167

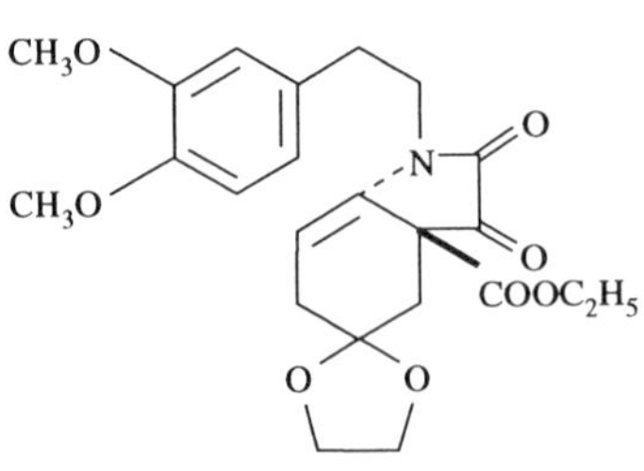

168

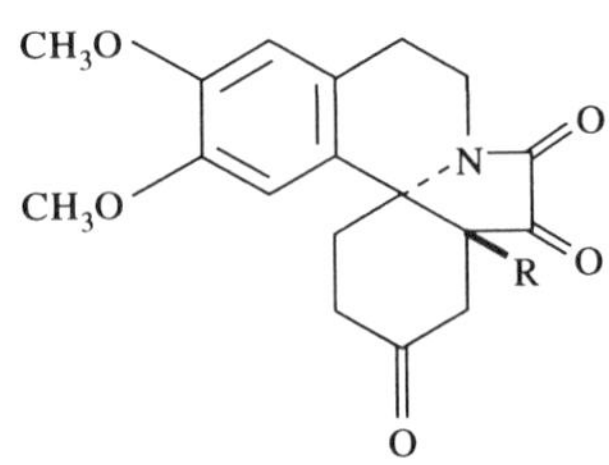

169 a R=COOC$_2$H$_5$
b R=H

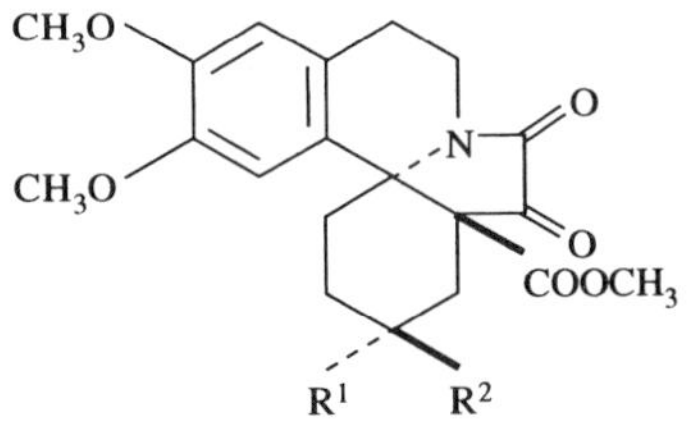

170 a R^1=OSi $(CH_3)_3$, R^2=H
b R^1=H, R^2=OSi $(CH_3)_3$

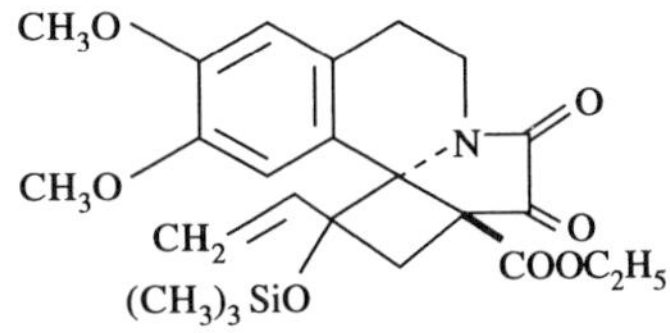

171

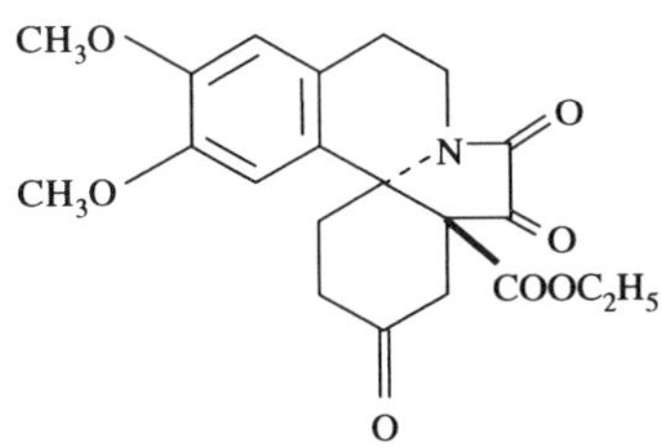

169 a

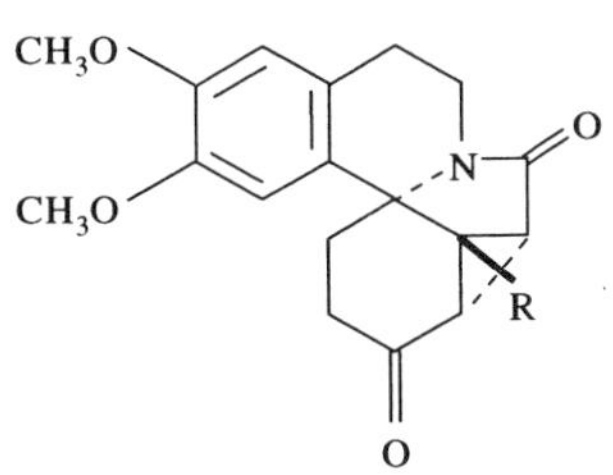

172 a R=H
b R=$COOC_2H_5$

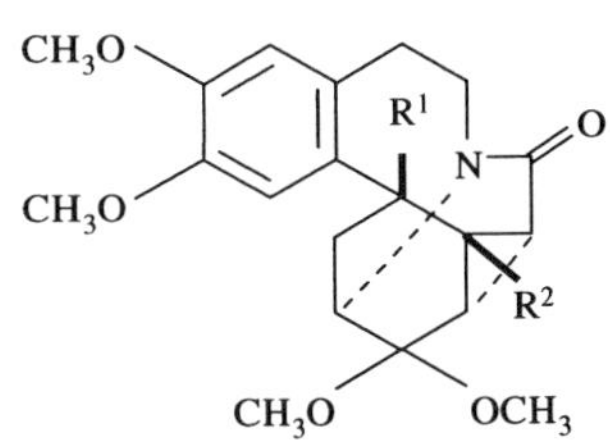

173 a R^1=OCH_3, R^2=H
b R^1=OCH_3, R^2=$COOC_2H_5$

Following the strategy based on the Diels-Alder reaction of activated butadienes to a dioxopyrroline as discussed above, Sano and co-workers [189] have achieved the total synthesis of erythrinan alkaloids. The reaction of isoquinolinopyrrolinedione **(174)** with 1,3-di-*O*- substituted butadienes proceeded in a regiospecific and regioselective manner to give erythrinan derivatives **(175)** and **(176)** (Fig. 11).

Reduction of either derivative with lithium borohydride in THF at -70°, followed by dehydration of the resulting product with hydrochloric acid, afforded the hydroxy-enone **(177)**. Mesylation of **177a** with methanesulphonyl chloride gave the mesylate **(178)**, which on demethoxycarbonylation in the presence of $MgCl_2$ in DMSO, yielded the dione **(179)**. Meerwein-Ponndorf reduction of this dione proceeded stereoselectively to give the epimeric alcohols **(180)** and **(181)** in 70 per cent and 25 per cent yields, respectively. Methylation of **180** with methyl iodide in the presence of a phase-transfer catalyst (KOH with $Et_4N^+Br^-$) furnished (±)-erysotramidine **(8)** (84%); this on reduction with aluminium hydride [190] (generated from $AlCl_3$ and $LiAlH_4$ in THF at room temperature), gave (±)-erysotrine **(7)**. The total synthesis of (±)-erythraline **(10)** (86%) was also achieved from the enedione **(177b)**, by the same sequence of reactions as described above. These results clearly show that the Diels-Alder strategy is an effective method for the synthesis of *Erythrina* alkaloids. These ten-step processes from commercially available arylethylamines represent the shortest route and the highest yield of any of the currently known methods.

The synthesis of **7** and **8** have been achieved by a different route which utilizes a new carbonyl-transposition reaction *via* phenylselenylation (Fig. 12) [191]. Treatment of the dioxoerythrinan **(172b)** with PhSeCl and $BF_3.EtO_2$ in THF afforded the phenylselenide **(182a)**; this on reaction with mercury (II) perchlorate (2 equiv.) in methanol, yielded α,α-dimethoxy ketone **(182b)** as a gum. This was characterised as the crystalline 2α-alcohol **(183a)** and its transesterification with

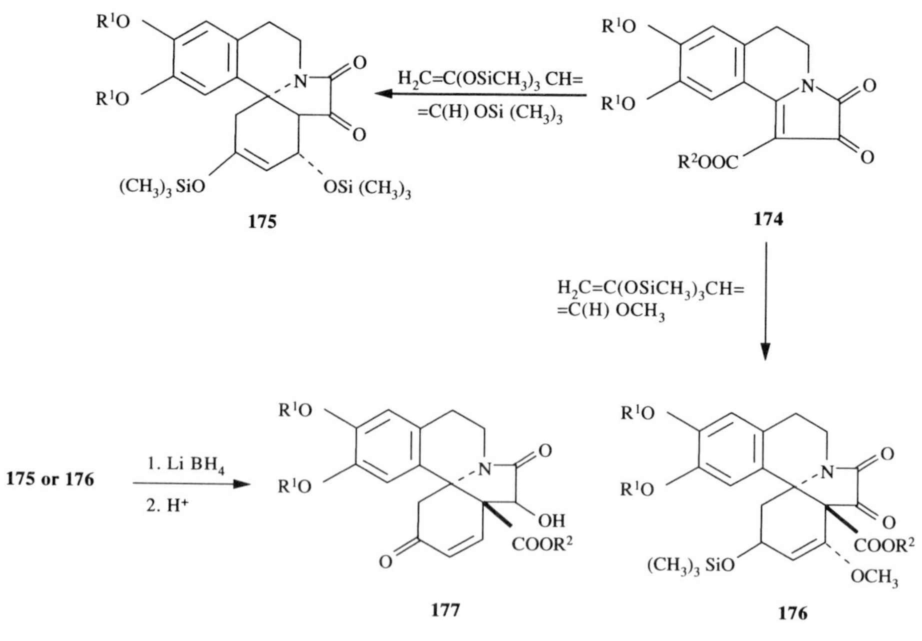

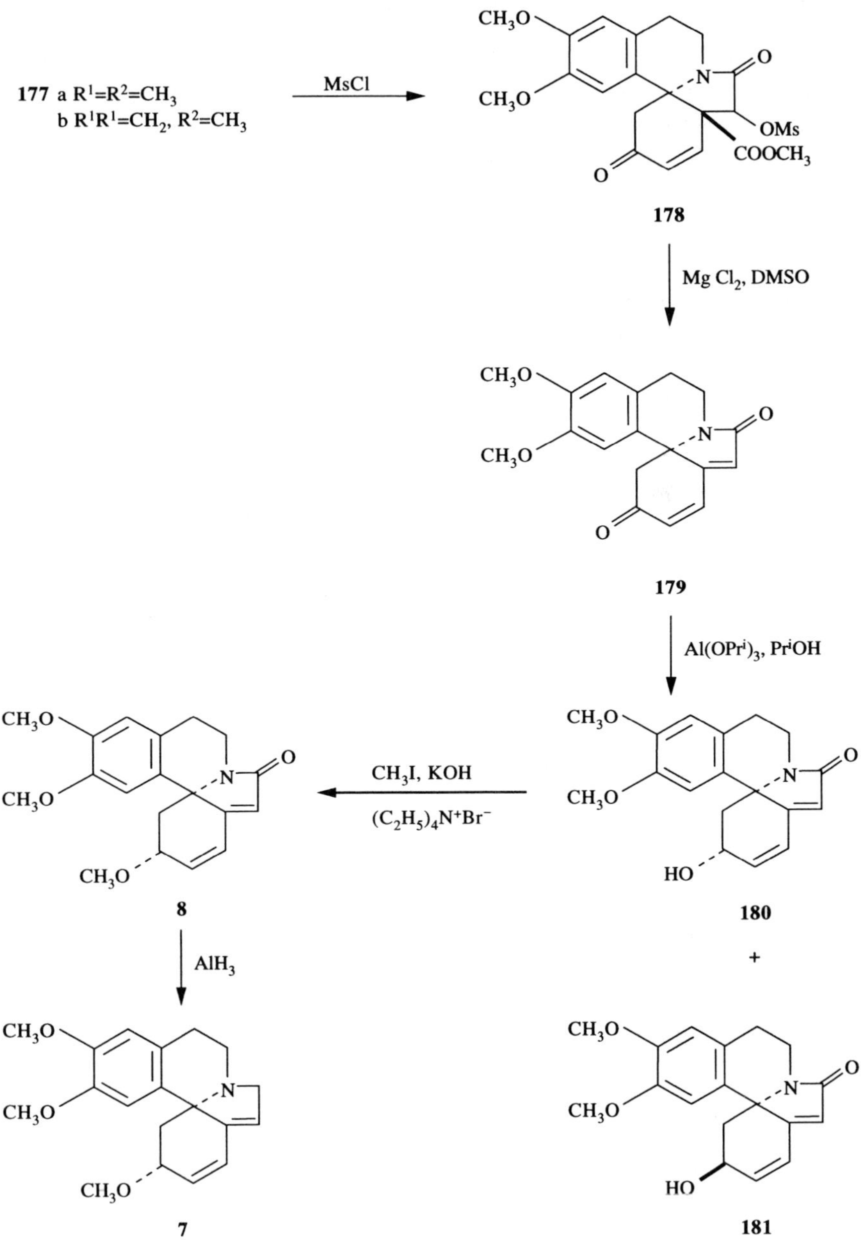
177 a R1=R2=CH3
b R1R1=CH2, R2=CH3
MsCl
178
Mg Cl2, DMSO
179
Al(OPri)3, PriOH
180
+
181
CH3I, KOH
(C2H5)4N+Br−
8
AlH3
7
CH3O
OMs
COOCH3
HO

Fig. 11

sodium methoxide in methanol gave the methyl ester **(183b)** (92%), which was then converted into the dithiocarbonate **(183c)** (81%). When this compound was treated with tributyltin hydride it yielded the deoxy- olefin **(184)** (97%) with concomitant opening of the cyclopropane ring. Acid hydrolysis of the latter gave the enone **(185)** (100%), which, if heated with calcium chloride in DMSO, underwent demethoxycarbonylation to yield the enones **(186a)** and **(186b)** in a ratio of 1:5 (total yield 83%); the reaction of **185** with magnesium chloride gave the

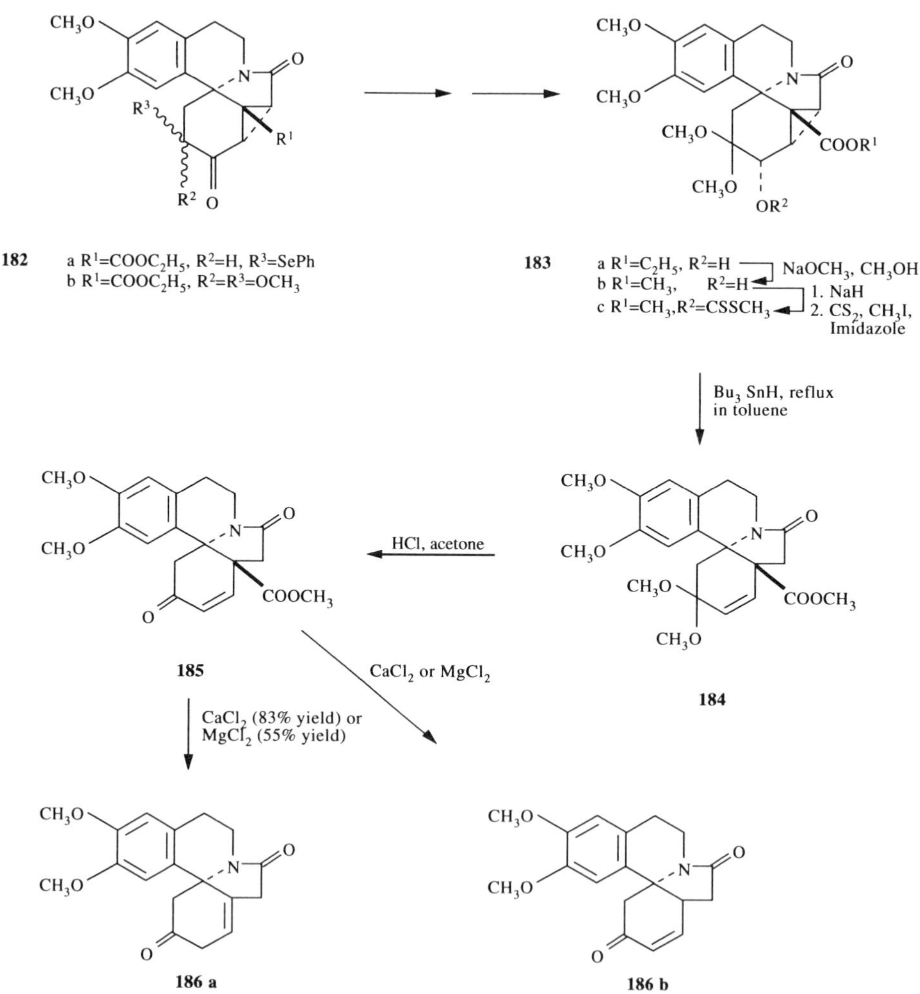

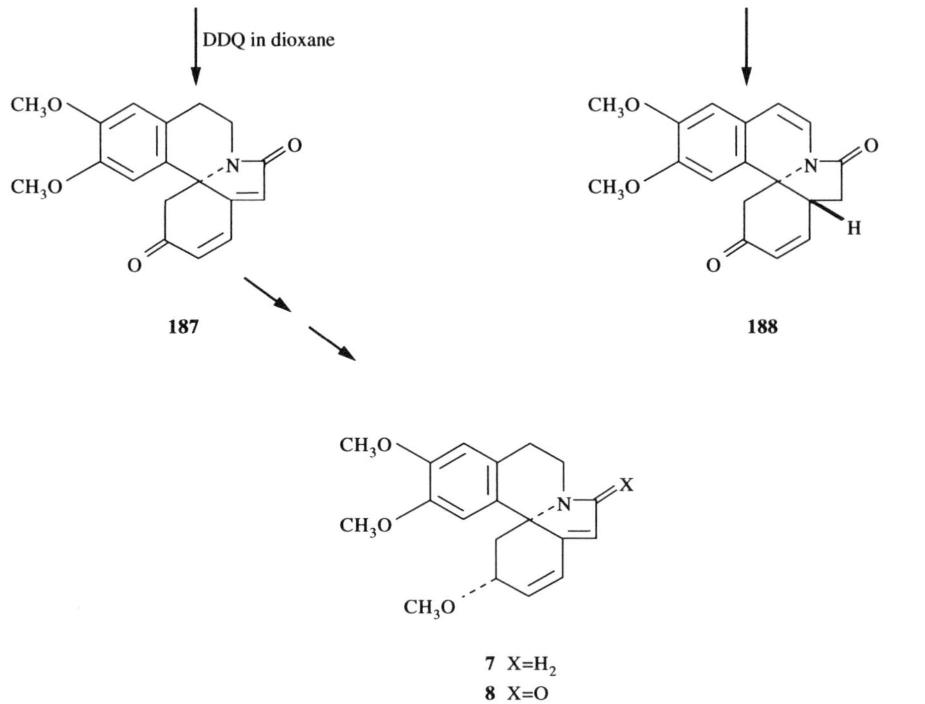

Fig. 12

same products in 1:1 ratio (55%). The compound **186a** readily isomerized to **186b** with 1,8-diazabicyclo[5.4.0]undec-7-ene (DBU) in benzene. The enone **186a** was oxidised by DDQ in dioxane to the dienone **(187)** (24%). A similar treatment of **186b** gave an isomeric dehydro compound **(188)**. The conversion of **187** into **7** and **8** has already been reported [185]. The enones **186a** and **186b** are potential intermediates in the preparation of alkenoid alkaloids such as erythramine **(45)** and isococculidine **(80)**.

Recently, the synthesis of **8** has been accomplished [192]. Its total synthesis involving the novel ring cleavage of the 15,16- dimethoxy- 2,8-dioxo- 1,7-cycloerythrinan with phenylselenyl chloride has earlier been reported [43].

The partial synthesis of 11-oxygenated *Erythrina* alkaloids has been achieved from the naturally occurring abundant alkaloids erysodine **(2)** and erysovine **(3)** [193]. Oxidation of **3** with lead tetraacetate in acetic acid afforded 11α-acetoxyerysovine **(190)** stereospecifically, in virtually quantitative yield, presumably *via* the intermediate **(189)**. The isomer **2**, when oxidized with the same reagent, afforded an *ortho*- quinone derivative **(191)** which was subsequently transformed into an 11β-acetoxyerysodine derivative **(193)** in 10 per cent yield, through the *para*-quinonoid intermediate **(192)**; the aromatic oxidation products **(194a)** and **(194b)** were also formed, by an alternative route (Fig. 13). The apparent regiospecificity of the reactions leading to the 11β−acetate may not be very significant, owing to the low overall yield obtained, but the high yield of the 11α-acetate that is produced in the oxidation of **3** may perhaps be due to co-ordination of acetic

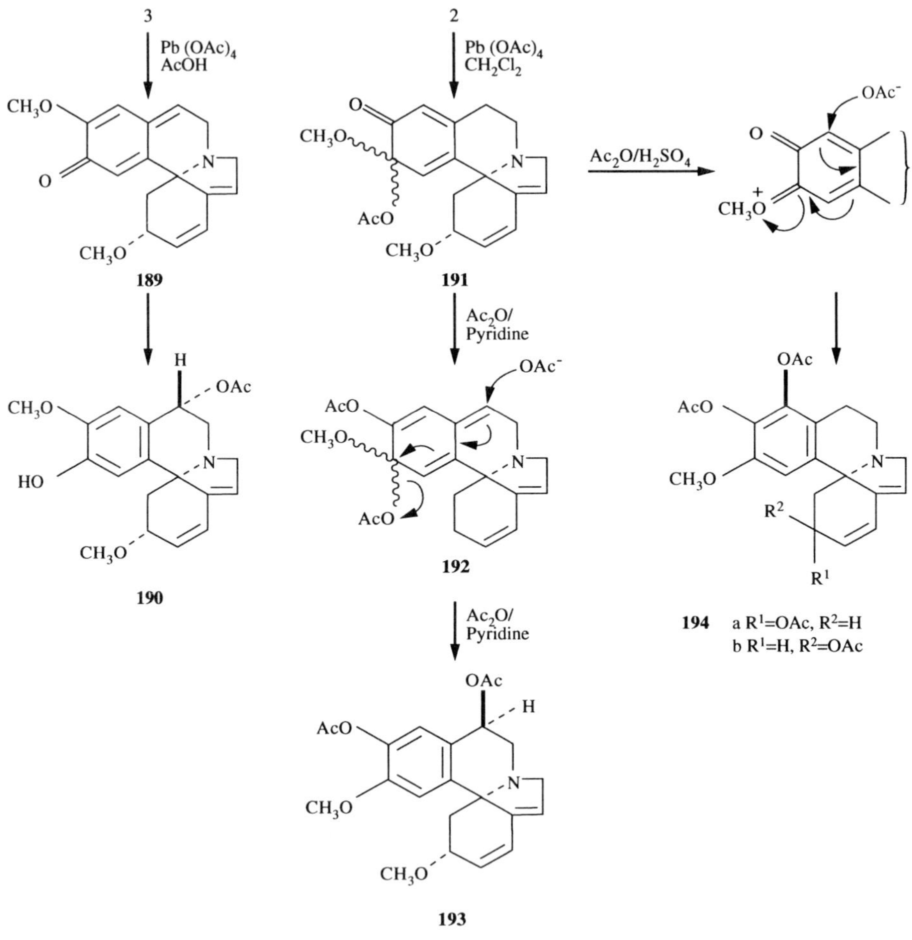

Fig. 13

acid with the tertiary nitrogen atom and delivery of the the acetate unit at position 11 from the α-face of the molecule. Other studies on the partial syntheses of 11-oxygenated alkaloids have been described [37,38,42,194,195].

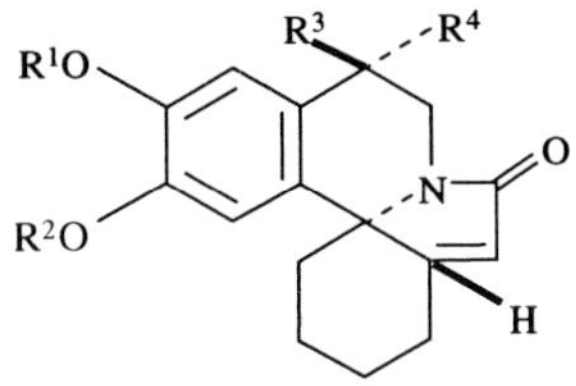

195 a R^1=R^2=CH_3, R^3=R^4=H
b R^1=R^2=CH_3, R^3=OCH_3,=R^4=H
c R^1=R^2=CH_3, R^3=H, R^4=OCH_3
d R^1=R^2=CH_3, R^3R^4=O
e R^1R^2= –CH_2–, R^3=R^4=H
f R^1R^2= –CH_2–, R^3=OCH_3, R^4=H

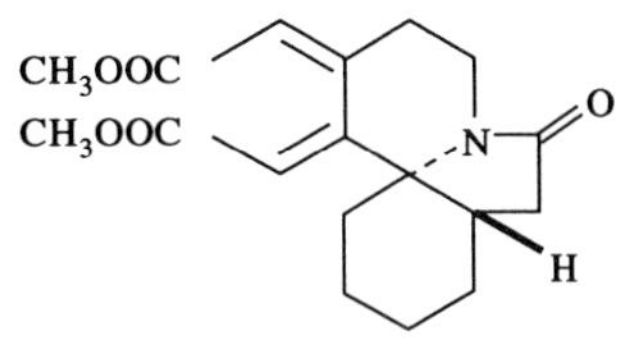

196

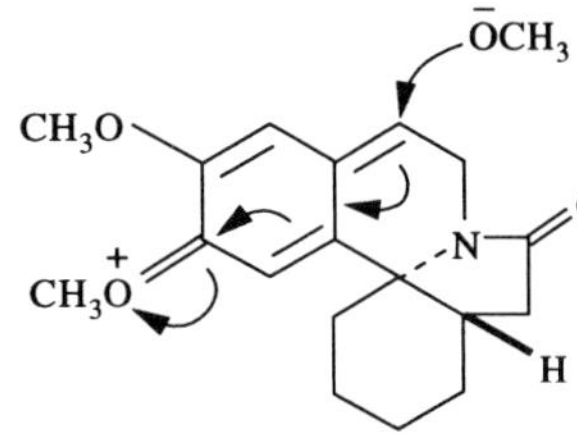

197

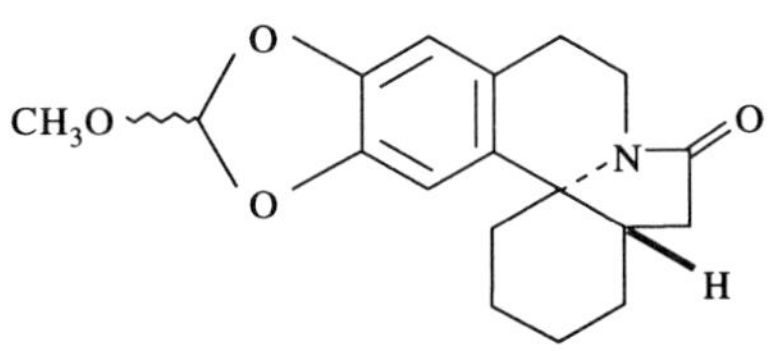

198

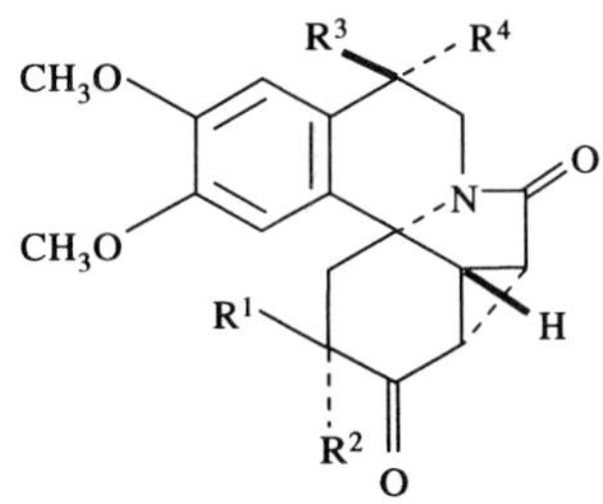

199 a R^1=R^2=R^3=R^4=H
b R^1=R^2=R^4=H, R^3=OCH_3
c R^1=R^2=H, R^3R^4=O
d R^1=R^2=OCH_3, R^3=R^4=H
e R^1=R^2=R^3=OCH_3, R^4=H
f R^1=R^2=OCH_3, R^3R^4=O

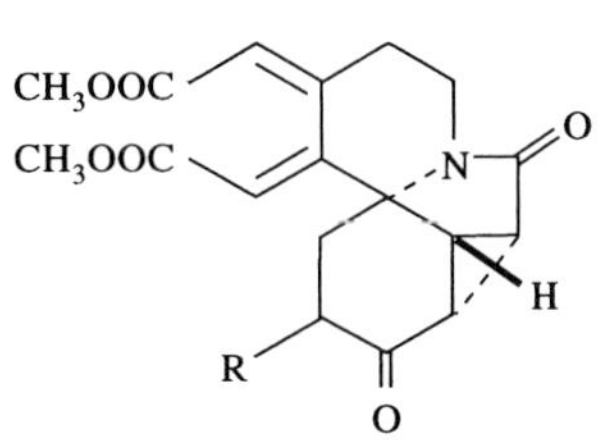

200 a R=H
b R=OCH_3

Isobe *et al.* [196] have reported the total synthesis of (±)-erythristemine **(15)**. Treatment of 15,16-dimethoxy-*cis*-erythrinan-8-one **(195a)** with ceric ammonium nitrate in methanol at room temperature gave the 11β-methoxy derivative **(195b)** (73%), the 11α-isomer **(195c)** (3%), the 11-ketone **(195d)** (6%) and the ring opened compound **(196)** (7%). The 11-oxygenated derivatives are considered to be formed *via* the intermediate **(197)**. The assigned β-configuration of the methoxy group in **195b** agrees well with the stereochemical consideration that the β-face of ring C in **195a** is less hindered for reagent approach. A similar oxidation of 15,16-methylenedioxy-*cis*-erythrinan-8-one (195e) also produced the 11β -methoxy derivative **(195f)**, though in lower yield (30%). The major product in this oxidation was an *ortho* ester **(198)**. Oxidation of 15,16-dimethoxy-1,7-cyclo-*cis*-erythrinan-2,8-dione **(199a)** with ceric ammonium nitrate gave the β-methoxy derivative **(199b)** (68%), the ketone **(199c)** (13%) and a ring opened product

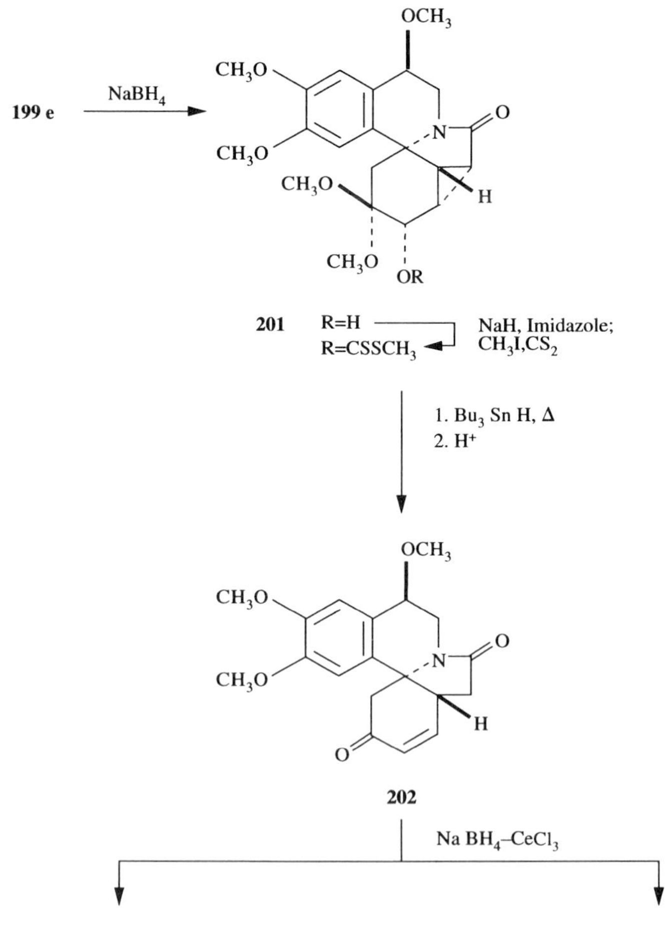

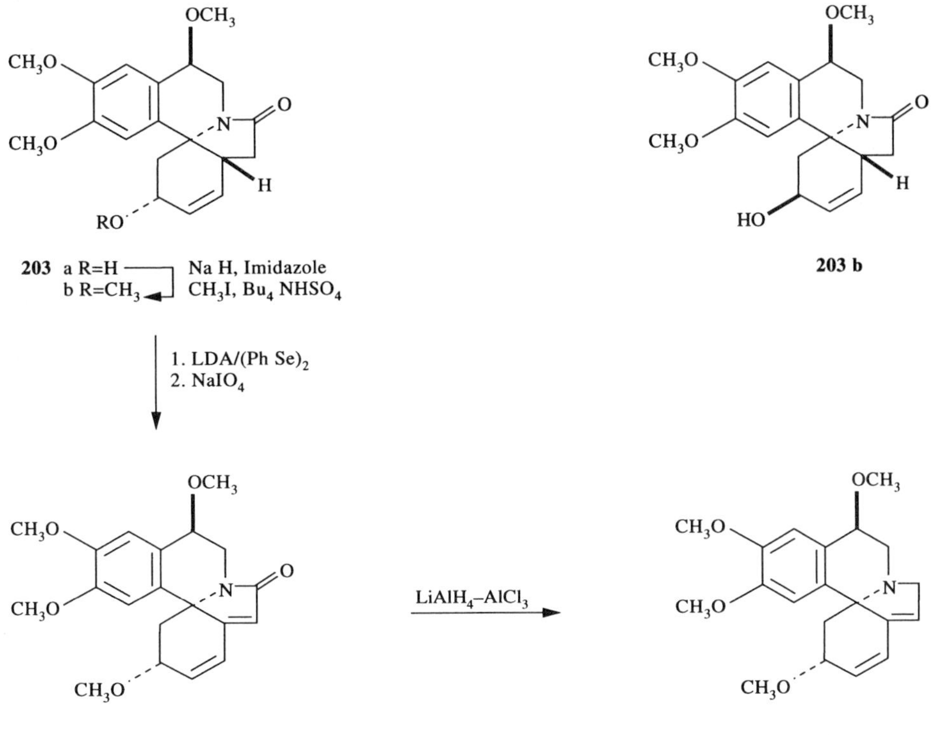

Fig. 14

(200a) (8%). The introduction of the methoxyl group at the 11β-position was stereoselective. Similarly, oxidation of **199d** with ceric ammonium nitrate produced the 11β-methoxy derivative **(199e)** together with the 11-oxo derivative **(199f)** and the *seco* compound **(200b)**. The 11β-methoxy compound **(199e)** was converted into **15** in the racemic form by the sequence of reactions given in Fig. 14.

Oxidation of **8** with ceric ammonium nitrate in AcOH–CH_3CN gave in moderate yield, the 11β-acetoxy derivative which was transformed into erythrartine **(14)** with $LiAlH_4$- $AlCl_3$(3:1) in THF [197].

3-Demethoxyerythratidinone **(77)** has been synthesized (in racemic form) by five different routes (Fig.15) [198]. These constitute the first total synthesis of an alkenoid *Erythrina* alkaloid. In method A, homoveratrylamine **(139)** was converted into the 2,8-dioxo-7α-hydroxyerythrinan ethylene acetal **(205a)** by a series of steps, in an overall yield of 65%. The latter was converted into the mesylate **(205b)**, which was transformed into the enone **(206)** in 90% yield. Reduction of the latter with $LiAlH_4$ and $AlCl_3$ in THF, followed by acid hydrolysis of the resulting amine

(207), furnished (±)-demethoxyerythratidinone **(77)**, identical with the natural product.

Method B provided an alternate route to **(206)**. The 2,8-dioxoerythrinan ethylene acetal **(209)** was prepared from 4-benzyloxy-3-methoxyphenylethylamine

Method A

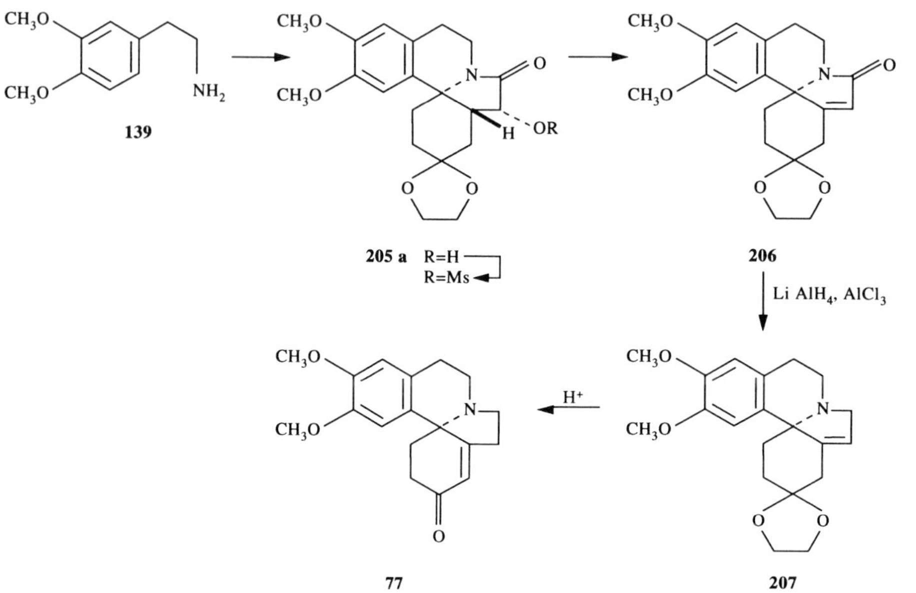

Method B

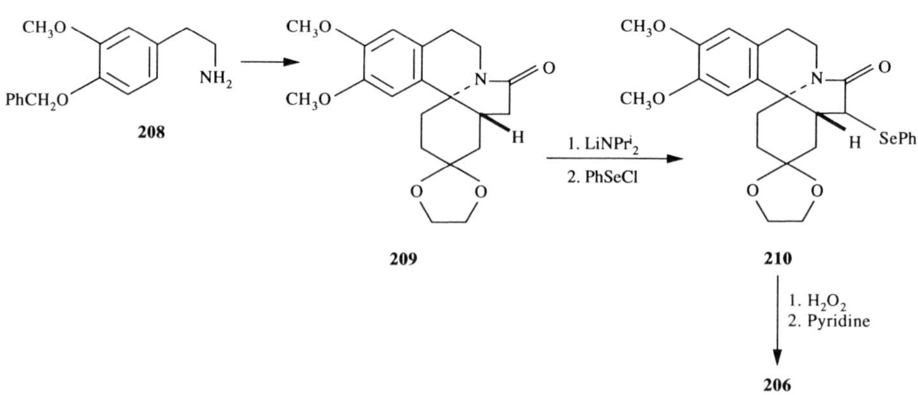

Method C

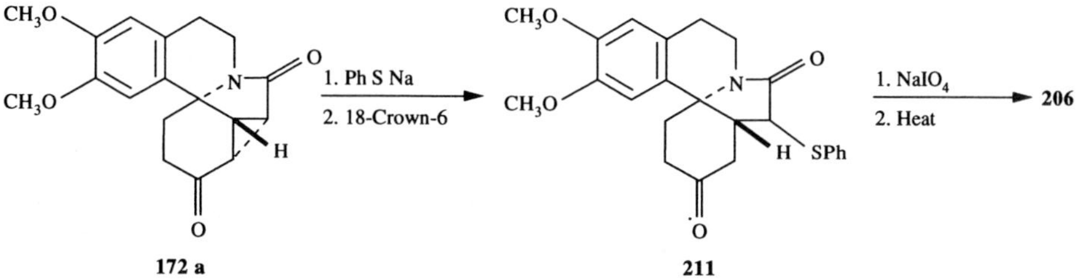

Method D

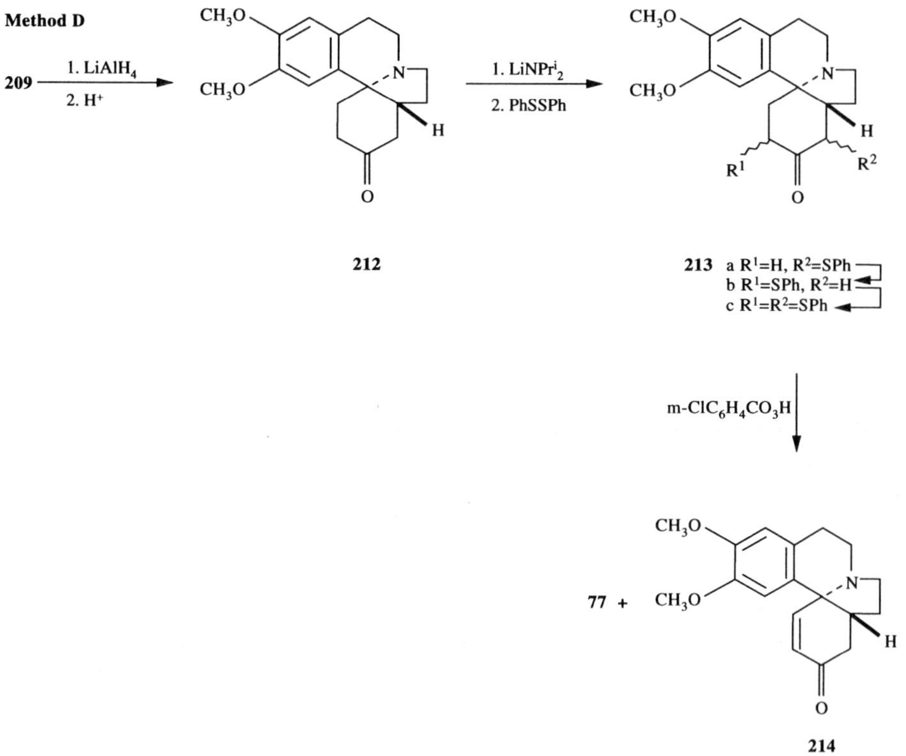

Method E

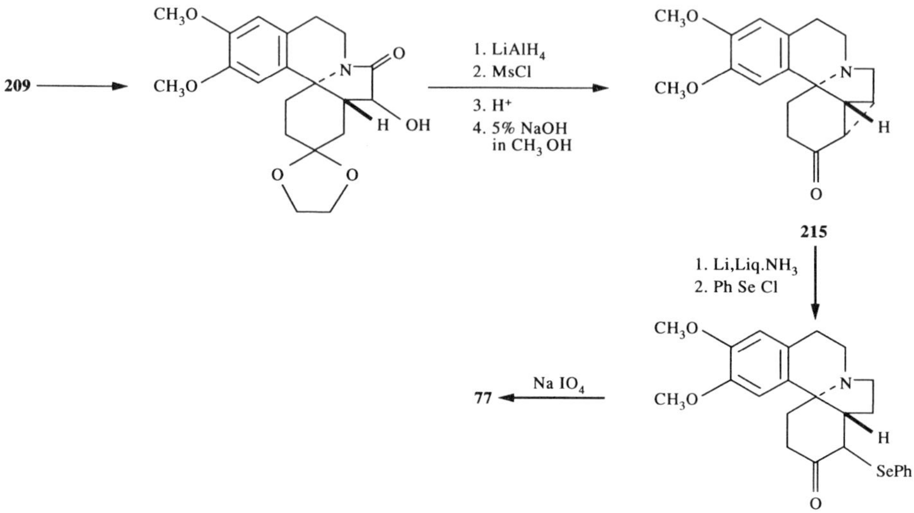

215

216

Fig. 15

(208). It was converted into the acetal **(206)** by phenylselenation [to form **(210)**] followed by treatment with hydrogen peroxide in pyridine. As a variation, in method C the intermediate **(206)** was synthesized from 2,8-dioxo-1,7-cycloerythrinan **(172a)** *via* the phenylsulphenylamide **(211)**.

In method D, 2,8-dioxoerythrinan ethylene acetal **(209)** was reduced with $LiAlH_4$; deacetalization of the product yielded the 2-oxoerythrinan derivative **(212)**. Treatment of the latter with $LiNPr^i_2$ followed by phenylsulphenylation with (PhSSPh) afforded a mixture of mono- {**(213a)** and **(213b)** [62.7%]} and di {**(213c)** [6%]}phenylsulphides. If the inseparable mixture of sulphides **(213a)** and **(213b)** was oxidized with *m*-chloroperbenzoic acid and then heated, it gave a mixture of the isomeric enones (**77**, 48%) and (**216**, 3.3%), which were separated by chromatography on silica gel. The identity of the former with demethoxy-erythratidinone was confirmed by spectral comparisons with an authentic sample.

In method E, the 2-oxo-1,7-cycloerythrinan **(215)** was obtained from the acetal **(209)** in five steps, in 62% yield. Birch reduction followed by phenylselenylation gave the α-phenylselenyl derivative **(216)**, which afforded the unsaturated ketone **(77)** when it was oxidized.

Of the five routes that are shown in Fig. 15, methods A and B seem to be the most practical ones, bearing in mind the availability of starting materials, the simplicity of the procedures, and the high overall yields.

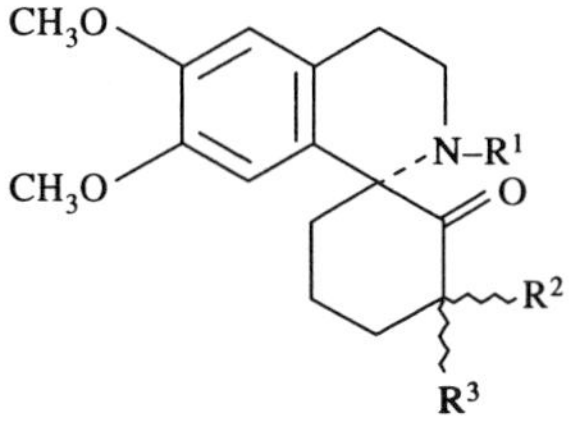

217a R^1=R^2=R^3=H
R^1=BOC, R^2=R^3=H
R^1=BOC, R^2=SePh, R^3=H

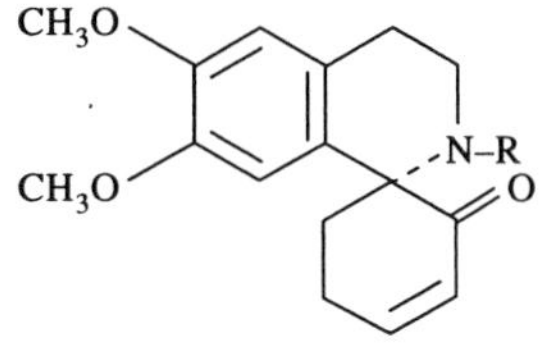

218 a R=BOC
b R=H
c R=$COCH_2P(O)(OCH_3)_2$

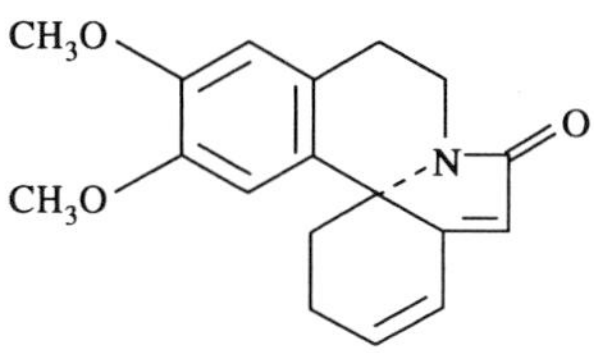

219

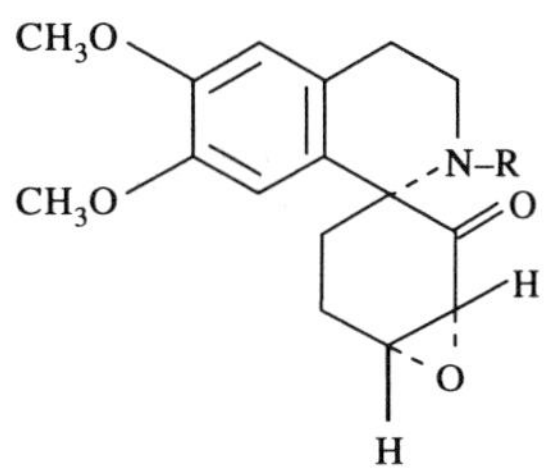

220 a R = BOC
b R = H
c R = $COCH_2P(O)(OCH_3)_2$

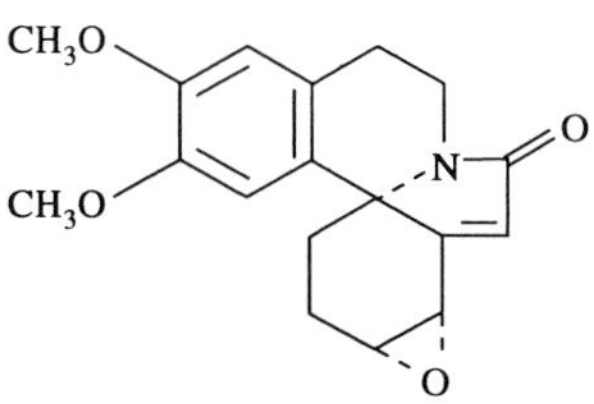

221

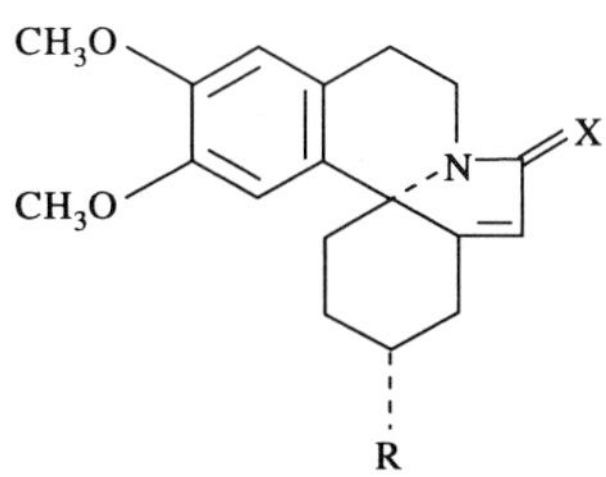

222 a R = OAc, X=O
b R = OH, X=H_2

3-Demethoxyerythratidinone **(77)** has also been synthesized *via* the dibenzazonine base [199], and from homoveratrylamine and cyclohexane-1,4-dione monoethylene acetal [200,201]. The synthesis of **77** has also been accomplished by application of an intramolecular Wittig reaction as the key step for constructing the α,β- unsaturated five-membered lactam [202]. Treatment of the spiro-amine **(217a)** with di-*tert*-butylcarbonate in chloroform gave the BOC amide **(217b)** which on reduction with LDA followed by phenylselenyl chloride gave the seleno derivative **(217c)**. Oxidation of **217c** with sodium periodate in ethanol furnished the conjugated ketone **(218a)** in good yield. Removal of the BOC group with TFA in methylene chloride then gave the amine **(218b)** which on acetylation with dimethylphos- phonoacetyl chloride in methylene chloride, in the presence of pyridine, afforded the phosphonate **(218c)**. Intramolecular Wittig reaction of **218c** with aqueous potassium hydroxide and benzene gave the erythrinan derivative **(219)**. Oxidation of **218b** with hydrogen peroxide and aqueous sodium hydroxide in methanol gave the epoxide **(220a)** in good yield. This after deprotection of the BOC group gave the amine **(220b)** which, on acetylation with phosphonoacetyl chloride, yielded the phosphonate **(220c)**. The latter was submitted to intramolecular Wittig reaction under the same conditions as described above to afford the lactam

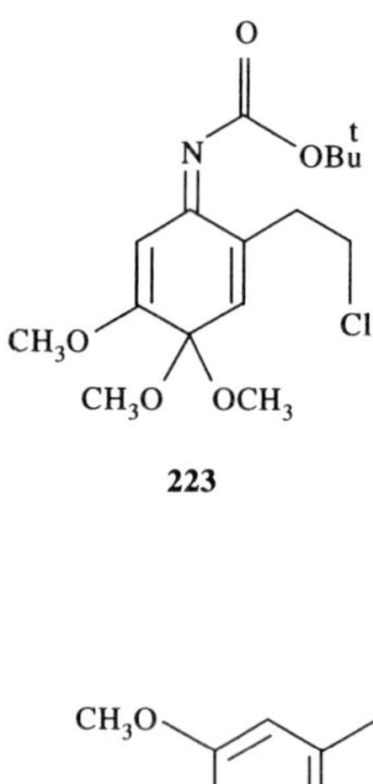

223

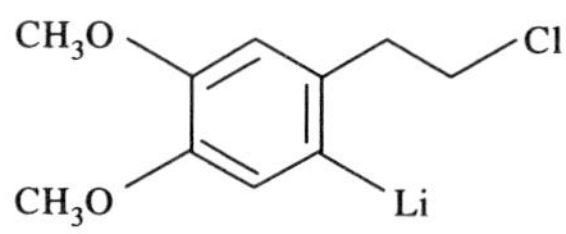

224

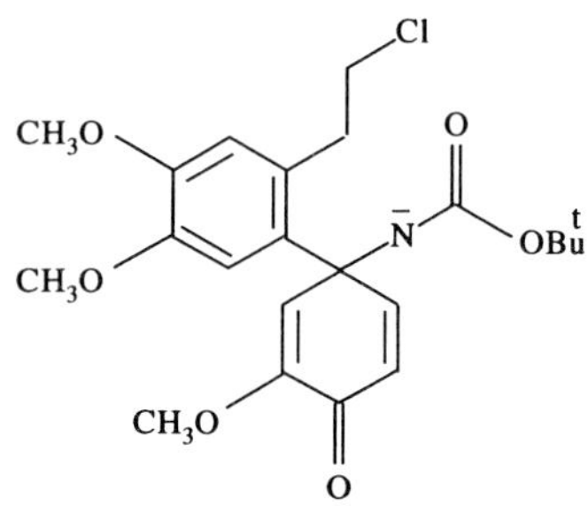

225

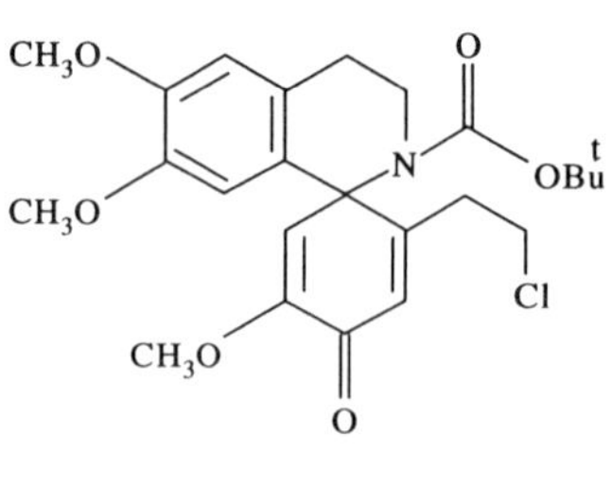

226

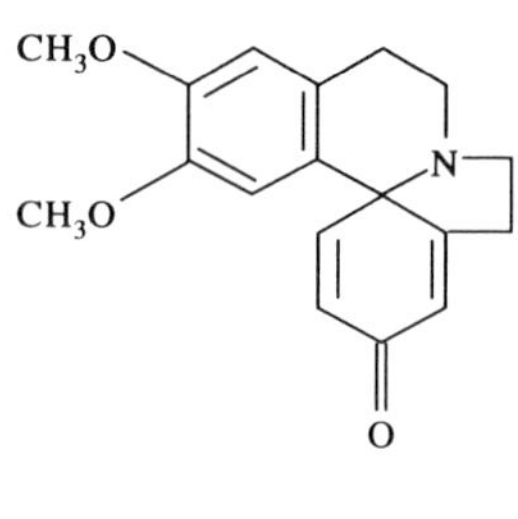

227

(221). Reduction of **221** with zinc in acetic acid and acetic anhydride gave the acetate **(222a)** and **219** in 67 per cent and 26 per cent yields, respectively. Treatment of **222a** with aluminium hydride (prepared from $LiAlH_4$ and $AlCl_3$ in THF and ether) furnished **222b**. This on Swern oxidtion [203] gave **77** in 45 per cent yield.

Recently, Tsuda and co-workers [204] have synthesized **77** starting from ethyl 5,5-ethylenedioxy-2-oxocyclohexane carboxylate. 6β-Ethoxycarbonyl-7β-hydroxy-15,16-dimethoxy- 2,8-dioxo-*cis*-erythrinan is the key intermediate in this synthesis.

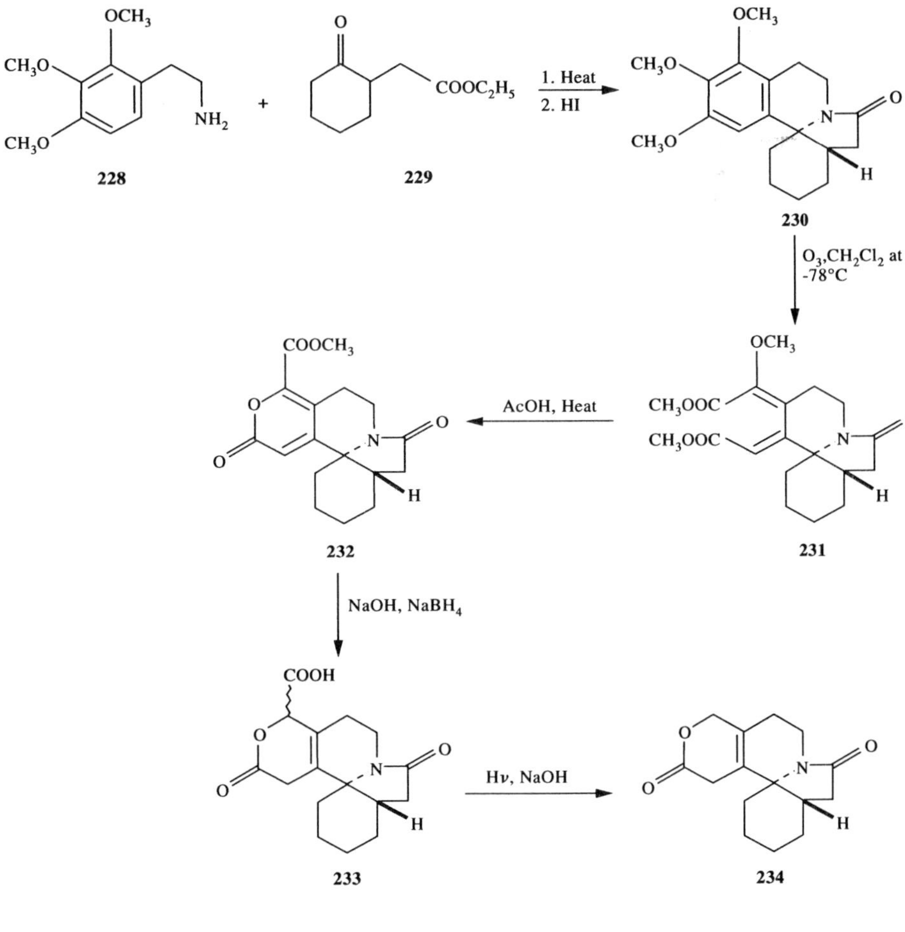

Fig. 16

Usefulness of such an intermediate in the synthesis of natural erythrinan alkaloids has been demonstrated [205]. A total synthesis of **7** has recently been achieved following this route [206, 207].

A new convergent approach to the synthesis of the *Erythrina* alkaloids has been reported [208]. The required precursor to the quinone imine **(223)** was

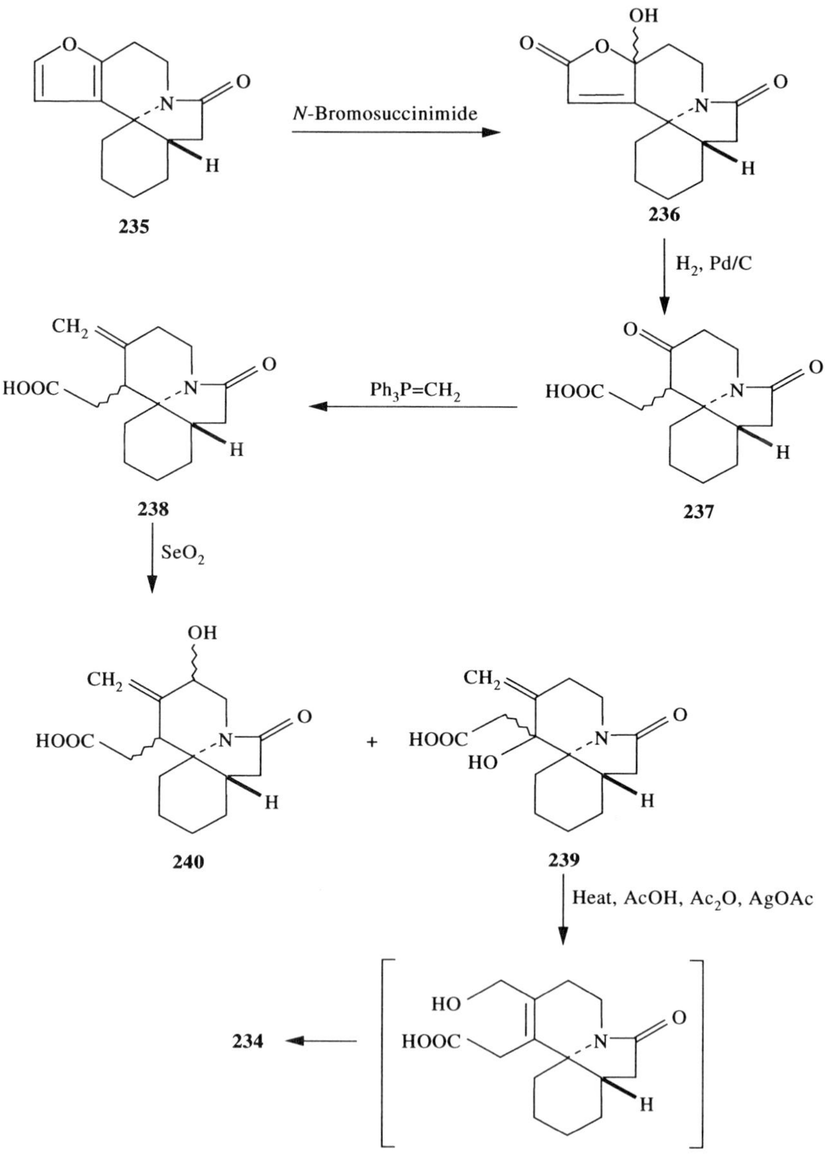

Fig. 17

prepared by reaction of 4,5-dimethoxy-2-(2-chloroethyl)aniline with phosgene to form the isocyanate followed by reaction with *tert*-butyl alcohol to give an intermediate urethane. Anodic oxidation of this urethane gave the quinone imine ketal (**223**, 90%). Reaction of the organolithium reagent (**224**) with **223** afforded a crystalline amide (**226**) *via* the intermediate anion (**225**). Reaction of **226** with trifluoroacetic acid/toluene *p*-sulphonic acid at room temperature hydrolysed the *tert*-butoxycarbonyl amide to give the crude amine, which underwent cyclization to give erysodienone methyl ether (**227**).

The synthesis of *β*-erythroidine skeleton (**234**) has been reported by two different routes [209]. One route started from 15,16,17-trimethoxy-*cis*-erythrinan-8-one (**230**), which had been prepared by the condensation of 2-(2,3,4,-trimethoxy- phenyl)ethylamine (**228**) and 2-(ethoxycarbonylmethyl) cyclohexanone (**229**) followed by cyclization with hydriodic acid. Selective ozonolytic cleavage of the aromatic ring of **230** gave the diester (**231**) (33%), which on heating with 70% acetic acid in a sealed tube yielded the pyrone derivative (**232**) (57%). Alkaline hydrolysis of the latter followed by treatment of the resulting keto-acid with sodium borohydride gave (after acidification) a mixture of epimeric *δ*-lactones (**233**) (66%). Photochemical removal of the carbonyl group then yielded 14,17-dihydro- 16(15*H*)-oxaerythrinane-8,15-dione (**234**) (18%) (Fig.16).

The second route involved the treatment of D-furanoerythrinane (**235**) with *N*-bromosuccinimide to yield the hydroxy-*γ*-lactone (**236**) (27%), which on hydrogenation over 10% Pd/C in ethanol afforded an epimeric mixture of keto-acids (**237**). Wittig reaction of the mixture with methylenetriphenylphosphorane yielded the oxomethylene derivative (**238**) (93%) as the sole product. Oxidation of the latter with selenium dioxide gave two hydroxy-derivatives (**239**) (50%) and (**240**) (49%). Heating the former in acetic acid and acetic anhydride with silver acetate (or with tetrabutylammonium acetate) in a sealed tube effected allylic rearrangement, and cyclization of the resultant primary alcohol gave the compound (**234**) (Fig.17).

The dibenzazonine (**242**), related to a biosynthetic precursor of the *Erythrina* alkaloids, has been prepared by intramolecular nickel promoted coupling of bis-(2-phenylethylamine) (**241**) which in turn was obtained from the commercially available 3-methoxyphenylacetic acid by a conventional series of reactions [210]. The other derivative (**243**) was obtained by intramolecular oxidative coupling of tetramethoxytrifluoroacetamide by thallium (III) trifluoroacetate in trifluoroacetic acid at 25° [211]. Employing the dibenzazonine base (**244**) a novel synthesis of *cis*-15,16-dimethoxyerythrinan-3-one (**246b**) has been reported [212]. Treatment of **244** with sodium in liquid ammonia produced the desired diene (**245**, 96%). The latter, when heated with 10 per cent sulphuric acid, underwent cyclization to give the erythrinane base (**246a**) as the sole product, and this was converted (by treatment with excess diazomethane) into the dimethoxy derivative (**246b**). This reaction seems to represent a useful route for the synthesis of *Erythrina* alkaloids Total synthesis of dibenz[*d,f*]azonine alkaloid bractazonine [213], laurifonine and laurifine [214] have been described.

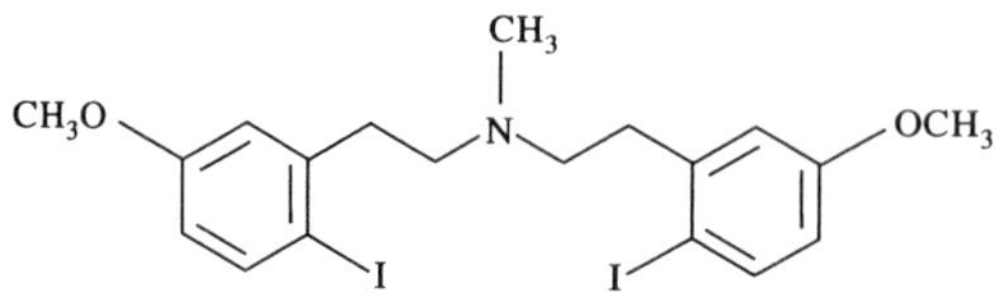

241

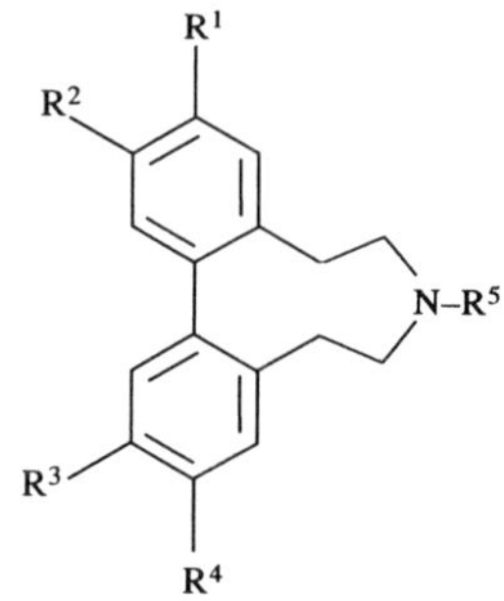

242 R^1=R^4=OCH_3, R^2=R^3=H, R^5=CH_3

243 R^1=R^2= R^3=R^4 =OCH_3, R^5=$COCF_3$

244 R^1=OH, R^2= R^3=OCH_3, R^4=R^5=H

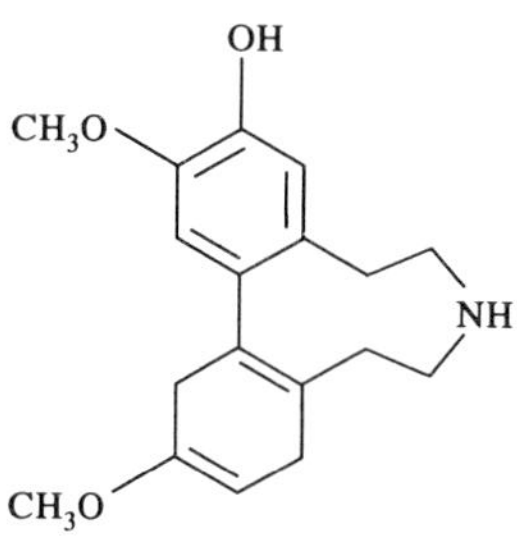

245

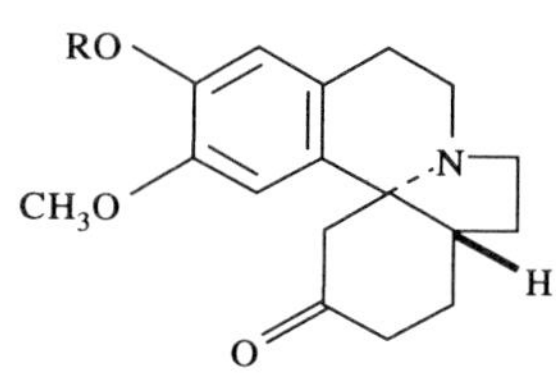

246 a R=H
b R=CH_3

The synthesis of cristadine **(97)** has been reported [215].

The stereochemistry of hydride reduction of erythrinan-7, 8-diones has recently been studied, and it has been found that formation of the product, which is a mixture of two stereoisomeric 7α- and 7β-alcohols, is affected by the bulkiness of the reagent and polarity of the solvent: sodium borohydride in ethanol-tetrahydrofuran gave the 7β-hydroxy isomer and tetrabutylammonium borohydride in methanol gave the 7α-hydroxy isomer, stereoselectively [216]. Conformational fluctuations and stereochemical pathway of reaction of phenylselenyl chloride with Δ^2-erythrinans in methanol have also been studied [217].

Aiming towards the total synthesis of non-aromatic *Erythrina* alkaloids, Tsuda and co-workers [218] have synthesized bicyclic unsaturated δ-lactones in a regioselective manner as a model of the C/D ring system of erythroidines.

6.2 Abnormal *Erythrina* Alkaloids

Ju-ichi *et al.* [219] have achieved the total synthesis of abnormal *Erythrina* alkaloids, e.g. (±)- coccolinine **(40)**, (±)-isococculidine **(80)**, (±)-coccuvinine **(38)**, and (±)-cocculidine **(63)**. 16-Ethoxycarbamido-2,15-dimethoxyerythrinan-7,8-dione, a key intermediate, was prepared by condensation of 3-ethoxycarbamido-4-methoxy-phenylethylamine **(247)** with ethyl 4-methoxycyclohexanone-2-glyoxalate **(248)** followed by treatment with 85 per cent phosphoric acid. The ethoxycarbamido

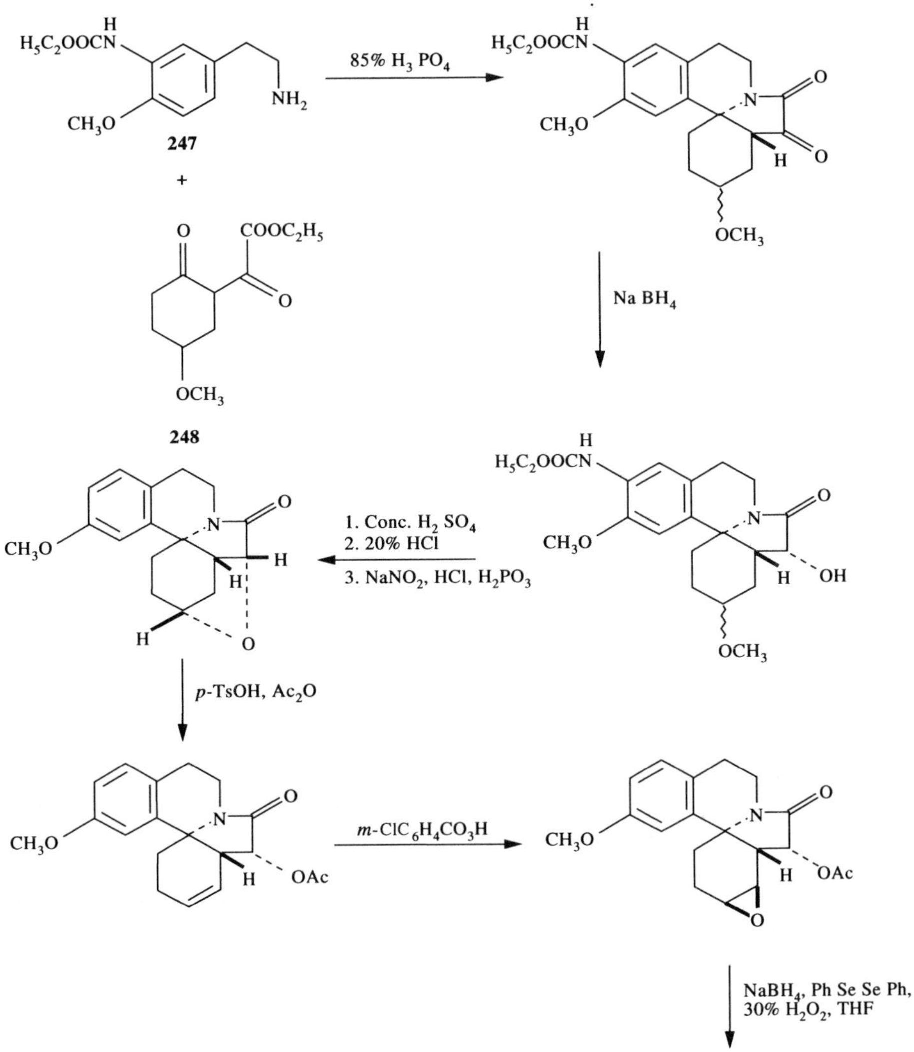

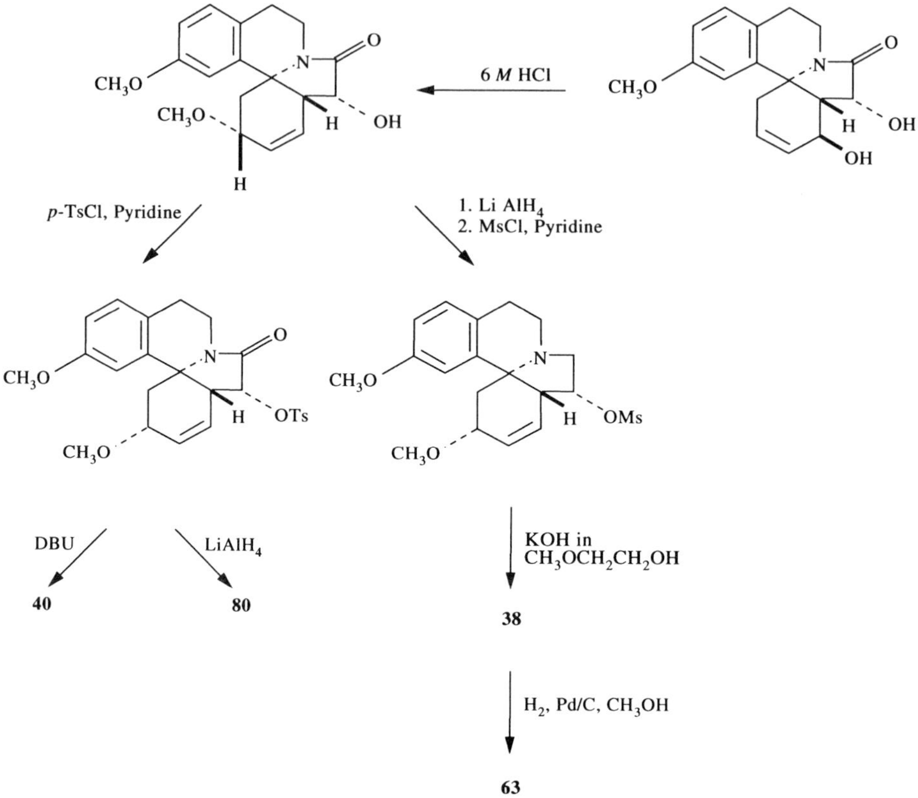

Fig. 18

group at C-16 position was effectively utilized as a regiospecific *para*- directing group in the ring closure of the isoquinoline (Fig. 18).

A new synthetic route (Fig. 19) for the preparation of **40** and **38** has been developed [220]. Condensation of 2-(4- methoxyphenyl)ethylamine **(249)** with methyl chloroformylacetate afforded the amide **(250)**, which underwent Bischler-Napieralski cyclization with polyphosphate ester to give 1,2,3,4-tetrahydro-7-methoxy-1-methoxycarbonylmethylideneisoquinoline **(251)**. This, on treatment with oxalyl chloride, yielded the isoquinolino- pyrrolinedione **(252)**. Diels-Alder reaction of the latter with 1,3-bistrimethylsilyloxybutadiene afforded the adduct **(253)**, which on reduction with lithium borohydride followed by treatment of the resulting alcohol with hydrochloric acid gave the enone **(254a)**. Formation of the mesylate **(254b)** and removal of the methoxycarbonyl group at C-6 (elimination caused by heating with $MgCl_2$) produced the dienone (**255**). Meerwein-Ponndorf reduction of

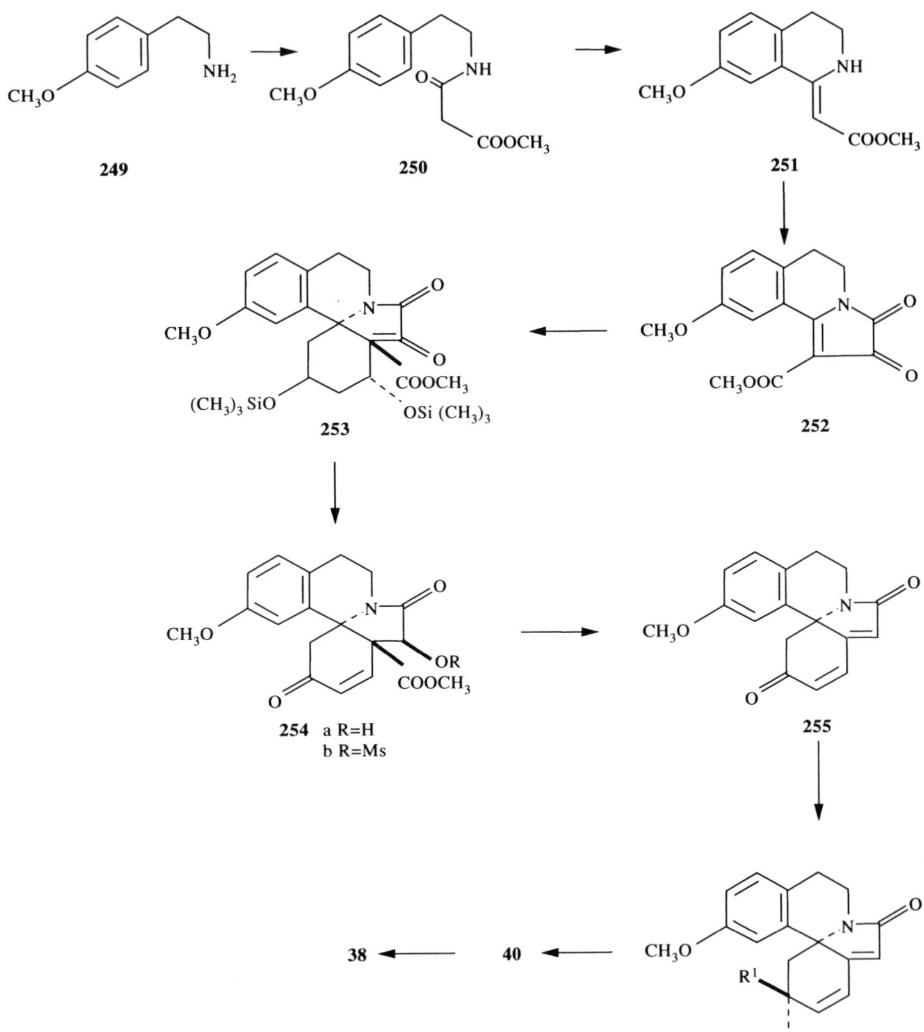

Fig. 19

255 gave the α-alcohol **(256a)** and the β-alcohol **(256b)**. Methylation of **256a** with methyl iodide in the presence of a phase-transfer catalyst afforded **40**. Reduction of this with aluminium hydride yielded **38**. The synthesis of 15-demethoxy-coccuvinine, which is an unnatural erythrinan compound, has also been achieved *via* the same route, starting from phenylethylamine.

Cocculolidine **(75)** is a lower homologue of β-erythroidine. The synthesis of the cocculolidine skeleton **(262)** has now been described (Fig. 20) [221]. Condensation of 2-(2,4,5-trimethoxyphenyl)ethylamine **(257)** and 2-(ethoxycarbonylmethyl)cyclohexanone **(229)**, followed by cyclization with an excess of polyphosphoric acid, gave 14,15,17-trimethoxyerythrinan-8-one**(258)**, which on ozonolysis at -78°, gave the bisnor-diester **(259)**. Hydrolysis of latter with 70 per cent acetic acid containing 10 per cent HCl yielded the diacid, which on heating with acetic anhydride afforded the anhydride **(260)**. Reduction of this with potassium tri-*sec*-butylborohydride gave **261** as the sole product.

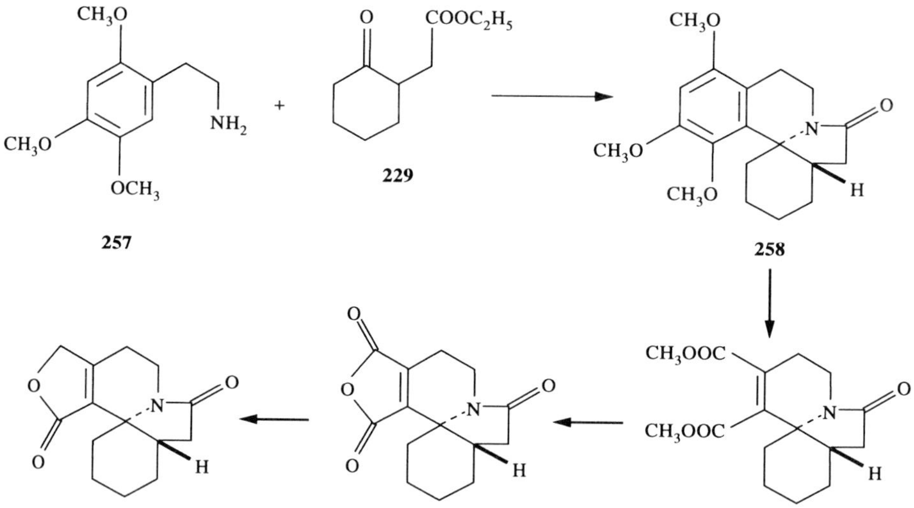

Fig. 20

7. PHARMACOLOGY

The *Erythrina* alkaloids and related synthetic analogues have been investigated for their pharmacological activity, particularly paralyzing potency [222,223]. These alkaloids, though tertiary bases, have relatively high paralyzing activity, and quaternization greatly diminishes the potency. The most active of the alkaloids is β-erythroidine **(73)**, and dihydro-β-erythroidine (obtained by hydrogenating β-erythroidine) is about five times more active than **73**. The erythroidines are active by mouth. Clinical trials indicate depression of blood pressure and respiration as the disadvantages.

The total alkaloids from the trunk bark of *E. variegata* showed several characteristic pharmacological effects [53,224] : neuromuscular blocking, smooth muscle relaxation, CNS depressant, anticonvulsant, potentiation of pentobarbital hypnosis, inhibition of acetylcholine-induced spasm, moderate negative inotropic and chronotropic effect. However, the alkaloids did not have any analgesic, antipyretic, antiinflammatory, laxative and diuretic effects. The ethanolic extract from *E. velutina* evidenced a spasmolytic activity [225]. Dhar *et al.* [226] observed the 50 per cent ethanol extractive of the leaves of *E. suberosa* to have antineoplastic activity against Sarcoma 180 in the mouse. Erysotrine **(7)** was found to exhibit effects (respiratory depression, decreased blood pressure and skeletal muscle relaxation) consistent with those of a competitive neuromuscular blocking agent in anaesthetized dogs [227]. The 1-azaerythrinan derivative **(262)** showed analgesic activity superior to that of codeine phosphate [228].

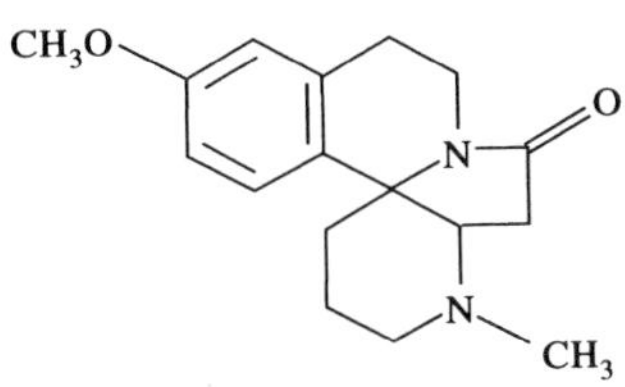

262

Bhakuni *et al.* [229] observed that 50 per cent aqueous ethanolic extract of the leaves of *Cocculus laurifolius*, a shrub from India, exhibited hypotensive and neuromuscular blocking activities. Cocculine **(41)** and cocculidine **(63)** nitrates exhibited hypotensive action in dogs; mainly due to their ganglionic blocking action. Neither of the alkaloids had a significant effect on the CNS system [230]. Isococculidine **(80)** showed weak neuromuscular blocking activity at the cholinoceptive site on the frog's rectus abdominus muscle [231]. Cocculolidine **(75)** was reported to be an insecticidal alkaloid [232].

8. REFERENCES

1. BA Krukoff and RC Barneby, Lloydia 37: 332 (1974).
2. BA Krukoff,Lloydia 40: 407 (1977).
3. BA Krukoff, Allertonia 3:121 (1982).
4. CR Gunn and DE Barnes, Lloydia 40:454 (1977).
5. WH Lewis, Lloydia 37:460 (1974).
6. P Rey, J Therapeutique 10:843 (1883).
7. AJ Lehman, J Pharmacol 60:69 (1937).

8. VH Cicardo and E Hug, C R Soc Biol 126:154 (1937).
9. I Simon, Arch Farmacol Sper Sci Affini 49:193 (1935).
10. K Folkers and K Unna, J Am Pharm Assoc 27:693 (1938).
11. K Folkers and K Unna, J Am Pharm Assoc 28:1019 (1939).
12. R Pichard and JV Luco, J Pharmacol 80:62 (1944).
13. L Marion, in : The Alkaloids, Vol 2, RHF Manske and HL Holmes, Eds., Academic Press, New York,1952, pp. 499-511.
14. V Boekelheide, in: The Alkaloids,Vol 7, RHF Manske, Ed., Academic Press, New York, 1960, pp. 201-227.
15. RK Hill, in: The Alkaloids, Vol 9, RHF Manske, Ed., Academic Press, New York, 1967, pp. 483-515.
16. A Mondon, in: Chemistry of the Alkaloids, SW Pelletier, Ed., Van Nostrand Reinhold Company, New York, 1970, pp. 173-198.
17. T Kametani and K Fukumoto, Synthesis 657 (1972).
18. SF Dyke and SN Quessy, in: The Alkaloids, Vol 18, RGA Rodrigo, Ed., Academic Press, New York, 1981, pp. 1- 98.
19. VA Snieckus, The Alkaloids (London) 1:145(1971); 2:199 (1972); 3:180 (1973); 4:273(1974); 5:176 (1975); 7:176 (1977).
20. SO De Silva and VA Snieckus, ibid 8:144 (1978).
21. AH Jackson, ibid 9:144 (1979).
22. AS Chawla and AH Jackson, ibid 11:137 (1981); 12:155 (1982); 13:196 (1983).
23. AS Chawla and AH Jackson, Nat Prod Reports 1:371 (1984); 3:555 (1986); 6:55 (1989); 7:565 (1990).
24. E Dagne and W Steglich, Tetrahedron Lett 24:5067 (1983).
25. GW Kenner, HG Khorana, and V Prelog, Helv Chim Acta 34:1969 (1951).
26. V Boekelheide and MF Grundon, J Am Chem Soc 75:2563 (1953).
27. V Boekelheide, J Weinstock, MF Grundon, GL Sauvaga, and EJ Agnello, J Am Chem Soc 75:2550 (1953).
28. DHR Barton, R James, GW Kriby, DW Turner, and DA Widdowson, Chem Commun 294 (1966).
29. DHR Barton, R James, GW Kriby, DW Turner, and DA Widdowson, J Chem Soc (C) 1529 (1968).
30. DHR Barton, PN Jenkins, R Letcher, DA Widdowson, E Hough, and D Rogers, Chem Commun 391 (1970).
31. DHR Barton, AAL Gunatilaka, RM Letcher, AMFT Lobo, and DA Widdowson, J Chem Soc Perkin Trans I 874 (1973).
32. K Ito, H Furukawa, and H Tanaka, Chem Commun 1076 (1970).
33. K Ito, H Furukawa, and H Tanaka, Chem Pharm Bull 19:1509 (1971).
34. K Ito, H Furukawa, and H Tanaka, Yakugaku Zasshi 93:1211 (1973); Chem Abstr 79: 146713 (1973).
35. K Ito, H Furukawa, and H Tanaka, Yakugaku Zasshi 93:1215 (1973); Chem Abstr 79: 146715 (1973).
36. K Ito, H Furukawa, H Tanaka, and T Rai, Yakugaku Zasshi 93:1218 (1973); Chem Abstr 79:146714 (1973).
37. K Ito, H Furukawa, and M Haruna, Yakugaku Zasshi 93:1611 (1973); Chem Abstr 80: 68387 (1974).
38. K Ito, H Furukawa, and M Haruna, Yakugaku Zasshi 93:1617 (1973); Chem Abstr 80: 48212 (1984).
39. K Ito, H Furukawa, M Haruna, and ST Lu, Yakugaku Zasshi 93:1671 (1973); Chem Abstr 80:68390 (1974).
40. K Ito, H Furukawa, M Haruna, and M Ito, Yakugaku Zasshi 93:1674 (1973); Chem Abstr 80:68391 (1974).
41. K Ito, M Haruna, and H Furukawa, Yakugaku Zasshi 95:358 (1975); Chem Abstr 82: 167515 (1975).

42. K Ito, M Haruna, Y Jinno, and H Furukawa, Chem Pharm Bull 24:52 (1976).
43. K Ito, F Suzuki, and M Haruna, J Chem Soc Chem Commun 733 (1978).
44. H Singh and AS Chawla, Experientia 25:785 (1969).
45. H Singh and AS Chawla, J Pharm Sci 59:1179 (1970).
46. H Singh and AS Chawla, Planta Med 19:71 (1970).
47. H Singh and AS Chawla, Planta Med 19:378 (1971).
48. H Singh, AS Chawla, AK Jindal, AH Conner, and JW Rowe, Lloydia 38:97 (1975).
49. H Singh, AS Chawla, VK Kapoor, N Kumar, DM Piatak, and W Nowicki, J Nat Prod 44: 526 (1981).
50. H Singh, AS Chawla, VK Kapoor,and J Kumar, Planta Med 41:101 (1981).
51. S Ghosal, DK Ghosh, and SK Dutta, Phytochemistry 9:2397 (1971).
52. S Ghosal, SK Majumdar, and A Chakraborti, Aust J Chem 24:2733 (1971).
53. S Ghosal, SK Dutta, and SK Bhattacharya, J Pharm Sci 61:1274 (1972).
54. S Ghosal, A Chakraborti, and RS Srivastava, Phytochemistry 11:2101 (1972).
55. S Ghosal and RS Srivastava, Phytochemistry 13:2603 (1974).
56. RT Hargreaves, RD Johnson, DS Millington, MH Mondal, W Beavers, L Becker, C Young, and KL Rinehart Jr., Lloydia 37:569 (1974).
57. DS Millington, DH Steinman, and KL Rinehart Jr., J Am Chem Soc 96: 1909 (1974).
58. DE Games, AH Jackson, NA Khan, and DS Millington, Lloydia 37:581 (1974).
59. I Barakat, AH Jackson, and MI Abdullah, Lloydia 40:471 (1977).
60. MI Abdullah, IE Barakat, DE Games, P Ludgate, VG Mavraganis, VU Ratnayake, and AH Jackson, Ann Missouri Bot Gard 66:533 (1979).
61. AH Jackson and AS Chawla, Allertonia 3:39 (1982).
62. AH Jackson, P Ludgate, V Mavraganis, and F Redha, Allertonia 3:47 (1982).
63. AS Chawla, AH Jackson, and P Ludgate, J Chem Soc Perkin I 2903 (1982).
64. AS Chawla, S Chunchatprasert, and AH Jackson, Org Magn Reson 21:39 (1983).
65. AS Chawla, FMJ Redha, and AH Jackson, Phytochemistry 24:1821 (1985).
66. AS Chawla, MP Gupta, and AH Jackson, J Nat Prod 50:1146 (1987).
67. AS Chawla, R Raja Reddy, and AH Jackson, Indian J Pharm Sci 51:189 (1989).
68. E Dagne and W Steglich, Phytochemistry 23:449 (1984).
89. PG Mantle, I Laws, and DA Widdowson, Phytochemistry 23:1336 (1984).
70. KP Tiwari and M Masood, Phytochemistry 18:704 (1979).
71. KP Tiwari and M Masood, Phytochemistry 18:2069 (1979).
72. M Masood and KP Tiwari, Phytochemistry 19:490 (1980).
73. SV Bhat, H Dornauer, and NJ De Souza, J Nat Prod 43:588 (1980).
74. H Ripperger, A Preiss, and M Diaz, Phytochemistry 22:2603 (1983).
75. MH Sarragiotto, HL Filho, and AJ Marsaioli, Can J Chem 59:2771 (1981).
76. AS Chawla, A Sood, M Kumar, and AH Jackson, Phytochemistry 31:372 (1992).
77. VU Ahmad, Q Najmus-Saqib, K Usmanghani, and GA Miana, J Chem Soc (Pakistan) 1:1 (1979).
78. ME Amer, S El-Masry, M Shamma, and AJ Freyer, J Nat Prod 54:161 (1991).
79. ME Amer, M Shamma, and AJ Freyer, J Nat Prod 54:329 (1991).
80. VU Ahmad, Q Najmus-Saqib, K Usmanghani, and GA Miana, Sci Pharm 48:169 (1980).
81. M Ju-ichi, Y Fujitani, and H Furukawa, Heterocycles 19:849 (1982).
82. MM El-Olmey, AA Ali, and MA El-Mottaleb, Lloydia 41:342 (1978).
83. DK Ghosh and DN Majumdar, Curr Sci 41:578 (1972).
84. K Folkers and F Koniuszy, J Am Chem Soc 62:436 (1940).
85. AS Chawla, TR Krishnan, AH Jackson, and DA Scalabrin, Planta Med 54:526 (1988).
86. RM Letcher, J Chem Soc (C) 652 (1971).
87. M Carmack, BC McKusick, and V Prelog, Helv Chim Acta 34:1601 (1951).
88. J Blake, JR Tretter, and H Rapoport, J Am Chem Soc 87:1397 (1965).
89. J Blake, JR Tretter, GJ Juhasz, W Bonthrone, and H Rapoport, J Am Chem Soc 88:4061 (1966).
90. B Belleau, J Am Chem Soc 75:5765 (1953).

91. MF Grundon and V Boekelheide, J Am Chem Soc 75:2537 (1953).
92. V Boekelheide and GC Morrison, J Am Chem Soc 80:3905 (1958).
93. W Nowacki and GF Bonsma, Z Krist 110:89 (1958).
94. V Prelog, BC McKusick, JR Merchant, S Julia, and M Wilhelm, Helv Chim Acta 39:498 (1956).
95. M Ju-ichi, Y Ando, Y Yoshida, K Kunitomo, T Shingu, and H Furukawa, Chem Pharm Bull 25:533 (1977).
96. M Ju-ichi, Y Ando, A Satoh, J Kunitomo, T Shingu, and H Furukawa, Chem Pharm Bull 26:563 (1978).
97. MI Aguilar, F Giral, and O Espejo, Phytochemistry 20:2061 (1981).
98. RB Boar and DA Widdowson, J Chem Soc (B) 1591 (1970).
99. AH Jackson, in:The Chemistry and Biology of Isoquinoline Alkaloids, JD Phillipson, MF Roberts, and MH Zenk, Eds., Springer-Verlag, Berlin, 1985, pp. 62-78.
100. Y Migron and ED Bergmann, Org Mass Spectrometry 12:500 (1977).
101. AF Beecham, Tetrahedron 27:5207 (1971).
102. R Razakov, SY Yonusov, SM Nasirov, AN Chekholov, VG Andrianov, and YT Struchkov, Chem Commun 150 (1974).
103. R Razakov, SY Yunusov, SM Nasirov, VG Andrianov, and YT Struchkov, Izv Akad Nauk SSSR Ser Khim 1:218 (1974); Chem Abstr 80:108727 (1974).
104. AW Hanson, Proc Chem Soc 52 (1963); Acta Cryst 16:939 (1963).
105. GR Wenzinger and V Boekelheide, Proc Chem Soc 53 (1963).
106. V Boekelheide and GR Wenzinger, J Org Chem 29:1307 (1964).
107. E Hough, Acta Crystallog, Sect B 32:1154 (1976).
108. AT McPhail and KD Onan, J Chem Soc Perkin II 115 (1977).
109. GA Miana, M Ikram, F Sultana, and MI Khan, Lloydia 35:92 (1972).
110. DS Bhakuni and S Jain, Tetrahedron 36:3107 (1980).
111. K Folkers and F Koniuszy, J Am Chem Soc 61:1232 (1939).
112. K Folkers and F Koniuszy , J Am Chem Soc 61:3053 (1939).
113. V Prelog, K Wiesner, HG Khorana, and GW Kenner, Helv Chim Acta 32:453 (1949).
114. JA Mills, J Chem Soc 4976 (1952).
115. A Mondon and M Ehrhardt, Tetrahedron Lett 2557 (1966).
116. DHR Barton, RB Boar, and DA Widdowson , J Chem Soc (C) 1213 (1970).
117. V Deulofeu, Chem Ber 85:620 (1952).
118. DHR Barton, RB Boar, and DA Widdowson,J Chem Soc (C) 1208 (1970).
119. DS Bhakuni, H Uprety, and DA Widdowson, Phytochemistry 15:739 (1976).
120. RS Singh, S Jain, and DS Bhakuni, Nat Acad Sci Letters 1:93 (1978).
121. O Prakash, R Roy, S Jain, and DS Bhakuni, J Nat Prod 51:603 (1988).
122. H Pande, NK Saxena, and DS Bhakuni, Indian J Chem 14B:366 (1976).
123. AN Singh, H Pande, and DS Bhakuni, Experientia 32:1368 (1976).
124. AN Singh and DS Bhakuni, Indian J Chem 15B:388 (1977).
125. Y Inubushi, H Furukawa, and M Ju-ichi, Chem Pharm Bull 18:1951 (1970).
126. AN Singh, H Pande, and DS Bhakuni, Lloydia 40:322 (1977).
127. M Tomita and H Yamaguchi, Pharm Bull (Tokyo) 4:225 (1956); Chem Abstr 51 :8115 (1957).
128. M Ju-ichi, Y Fujitani, T Shingu, and H Furukawa, Heterocycles 16:555 (1981).
129. AT McPhail, KD Onan, H Furukawa, and MJu-ichi, Tetrahedron Lett 485 (1976).
130. K Wada, S Marumo, and K Munakata, Tetrahedron Lett 5179 (1966).
131. MA Elsohly, JE Knapp, PL Schiff Jr, and DL Slatkin J Pharm Sci 65:132 (1976).
132. KP Madhusudanan, S Gupta, and DS Bhakuni, Indian J Chem 22B :907 (1983).
133. S Jain, KP Madhusudanan, and DS Bhakuni, Indian J Chem 26B:308 (1987).
134. E Leete and A Ahmad, J Am Chem Soc 88:4722 (1966).
135. DHR Barton and T Cohen, Festschrift A Stoll, Birkhauser, Basel, 1957, p.117.
136. DHR Barton, CJ Potter, and DA Widdowson, J Chem Soc Perkin I 346(1974).
137. DHR Barton, RD Bracho, CJ Potter, and DA Widdowson, J Chem Soc Perkin I 2278

(1974).
138. DS Bhakuni, AN Singh, and RS Kapil, J Chem Soc Chem Commun 211 (1977).
139. DS Bhakuni and AN Singh, J Chem Soc Perkin I 618 (1978).
140. DS Bhakuni and S Jain, Tetrahedron 36:2153 (1980).
141. H Pande and DS Bhakuni, J Chem Soc Perkin I 2197 (1976).
142. B Belleau, Chem Ind (London) 410 (1956).
143. B Belleau, Can J Chem 35:651 (1957).
144. A Mondon, Chem Ber 92:1461 (1959).
145. A Mondon, Chem Ber 92:1472 (1959).
146. V Boekelheide, M Müller, J Jack, TT Grossnickle, and MY Chang, J Am Chem Soc 81: 3955 (1959).
147. A Mondon, G Hasselmeier, and J Zander, Chem Ber 92:2543 (1959).
148. A Mondon, J Zander , and HU Menz, Liebigs Ann Chem. 667:126 (1963).
149. V Prelog, A Langemann, G Rodig, and M Ternbah, Helv Chim Acta 42:1301 (1959).
150. RV Stevens and MP Wentland, Chem Commun 1104 (1968).
151. HJ Wilkens and F Troxler, Helv Chim Acta 58:1512 (1975).
152. H Iida, S Aoyagi, K Kohno, N Sesaki, and C Kibayashi, Heterocycles 4:1771 (1976).
153. A Mondon, HG Vilhuber, C Fischer, M Epe, B Epe, and C Wolff, Chem Ber 112:1110 (1979).
154. A Mondon, M Epe, C Wolff, T Clausen, and HG Vilhuber, Chem Ber 112:1126 (1979).
155. Y Tamura, H Maeda, S Akai, and H Ishibashi, Tetrahedron Lett 23:2209 (1982).
156. T Kametani, K Higashiyama, T Honda, and H Otomasu, Heterocycles 22:569 (1984).
157. H Ishibashi, K Sato, M Ikeda, H Maeda, S Akai, and Y Tamura, J Chem Soc Perkin I 605 (1985).
158. R Ahmed-Schofield and PS Mariano, J Org Chem 50:5667 (1985).
159. M Westling, R Smith, and T Livinghouse, J Org Chem 51:1159 (1986).
160. R Ahmed-Schofield and PS Mariano, J Org Chem 52:1478 (1987).
161. H Ishibashi, S Harada, K Sato, M Ikeda, S Akai, and Y Tamura, Chem Pharm Bull 33: 5278 (1985).
162. Y Zhang, S Takeda, T Kitagawa, and H Irie, Heterocycles 24:2151 (1986).
163. K Isobe, K Mohri, H Sano, J Taga, and Y Tsuda, Chem Pharm Bull 34:3029 (1986).
164. M Ikeda, T Uchino, K Maruyama, and A Sato, Heterocycles 27:2349 (1988).
165. A Mondon, KF Hansen, K Boehme, HP Faro, HJ Nestler, HG Vilhuber, and K Böttcher, Chem Ber 103:615 (1970).
166. A Mondon, HP Faro, K Boehme, KF Hansen, and PR Seidel, Chem Ber 103:1286 (1970).
167. A Mondon and PR Seidel, Chem Ber 103:1298 (1970).
168. A Mondon and K Bottcher, Chem Ber 103:1512 (1970).
169. A Mondon and H Witt, Chem Ber 103:1522 (1970).
170. K Ito, M Haruna, and H Furukawa, J Chem Soc Chem Commun 681 (1975).
171. M Haruna and K Ito, J Chem Soc Chem Commun 345 (1976).
172. RB Boar, DA Hawkins, JF McGhio, and DHR Barton, J Chem Soc Perkin I 654 (1973).
173. A Mondon and HJ Nestler, Angew Chem 76:651 (1964).
174. A Mondon and HJ Nestler, Chem Ber 112:1329 (1979).
175. A Mondon, S Mohr, C Fischer, and HG Vilhuber, Chem Ber 112:2472 (1979).
176. S Mohr, C Fischer, T Clausen, and A Mondon, Chem Ber 112:3110 (1979).
177. S Mohr, T Clausen, B Epe, C Wolff, and A Mondon, Chem Ber 112:3795 (1979).
178. Y Tsuda, Y Sakai, N Kashiwaba, T Sano, J Toda, and K Isobe, Heterocycles 16:189 (1981).
179. Y Tsuda, Y Sakai, M Kaneko, Y Ishiguro, K Isobe, J Taga, and T Sano, Heterocycles 15:431 (1981).
180. Y Tsuda, Y Sakai, and T Sano, Heterocycles 15:1097 (1981).
181. Y Tsuda and Y Sakai, Synthesis 119 (1981).

182. Y Tsuda, Y Sakai, M Kaneko, K Akiyama, and K Isobe, Heterocycles 16:921 (1981).
183. T Sano, J Toda, N Kashiwaba, Y Tsuda, and Y Iitaka, Heterocycles 16:1151 (1981).
184. T Sano, J Toda, Y Horiguchi, K Imakufu, and Y Tsuda, Heterocycles 16:1463 (1981).
185. T Sano, J Toda, and Y Tsuda, Heterocycles 18:229 (1982).
186. Y Tsuda, T Ohshima, T Sano, and J Toda, Heterocycles 9:2027 (1982).
187. T Sano, J Toda, Y Tsuda, and T Ohshima, Heterocycles 22:49 (1984).
188. Y Tsuda and A Nakai, Heterocycles 20:1259 (1983).
189. T Sano, J Toda, N Kashiwaba, T Ohshima, and Y Tsuda, Chem Pharm Bull 35:479 (1987).
190. MJ Jorgensen, Tetrah edron Lett 559 (1962).
191. Y Tsuda, S Hosoi, A Nakai, T Ohshima, Y Sakai, and F Kiuchi, J Chem Soc Chem Commun 1216 (1984).
192. Y Tsuda, S Hosoi, and M Murata, Heterocycles 30:311 (1990).
193. MI Abdullah, AS Chawla, and AH Jackson, J Chem Soc Chem Commun 904 (1982).
194. MH Sarragiotto, PA da Costa, and AJ Marsaioli, Heterocycles 22:453 (1984).
195. Y Tsuda, S Hosoi, F Kiuchi, T Sano, J Toda, and R Yamamoto, Heterocycles 22:2255 (1984).
196. K Isobe, K Mohri, N Takeda, S Hosoi, and Y Tsuda, J Chem Soc Perkin I 1357 (1989).
197. K Isobe, K Mohri, K Suzuki, M Haruna, K Ito, S Hosoi, and Y Tsuda, Heterocycles 32: 1195 (1991).
198. Y Tsuda, A Nakai, K Ito, F Suzuki, and M Haruna, Heterocycles 22:1817 (1984).
199. H Tanaka, M Shibata, and K Ito, Chem Pharm Bull 32:1578 (1984).
200. H Ishibashi, T Sato, M Takahashi, M Hayashi, and M Ikeda, Heterocycles 27:2787 (1988).
201. H Ishibashi, T Sato, M Takahashi, M Hayashi, K Ishikawa, and M Ikeda, Chem Pharm Bull 38:907(1990).
202. H Irie, K Shibata, K Matsuna, and Y Zhang, Heterocycles 29:1033 (1989).
203. AJ Mancuso, DS Brownfain, and D Swern, J Org Chem 23:4148 (1979).
204. Y Tsuda, Y Sakai, A Nakai, M Kaneko, Y Ishiguro, K Isobe, J Taga, and T sano, Chem Pharm Bull 38:1462 (1990).
205. Y Tsuda, Y Sakai, A Nakai, T Ohshima, S Hosoi, K Isobe, and T Sano, Chem Pharm Bull 38:2136 (1990).
206. Y Tsuda, S Hosoi, A Nakai, Y Sakai,T Abe, Y Ishi, F Kiuchi,and T Sano, Chem Pharm Bull 39:1365 (1991).
207. Y Tsuda, Y Sakai, T Sano, and J Toda, Chem Pharm Bull 39:1402 (1991).
208. Chun-Tzer Chou and JS Swenton, J Am Chem Soc 109:6898 (1987).
209. K Isobe, K Mohri, Y Itoh, Y Toyokawa, N Takeda, J Taga, and Y Tsuda, Chem Pharm Bull 35:2618 (1987).
210. S Brandt, A Marfat, and P Helquist, Tetrahedron Lett 2193 (1979).
211. E McDonald and RD Wylie, J Chem Soc Perkin I 1104 (1980).
212. H Tanaka, M Shibata, and K Ito, Chem Pharm Bull 32:3271 (1984).
213. HG Theuns, HBM Lenting, CA Salemink, H Tanaka, M Shibata, K Ito, and RJJCh Lousberg, Heterocycles 22:2007 (1984).
214. JB Bremner and C Drager, Heterocycles 23:1451 (1985).
215. L Kovacs and P Kerekes, Sci Kommun 11:8 (1986); Chem Abstr 108:6245 (1988).
216. Y Tsuda, Y Sakai, K Ariyama, and K Isobe, Chem Pharm Bull 39:2120 (1991).
217. Y Tsuda, A Ishiura, Y Sakai, and S Hosoi, Chem Pharm Bull 40:24 (1992).
218. Y Tsuda, A Ishiura, S Takamura, S Hosoi, K Isobe, and K Mohri, Chem Pharm Bull 39: 2797 (1991).
219. M Ju-ichi, Y Fujitani, and Y Ando, Chem Pharm Bull 29:396 (1981).
220. T Sano, J Toda, N Maehara, and Y Tsuda, Can J Chem 65:94 (1987).
221. K Isobe, K Mohri, M Maeda, T Takeda, and Y Tsuda, Chem Pharm Bull 35:2602 (1987).
222. WDM Paton, J Pharm Pharmacol 1:273 (1949).
223. LE Craig,in:The Alkaloids, Vol 5, RHF Manske, Ed., Academic Press, New York, 1955, pp. 265-293.
224. SK Bhattacharya, PK Debnath, AK Sanyal, and S Ghosal, J Res Indian Medicine 6:135

(1971).
225. SG Barros, FJA Matos, JEV Vieira, MP Sousa, and MC Medeiros, J Pharm Pharmacol 22 :116 (1970).
226. ML Dhar, MM Dhar, BN Dhawan, BN Mehrotra, and C Ray, Indian J Exp Biol 6:232 (1968).
227. A Qayum, K Khanum, and GA Miana, Pak Med Forum 6:35 (1971); Chem Abstr 77: 148526 (1972).
228. Nippon Shinyaku Co Ltd Jpn Kokai Tokkyo Koho J P 81 100 785; Chem Abstr 96: 6927 (1982).
229. DS Bhakuni, ML Dhar, MM Dhar, BN Dhawan, and BN Mehrotra, Indian J Exp Biol 7: 250 (1969).
230. UB Zakirov, Kh U Aliev,and NV Abdumalikova, Farmakol Alkaloidov Serdech Glikozidov 197 (1971); Chem Abstr 77:135092 (1972).
231. K Kar, KC Mukherjee, and BN Dhawan, Indian J Exp Biol 15:547 (1977).
232. K Wada and K Munakata, Agr Biol Chem 31:336 (1967).

Chapter Four

Chemistry, Biology and Chemoecology of the Pyrrolizidine Alkaloids

Thomas Hartmann and Ludger Witte
Institut für Pharmazeutische Biologie der Technischen Universität Braunschweig, Mendelssohnstrasse 1, D-38106 Braunschweig, Germany

CONTENTS

We are grateful to the Deutsche Forschungsgemeinschaft and Fonds der chemischen Industrie who have supported our work on the biosynthesis, metabolism and chemical ecology of PAs during the last eight years by grants to T.H.

1. INTRODUCTION

As in other classes of natural products the number of new pyrrolizidine alkaloids (PAs) has increased dramatically during the past decades. The first review by Leonard in 1950 [4] counted 27 PAs, this number increased to about 100 in 1966/68 [5, 6], to approximately 200 from 350 plant species in 1980 [2, 7] and stands now in excess of 370 structures isolated from more than 560 species (see Appendix 1 and 2).

Fortunately, there is not only an increasing number of chemical structures, but also growing understanding of the mechanisms involved in biosynthesis of PAs and their biological activities and functions. In particular two features of the PAs received great attention and stimulated extensive and often interdisciplinary studies:

- the health hazard for humans and domestic animals of potentially hepatotoxic PAs with a 1,2-unsaturated necine base moiety and their mode of action.

- the unique role of PAs in plant-insect relationships.

Another unique property of PAs is that in most (if not all) plants they are synthesized, translocated and stored as *N*-oxides and that even adapted insects keep sequestered PAs as *N*-oxides. The occurrence of salt-like, polar *N*-oxides has incidentally been observed in other classes of alkaloids but in no instance as the dominating if not exclusive form. As a matter of convenience in the following chapter, alkaloids are generally called PAs leaving out the suffix *N*-oxide. The role of the *N*-oxide form will be discussed in section 6.

In this chapter we intend to combine and discuss results from different areas of chemical and biological research in the field of PAs, to draw a more complete general picture of this fascinating class of secondary compounds. Not all aspects could be covered comprehensively to the same extent. For more detailed information the reader may be referred to excellent monographs or recent reviews dealing with the various aspects of PA research: chemistry and analytics [3, 8]; biosynthesis [9, 10]; physiology and *in vitro* cultures [11-13]; pharmacology and toxicology [3, 14-18]; chemical ecology [19-22]. Furthermore, there are annual reports with chapters on pyrrolizidine alkaloids [23] and their role in chemical ecology [24].

2. SURVEY AND CLASSIFICATION OF PLANT PAs

An updated list of all PA containing plant species is given in Appendix I. Only references which are not recorded in one of the three previous comprehensive compilations [1, 2, 3] are quoted; for older references the reader is referred to these reference sources. All structures are illustrated in Appendix II.

PAs are most commonly found in the plant families Asteraceae (tribes Eupatorieae and Senecioneae), Boraginaceae (most genera), Fabaceae (mainly the genus *Crotalaria*) and Orchidaceae (ten genera). More than 95% of the PA containing species so far investigated belong to these four families (Appendix 1). In addition PAs have been found scattered in a number of further families, i.e Apocynaceae (4 genera), Celastraceae (single report), Rhizophoraceae (single report; *S*-containing alkaloid, **362**), Ranunculaceae (single report), Santalaceae (single report), and Sapotaceae (2 genera). Very recently four related PAs have been isolated for the first time from the Convolvulaceae (genus *Ipomoea*) [146]. Two plant families previously recorded to contain PA carrying species [2, 3, 8] have been omitted from our list: (i) Euphorbiaceae (genus *Phyllanthus*, *Securinega*), because the alkaloids of the norsecurinine type are biosynthetically closely related to the *Securinega*-alkaloids [161] but unrelated to the other PAs; (ii) Scrophulariaceae (genus *Castilleja*), because it is clear now that this root parasite does not synthesize PAs *de novo* but obtains them from its *Senecio* host (see section 5.4).

PAs generally occur as ester alkaloids composed of a necine base and a necic acid. The two building blocks may form monoesters, open-chain diesters or even triesters and macrocyclic diesters. By chemical modification of both the necine and the necic acid moiety a great diversity of structures can be created (Appendix II). In the past PAs have preferably been classified according to the chemical structure of the necine base moiety [2] or just listed alphabetically [3, 8]. Culvenor [25] analyzed the implications for taxonomy and phylogeny of PA occurrences at the genus level and higher. He defined three alkaloid types: (a) aliphatic monocarboxylic esters, characteristic for the Boraginaceae; (b) macrocyclic diesters, common in the Senecioneae (Asteraceae) and the genus *Crotalaria* (Fabaceae); (c) esters of aryl and aralkyl acids, typical for the Orchidaceae. Following and extending Culvenor's suggestion and including biogenetic implications we suggest a classification into alkaloid types as described below (Fig 1).

Senecionine type (A) - 104 structures
Mostly twelve-membered macrocyclic diesters that are derived from or structurally closely related to senecionine (**1**).

- **Senecionine group (A1)** - 83 structures (Appendix II **1-83**)
  Structures that can be regarded as derivatives of **1**, formed by chemical modification of both the necine base and the necic acid moiety. There is evidence that in this group **1** is the primary product of biosynthesis (see section 5.3). The necic acids are biogenetically derived from isoleucine (see section 4). The otonecine esters (i.e. **20-26, 33, 34, 38-45, 47-49, 51, 52, 60-66**) which are often regarded as a separate group have been shown to be transformation products of the respective retronecine esters (e.g. **1** is the precursor of **20**; see section 3.1). The acylpyrrols **(78-83)** with necic acid moieties corresponding to the respective retronecine esters are also included in this group.

- **Senecivernine group (A2)** - 8 structures (Appendix II **84-91**)
  Small group often co-occuring with A1 from which it is distinguished by a different C-C-linkage of the necic acid part.

- **Nemorensine group (A3)** - 4 structures (Appendix II **92-95**)
  Thirteen-membered macrocycle; necic acid part presumably derived from isoleucine and leucine.

- **Rosmarinine group (A4)** - 9 structures (Appendix II **96-104**)
  C1- or C2-hydroxylated alkaloids. Biosynthesis of rosmarinine (**96**) differs from that of A1 type alkaloids in respect to stereochemistry (see section 3.1; Fig. 3).

Triangularine type (B) - 56 structures
Monoesters or diesters with C_5-acids (mostly angeloyl, tigloyl or senecioyl residues) and their hydroxylated derivatives.

- **Triangularine group (B1)** - 36 structures (Appendix II **105-144**)
  Monoesters or diesters of C7- and C9-hydroxylated necines.

- **Macrophylline group (B2)** - 11 structures (Appendix II **145-155**)
  Monoesters or diesters with C2- and C9-hydroxylated necines, and complex structures derived thereof.

- **Senampeline group (B3)** - 8 structures (Appendix II **156-163**)
  Related to B1 type diesters. The necine base is an acylpyrrol or its derivative with the C5-oxo-group reduced and esterified with acetate.

Lycopsamine type (C) -109 structures
Monoesters or diesters containing as necic acid a hydroxylated 2-isopropylbutyric acid. This branched C_7-acid is uniquely found in PAs. The basic structures are the four stereoisomers (+)- and (-)-trachelanthic acid and (+)- and (-)-viridifloric acid as well as their derivatives.

- **Lycopsamine group (C1)** - 94 structures (Appendix II **164-254**)
  Monoesters or diesters carrying a C_7-acid esterified exclusively to the C9-OH of the necine base. Diesters have a second necic acid which can be acetate or one of the C_5-acids (see type B) attached to the C7-OH of the necine base.

- **Isolycopsamine group (C2)** - 6 structures (Appendix II **256-260**)
  Monoesters or diesters carrying a C_7-acid esterified exclusively to the C7-OH of the necine base. Diesters have an acetate attached to the C9-OH of the necine.

- **Latifoline group (C3)** - 6 structures (Appendix II **261-266**)
  Like C1 but the C_7-acid moiety forms a five-membered lactone ring.

- **Parsonsine group (C4)** - 11 structures (Appendix II **267-277**)
  Macrocyclic PAs where the macrocyclic structure is obtained by an ester linkage between two necic acids attached to the C7-OH and C9-OH of retronecine; at least one of the acids is a C_7-acid.

Monocrotaline type (D) - 33 structures (Appendix II **278-309b**)
Eleven-membered macrocyclic diesters with retronecine as necine base and a variable substitution pattern at the necic acid moiety.

Phalaenopsine type (E) - 21 structures (Appendix II **310-331**)
Monoesters of (-)-isoretronecanol (**164**), lindelofidine (**167**), laburnine (**171**) or (-)-trachelanthamidine (**173**) with aryl or aralkyl, rarely alkyl, necic acid.

Miscellaneous PAs (M) - 32 structures (Appendix II **332-362**)

- **Unusual necine esters (M1)** - 14 structures (Appendix II **332-343**)
  Simple esters or even a PA-glycoside (**338**) which do not fit the types A to E. The novel ipangulines (**340-343**) isolated from *Ipomoea* (Convolvulaceae) may be considered as a distinctive group; they are diesters of the necine base turneforcidine; one necic acid is an alkyl acid, the other benzoic or phenylacetic acid.

- **Simple necine derivatives (M2)** - 18 structures (Appendix II **344-362**)
  Simple necine base derivatives (not esterified) which include the sulfur-containing cassipurine (**362**), the only PA derivative isolated from the Rhizophoraceae as well as the tussilagines (**354, 355**) found in *Tussilago* and *Arnica* (Asteraceae). Some other structures may be artifacts or postmortal degradation products.

Loline type (L) - 11 structures (Appendix II **363-373**)
The lolines are 1-aminopyrrolizidine derivatives with an ether-bridge linking C2 and C7. They are biogenetically unrelated to the 1-hydroxymethylpyrrolizidines (see section 9).

The different alkaloid types and subtypes (groups) are illustrated in Fig. 1; the structures of the name-giving representatives are shown. The distribution of the alkaloid types within taxa that contain PA producing species is summarized in Table 1. Interestingly, the occurrence of the two dominating alkaloid types, A (senecionine type) and C (lycopsamine type) each with more than 100 structures, obviously exclude each other. Type A is most characteristic for the tribe Senecioneae of the Asteraceae and also well represented in the Fabaceae. By contrast, type C seems to be absent from these two taxa but dominates in the tribe Eupatorieae of the Asteraceae, the Boraginaceae and the few so far studied species of the Apocynaceae which again do not contain type A. The single reports indicating the occurrence of type A alkaloids in the Eupatorieae and the Boraginaceae appear to be exceptions.

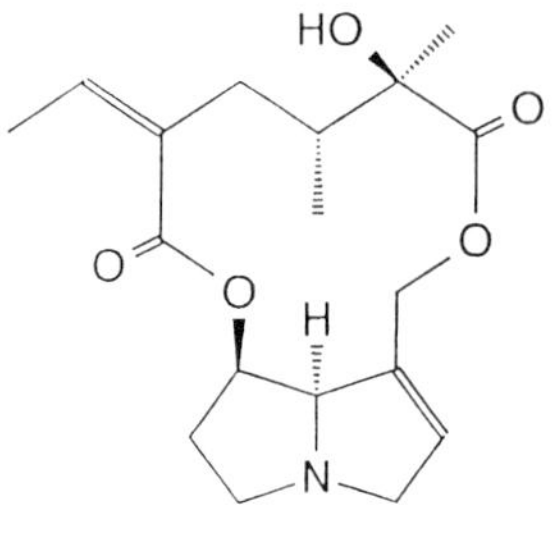
group A1 (**1**)

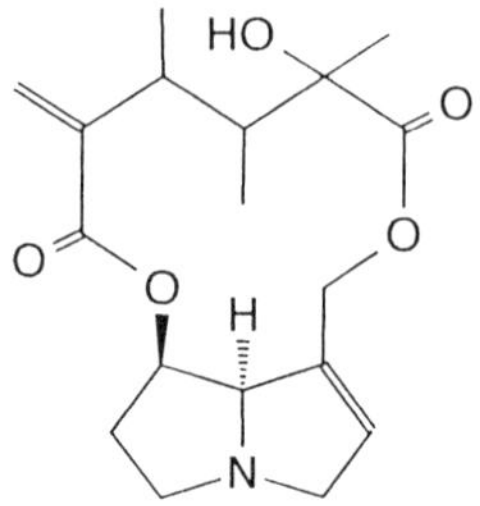
group A2 (**84**)

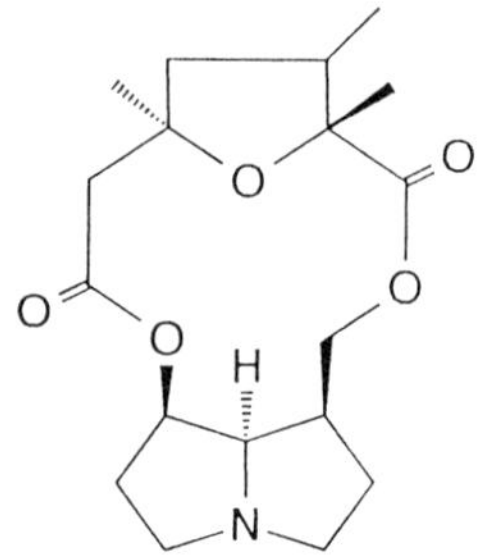
group A3 (**92**)

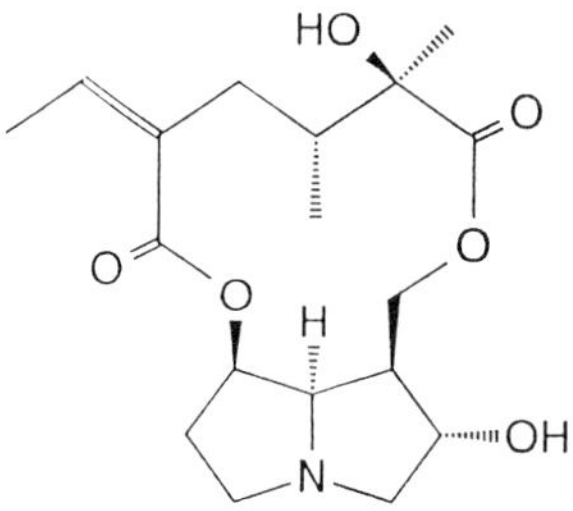
group A4 (**96**)

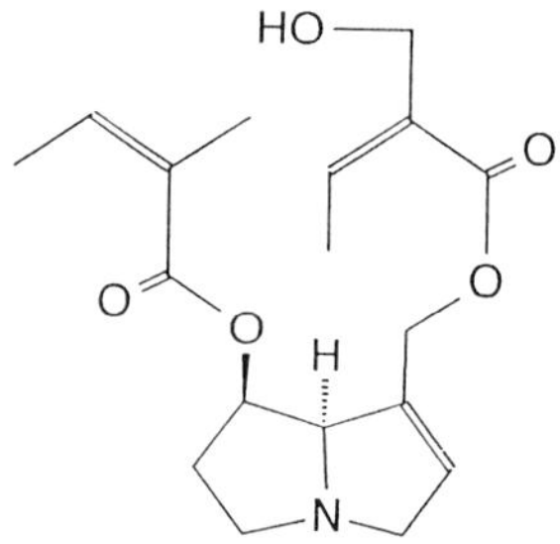
group B1 (**112**)

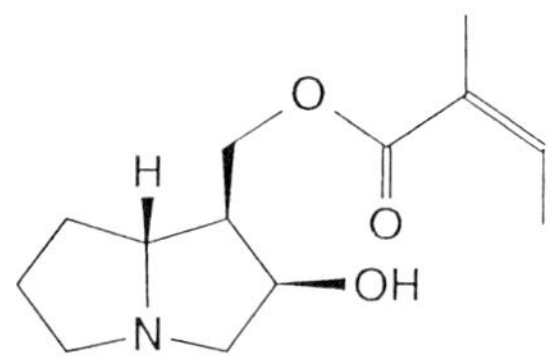
group B2 (**146**)

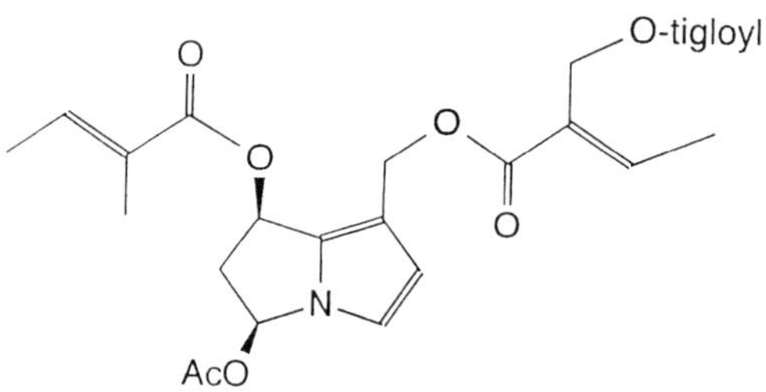

group B3 (**158**)

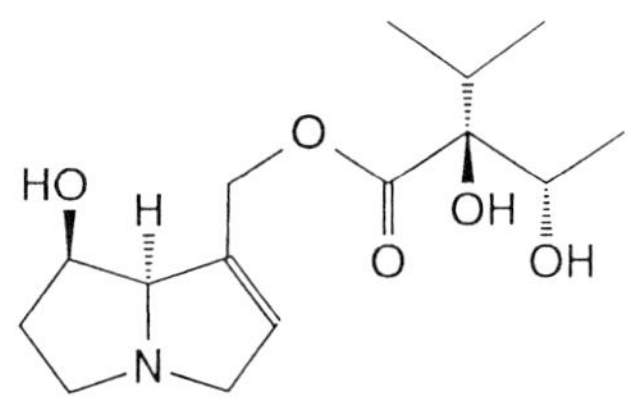
group C1 (**215**)

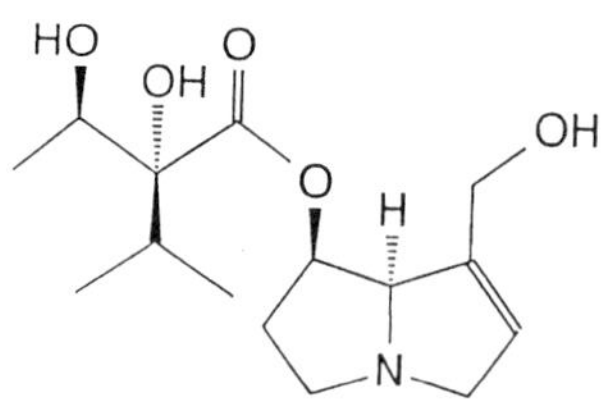
group C2 (**257**)

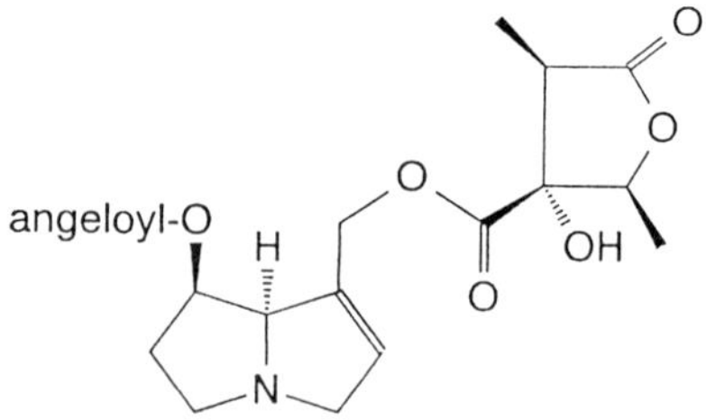

group C3 (**262**)

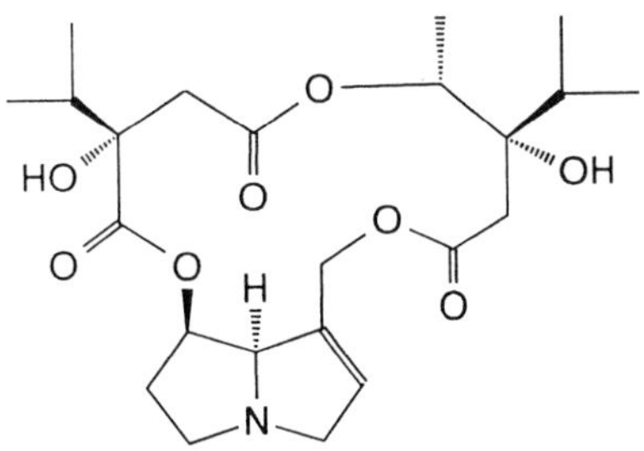

group C4 (**269**)

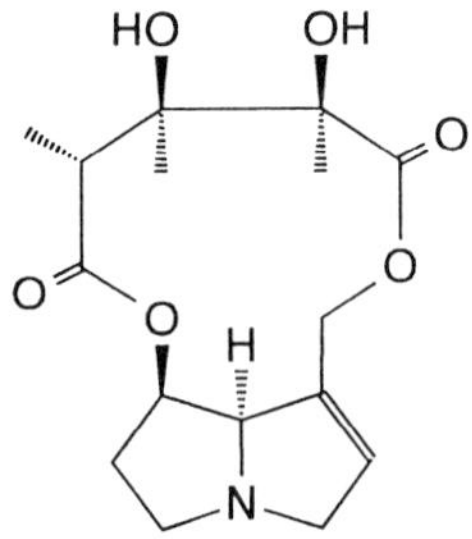

group D (**284**)

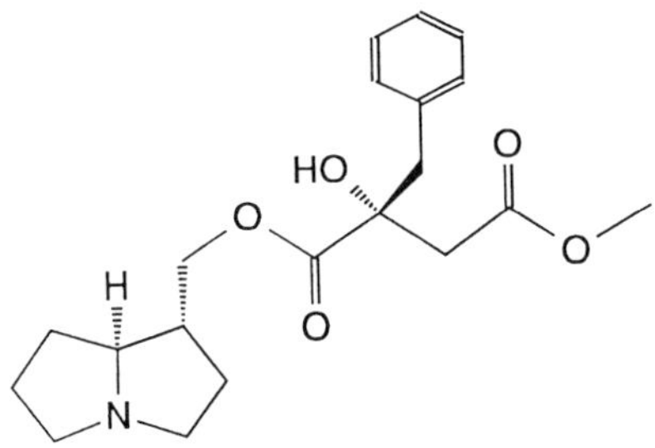

group E (**320**)

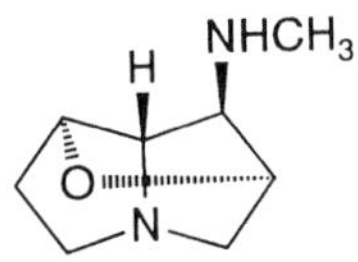

group L (**364**)

Figure 1. Classification of the pyrrolizidine alkaloids. A1, senecionine group (**1** senecionine); A2, senecivernine group (**84** senecivernine); A3, nemorensine group (**92** nemorensine); A4, rosmarinine group (**96** rosmarinine); B1, triangularine group (**112** triangularine; B2, macrophylline group (**146** macrophylline); B3, senampeline group (**158** senampeline B); C1, lycopsamine group (**215** lycopsamine); C2, isolycopsamine group (**257** isolycopsamine); C3, latifoline group (**262** latifoline); C4, parsonsine group (**269** parsonsine); D, monocrotaline group (**284** monocrotaline); E, phalaenopsine group (**320** phalaenopsine T); L, loline group (**364** loline).

Type B alkaloids are well represented in the Senecioneae. This seems reasonable because in many instances (particularly within the genus *Senecio*) the diesters are structurally related to the respective macrocyclic type A alkaloids; they represent the open-chain diester analoges of the macrocyclic diesters (Fig. 1, compare A1 and B1). In a similar way type B monoesters frequently co-occur with type C alkaloids in the Eupatorieae and Boraginaceae.

Type D alkaloids are most characteristic for the genus *Crotalaria* but have also been reported from some species of the Boraginaceae. The Orchidaceae exclusively produce type E alkaloids with an 1,2-saturated necine base. Thus, orchids are nontoxic as far as liver toxicity of PAs is concerned (see section 7.1).

Table 1. Distribution of the alkaloid types A to L within the taxa which contain PA producing species. The number of species containing the respective alkaloid type are given.

Taxa	Total No of species with PAs*	Number of species						
		A (104)**	B (56)**	C (109)**	D (33)**	E (21)**	M (32)**	L (11)**
Apocynaceae	8	-	-	5	-	1	3	-
Asteraceae								
Eupatorieae	23	1	2	23	-	-	-	-
Senecioneae	231	204	31	-	1	-	7	-
Boraginaceae	145	2	16	128	5	3	8	-
Fabaceae								
Crotalaria	81	19	4	-	58	-	15	-
other genera	37	22	-	-	-	-	8	8
Orchidaceae	35	-	-	-	-	33	2	
Poaceae	4	1	-	-	-	-	-	3

* According to Appendix I
** Number of structures (see Appendix II)

The occurrence of loline-type alkaloids (1-aminopyrrolizidines) in fungus-infected grasses (see section 9), as well as in the genus *Adenocarpus* (Fabaceae, tribe Genisteae), is intriguing. Until now a co-occurrence of lolines with other PAs has not been found.

3. BIOSYNTHESIS OF THE NECINE BASES

3.1 Evidence from Tracer Studies

First biogenetic studies were carried out by Nowacki and Byerrum [162] who fed ^{14}C-labeled ornithine, acetate, and propionate to *Crotalaria spectabilis*. The authors showed that labeled ornithine is exclusively incorporated into the necine base moiety of monocrotaline whereas the label from the two acids occurs preferentially in the necic acid moiety. Extension of these early studies with radioactively labeled tracers established the view that the necine base is formed from two molecules of L-ornithine or L-arginine by way of putrescine [163-168]. More detailed studies with radioactively labeled precursors were hampered by the lack of good procedures for a selective chemical degradation of the necine base.

Further insight into the mechanism of necine base biosynthesis came from tracer studies, chiefly using precursors labeled with stable isotopes. These studies were mainly performed by D.J.Robins and by I.D.Spenser and their co-workers. The authors used as experimental plants *Senecio isatideus* and *Senecio vulgaris*. These two related species produce retronecine esters, i.e. retrorsine **(14)**, senecionine **(1)**, and seneciphylline **(8)**. Feeding experiments with ^{13}C-labelled putrescines and analysis of the complete labeling pattern by ^{13}C n.m.r. spectroscopy clearly established the formation of retronecine from two putrescine units [169, 170]. Experiments with a putrescine, double labeled with ^{15}N and ^{13}C at the adjacent C-atom, demonstrated that a compound with C_{2v} symmetry serves as a precursor of retronecine [171-174]. As the best candidate homospermidine was suggested to be the symmetrical C4-N-C4 intermediate [172]. Tracer studies with ^{14}C-labeled and ^{13}C-labeled homospermidines showed this precursor to be incorporated intact into retronecine [175, 176]. In addition, [^{14}C]homospermidine could be identified as a free intermediate in a trapping experiment where [^{14}C]ornithine was fed to a *S. isatideus* plant [176]. Further support for homospermidine as an intermediate in necine biosynthesis was obtained by a biomimetic experiment in which homospermidine was incubated under physiological conditions with diamine oxidase from pea seedlings. The polyamine was oxidized to the corresponding dialdehyde in equilibrium with an iminium ion, which in turn underwent non-enzymatic cyclization to give 1-formylpyrrolizidine. After addition of alcohol dehydrogenase, (±)-trachelanthamidine **(171, 173)** was formed and could be isolated with an overall yield of 22% after one week [177]. The intermediacy of an iminium ion in necine biosynthesis could be proved by incorporation of the synthesized radioactively labeled derivative into PAs [178].

A more complete picture of the presumed stereospecific enzymatic steps leading from putrescine to the necine moiety (e.g. retronecine) comes from rigorous experiments with chiral deuterated putrescines, carried out independently in Spenser's and Robins' laboratories. (*R*)[1-D]putrescine and (*S*)[1-D]putrescine were fed to *S. vulgaris* [174, 179] and *S. isatideus* [180] and the deuterium label was localized by ^{2}H n.m.r. spectroscopy in the retronecine moiety of the synthesized PAs. The labeling patterns established in the retronecine moiety showed that the two oxidation steps of the primary amino groups of homospermidine (i.e. carbons 8 and 9 of retronecine) occur with stereospecific loss of the *pro-S*-hydrogen (Fig. 2).

This behavior is consistent with the known stereochemistry of diamine oxidase in which the *pro-R*-hydrogen is retained in the resulting aldehyde [181, 182]. The stereochemistry of the third oxidation step affecting the primary amino group of one putrescine moiety during homospermidine formation (i.e. carbons 3 and 5 of retronecine) is not as clear. In the two laboratories deuterium retentions of 69% and 68%, respectively, were established in experiments with (*S*)[1-D]putrescine instead of 50% as expected for the assumed involvement of diamine oxidase in homospermidine formation (Fig. 2). The discrepancy has been explained by intramolecular deuterium isotope effects in the enzymatic oxidation involved [179, 180].

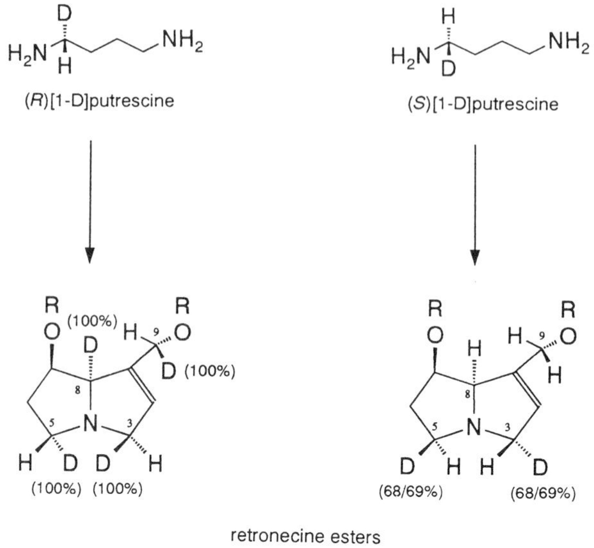

Figure 2. Incorporation of chiral deuterated putrescines into retronecine.

The common pathway of necine biosynthesis from putrescine via homospermidine and intermediate iminium ions may lead to one of the four stereoisomers of 1-hydroxymethyl-pyrrolizidine. All four isomers are known as necines of plant PAs (**164, 167, 171, 173**); three are given in Fig. 3. Isotope tracer experiments mainly carried out by Robins showed that (-)-trachelanthamidine (**173**) is specifically incorporated into retronecine, the most widespread necine base of plant PAs [183, 184], whereas (-)-isoretronecanol (**164**) is the preferred precursor of the 2-hydroxylated PAs such as rosmarinecine [184-187]. Cynaustraline (**170**) and cynaustine (**209**), two examples of the minority of necines with 8ß-stereochemistry known from *Cynoglosum officinale*, were shown to be formed via lindelofidine ((+)-isoretronecanol) (**167**) [188]. Epimerization of either trachelanthamidine or lindelofidine does not appear to take place during PA biosynthesis. Thus the pathways probably diverge prior to the formation of the C9-alcohols during cyclization of the iminium ions, which in all the examples shown in Fig. 3 are efficiently incorporated into the species specific PAs [189].

A further important question was the sequence by which the different co-occurring necines are formed. *Heliotropium spathulatum* produces lycopsamine-type PAs containing the necines (-)trachelanthamidine (**173**), (-)-supinidine (**180**) and retronecine. Pulse-chase labeling experiments with $^{14}CO_2$ applied to the plant and isolation of the three necines after

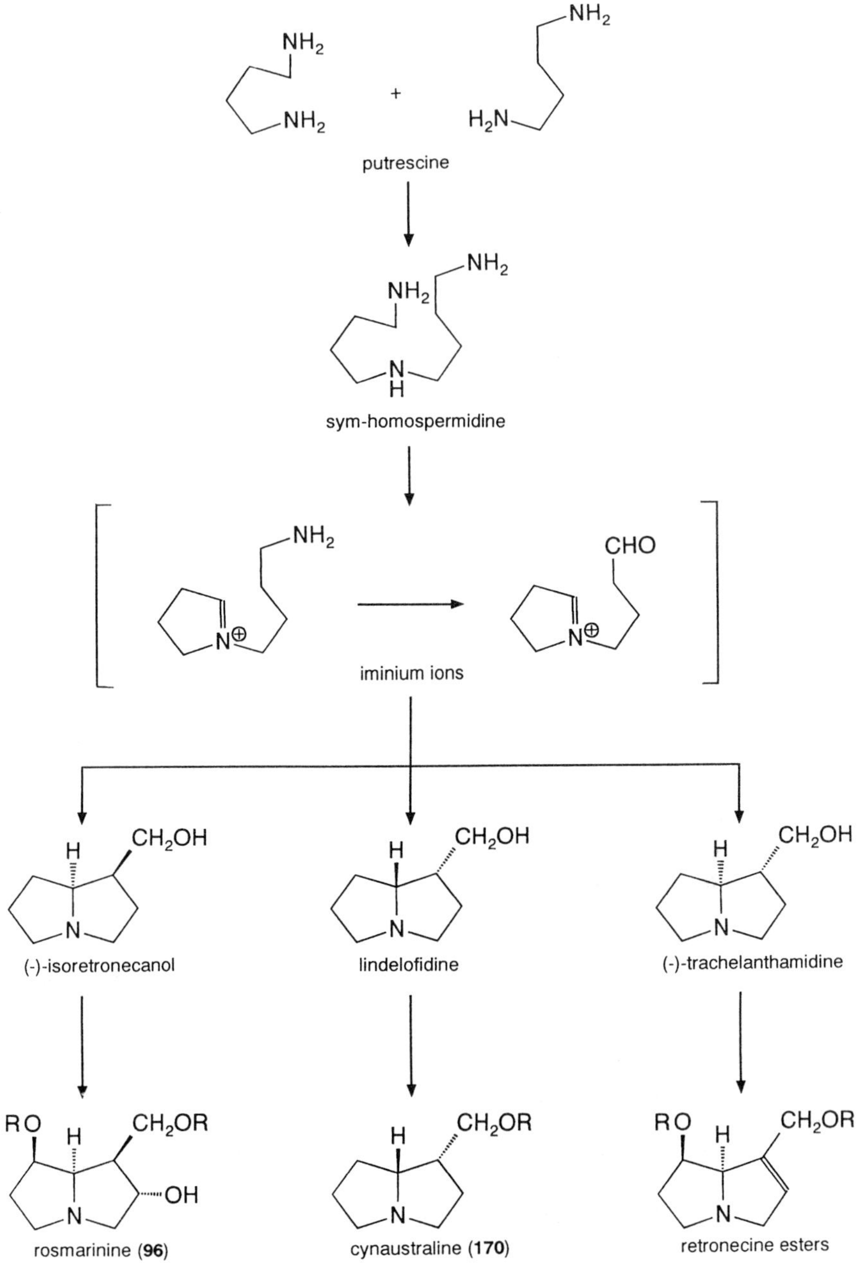

Figure 3. Biosynthesis of the isomeric 1-hydroxymethylpyrrolizidines and their role as specific precursors of necine esters.

different lengths of time showed that the changes in the the specific radioactivities are consistent with the biosynthetic sequence [190]:

(-)-trachelanthamidine ---------> (-)-supinidine ---------> retronecine

The same sequence was established in *Eupatorium cannabinum* root cultures which also produce lycopsamine-type PAs. In tracer studies with labeled precursors (e.g. [^{14}C]putrescine) no free necines were detectable. The conversion of (-)-trachelanthamidine via (-)-supinidine into retronecine proceeds at the stage of the respective O^{9}-esters [191]. Tracer experiments with chirally C-2 deuterated putrescines revealed that the hydroxylation at the C-7 of the retronecine moiety of retrorsine proceeds with retention of configuration whereas the formation of the 1,2-double bond of retronecine involves removal of the *pro-S*-hydrogen and retention of the *pro-R*-hydrogen at C-2 of retronecine [192, 193].

Retronecine esters, particularly of the senecionine type, are often accompanied by the respective otonecine esters. A direct conversion of retronecine into otonecine was demonstrated in root cultures of *Senecio vernalis* which convert senecionine (**1**) *N*-oxide into senkirkine (**20**) [95]. Tracer experiments with various necine precursors fed to *Emilia flammea* led to the same conclusion and established that the otonecine moiety of emiline (**51**) was formed at a late stage in the biosynthetic pathway, probably by hydroxylation at C-8 of retronecine and *N*-methylation, followed by ring cleavage [194].

3.2 Enzymatic Evidence

The results of the tracer studies and the elucidation of the stereochemistry involved in necine biosynthesis represent an excellent basis for the next stage of biosynthetic research, the characterization of the individual enzymes involved in the pathway. Recently root cultures of *Senecio* species were found to be excellent systems for biochemical and enzymological studies of PA biosynthesis [95, 195].

The formation of putrescine in root cultures of *Senecio vulgaris* [196] has been elucidated. Experiments using the suicide-inhibitors of ornithine decarboxylase (α-difluoromethylornithine, DFMO) and of arginine decarboxylase (α-difluoromethylarginine, DFMA) revealed that putrescine is exclusively synthesized via the arginine-agmatine pathway (Fig. 4). Ornithine is rapidly converted into arginine. This rapid conversion is in accordance with the observation that the two amino acids are equally well incorporated into PAs [165, 168, 195]. In *Heliotropium* species arginine was also found to be the specific endogenous precursor of PA biosynthesis [197]. The three enzymes of the arginine-agmatine pathway have been characterized from *Senecio* root cultures and ornithine decarboxylase was shown to be absent

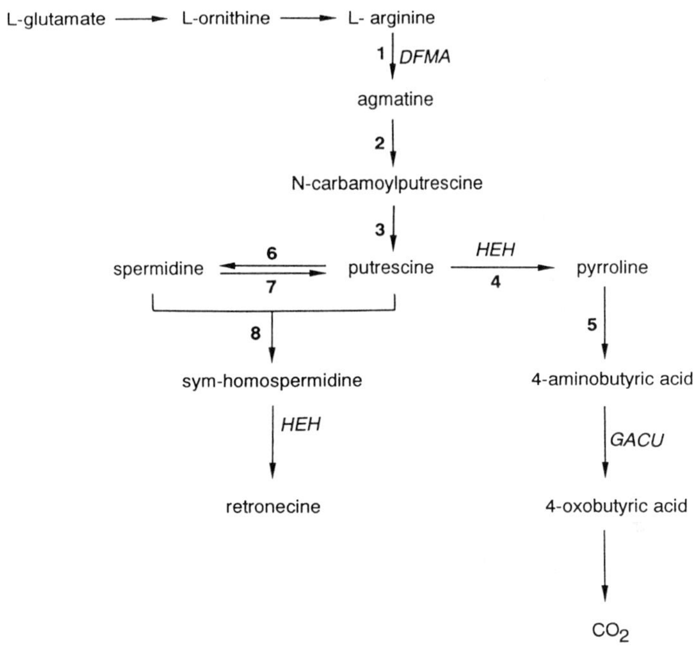

Fig. 4. Pathways of putrescine metabolism and alkaloid formation in PA producing root cultures of *Senecio* and *Eupatorium* root cultures. The numbered reactions have been characterized on the enzymatic level. Enzymes: 1 = arginine decarboxylase (EC 4.1.1.19), 2 = agmatine iminohydrolase (EC 3.5.3.12), 3 = *N*-carbamoylputrescine amidohydrolase (EC 3.5.1.53), 4 = diamine oxidase (EC 1.1.3.6), 5 = 4-aminobutyraldehyde dehydrogenase (EC 1.2.1.19), 6 = spermidine synthase (EC 2.5.1.16), 7 = a putrescine producing polyamine oxidase, 8 = homospermidine synthase (new enzyme). Inhibitors: *DFMA* = α-difluoromethylarginine, *HEH* = ß-hydroxyethylhydrazine, *GACU* = gabaculine.

[198]. Putrescine is an important metabolic branching point (Fig. 4). It is (i) converted into spermidine via spermidine synthase, (ii) oxidized via pyrroline and aminobutyric acid and (iii) channeled into the alkaloid specific pathway.

Homospermidine has been identified as the first intermediate of the alkaloid specific pathway in the tracer experiments described in the preceeding section. It also accumulates in *Senecio* root cultures in the presence of ß-hydroxyethylhydrazine (HEH), an inhibitor of diamine and polyamine oxidases (Fig. 4) [199]. This is consistent with the suggestion that a polyamine oxidase is involved in the oxidation of the primary amino groups of homospermidine, but makes a participation of a diamino oxidase in homospermidine formation unlikely. In addition, radioactively labeled pyrroline is not incorporated into homospermidine [199]. Finally a homospermidine synthase (HSS) was detected as the enzyme responsible for the conversion of putrescine into homospermidine. HSS is already known from two bacterial sources [200, 201] and seedlings of *Lathyrus sativus* [202], all organisms that do not produce PAs. The enzyme has been suggested to catalyze the following reaction:

$$\text{putrescine} + \text{putrescine} \xrightarrow{NAD^+} \text{sym-homospermidine} + NH_3$$

HSS has been partially purified and characterized from root cultures of *Eupatorium cannabinum* and has also been detected in five other PA producing species of Asteraceae [199]. Surprisingly, in addition to the above written reaction, HSS was found to accept spermidine as substrate instead of the first putrescine unit [203]. Spermidine is oxidized to aminobutanal which is combined with putrescine and 1,3-diaminopropane is released. Spermidine is also a substrate of the bacterial enzyme, but in comparison with the *Eupatorium* enzyme is less efficient. This reaction has obviously been overlooked because spermidine is a substrate only in the presence of putrescine. Plant HSS is essentially inactive with spermidine alone. Spermidine (K_m 15 μM) is incorporated into homospermidine with the same affinity and activity as putrescine (K_m 13 μM). Taking into account the steady-state concentrations of putrescine (70 - 120 μM) and spermidine (30 - 100 μM) it can be assumed that the ratio of incorporation of the two substrates is roughly 1 : 1 [203]. The overall reaction catalyzed by plant HSS is shown in Fig. 5. The role of spermidine as a direct substrate of enzymatic homospermidine formation is well in accordance with the observation that in tracer experiments spermidine and putrescine are equally efficient as precursors of PA biosynthesis [166, 167, 195, 196]. The enzymatic evidence is in agreement with the tracer experiments described in section 3.1; tracer incorporations cannot distinguish whether a putrescine moiety is incorporated directly or via spermidine into homospermidine (see Fig 4). There is only one discrepancy: in the HSS catalyzed reaction NAD^+ was shown to function as a hydrid acceptor in the first part of the reaction and subsequently as a hydrid donor in the second part; i.e. both (*R*)-[1-^{2}H]putrescine and (*S*)-[1-^{2}H]putrescine were incorporated into homospermidine with 100% of retention of deuterium [203]. As already discussed, Robins and Spenser found only 69 and 68% for this step when they fed (*S*)-[1-^{2}H]putrescine (Fig. 2). Further studies are needed to solve this discrepancy; isotope effects might be involved in the formation of spermidine or/and its incorporation into homospermidine.

4. BIOSYNTHESIS OF THE NECIC ACIDS

By contrast to the necine bases our knowlege concerning the biosynthesis of the necic acids is very scanty. All aliphatic necic acids so far studied are derived from branched-chain amino acids such as valine, isoleucine, leucine and their precursors. Most studies were carried out by Crout 10 to 30 years ago. The two necic acids of the diester PA heliosupine (**239**) in *Cynoglossum officinale* are derived from isoleucine (angelic acid) and valine (echimidinic acid, presumably five carbon atoms) [204, 205]. The necic acid moiety of senecionine (**1**) in *Senecio* species is derived from two molecules of L-isoleucine which after (oxidative?)

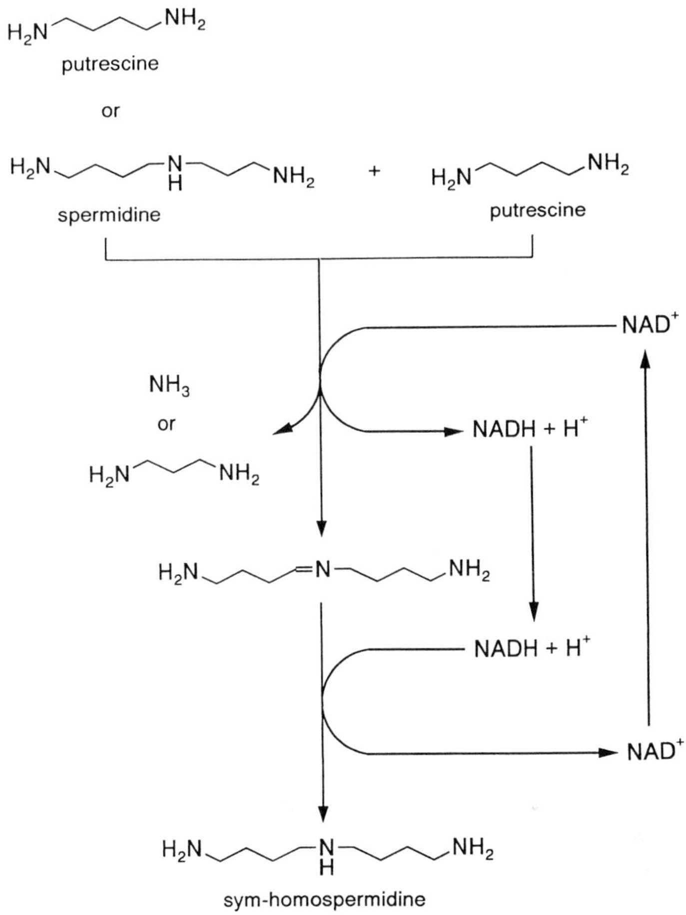

Fig. 5. Reaction catalyzed by plant homospermidine synthase.

decarboxylation supply the two five-carbon precursors [206-210]. The intermediates are still unknown. The necic acid (eight carbon acid) of monocrotaline (**284**) in *Crotalaria retusa* contains a C_5 unit from isoleucine, while the origin of the remaining three carbon atoms remains uncertain [211]. One half of the C_{10} necic acid of trichodesmine (**294**), a macrocyclic PA from *Crotalaria globifera*, was shown to be derived from isoleucine, while valine and leucine are precursors of the other half [212].

5. TRANSPORT, STORAGE AND METABOLISM OF PAs IN PLANTS

5.1 Tissue Specificity of PA Synthesis

PAs are synthesized either in roots, shoots or even at both sites depending on the plant family from which the species is derived.

In Asteraceae (e.g. *Senecio* and *Eupatorium*), roots were shown to be the major, if not exclusive, sites of PA biosynthesis [213]. In *Senecio vulgaris* total incorporation of ^{14}C-labeled putrescine into PAs was 10 to 35% in excised roots of intact plants and in vitro cultivated roots, but below the analytical detection limit (< 0.01%) in detached shoots. Birecka and co-workers, who also were unsuccessful in finding incorporation of radioactivity into PAs of detached shoots of *S. vulgaris* exposed to pulse labeling with $^{14}CO_2$ [197], claimed low but significant synthesis of labeled PAs in detached shoots of *S. riddellii* and *S. longilobus* [214]. All root cultures so far established from PA-producing Asterceae were found to synthesize PAs in quantities comparable to those found in intact plants, while shoot cultures and dedifferentiated cell cultures were free of PAs and unable to incorporate labeled tracers into PAs [11, 12, 95]. PA biosynthesis in roots seems to be a general feature of PA-producing species of the Asteraceae; synthesis occurs preferentially in the root apex and thus coincides with the site of cell division and enhanced protein synthesis [215].

By contrast, in detached shoots of *Heliotropium spathulatum* (Boraginaceae) exposed to pulse labeling with $^{14}CO_2$, leaves were the main sites of PA synthesis [190]. In *H. indicum* all parts of the shoot and again preferentially young leaves were found to incorporate labeled putrescine into PAs, while the roots were ineffective [217]. Interestingly, young leaves of *H. indicum* were the only organs of the plant containing detectable amounts of free sym-homospermidine [216]. On the other hand, *Cynoglossum officinale* actively synthesizes PAs in leaves as well as in roots [218]. Root cultures of two *Heliotropium* species were unable to produce PAs while the culture of *C. officinale* does [12]. Obviously in species of the Boraginaceae, leaves as well as roots may possess the ability to synthesize PAs. Again undifferentiated cell cultures do not produce any PA (e.g. *Symphytum officinale* [12, 219], *Lithospermum erythrorhizon* [220]).

In the first successful tracer study of PA biosynthesis carried out by Nowacki and Byerrum [162], excised shoots of *Crotalaria spectabilis* were used. Root and cell cultures of different *Crotalaria* species were ineffective [11, 12, 213]. Thus, in the family Fabaceae the shoots may be the preferred organs of PA formation.

5.2 Tissue Distribution, Translocation and Vacuolar Storage

Unfortunately most phytochemical reports do not contain any quantitative information on PA levels or even a specification of the plant organ that was analyzed. Although PAs are generally found in all plant parts, there are often considerable differences in quality and quantity between different tissues and during the ontogenesis of the plant. In *Heliotropium* species young leaves, inflorescences and roots contain the highest levels of alkaloids [221]. In the monocarpic species *Senecio vulgaris* and *S. vernalis* up to 90% of total PAs may be found in the inflorescences of flowering specimens [222]. PA concentrations in the flower heads are about 10 to 30-fold higher than in the leaves. In stems the highest PA concentrations are found in the epidermal and subepidermal tissues [213]. In *S. vernalis* senecionine (**1**) dominates in the inflorescences while senkirkine (**20**) is the major PA in leaves [222]. In *Chromolaena odorata* (Eupatorieae) leaves were almost devoid of PAs while the inflorescences (i.e. tubular flowers) were a rich PA source [37]. On the other hand, leaves of *Senecio riddellii* contain up to 18% PAs (on a dry weight basis), which is the highest PA content ever recorded for a plant [223].

Considerable inter- and intraspecific variations in total PA content were found in *Senecio, Amsinckia,* and *Crotalaria* species which were analyzed during growing seasons for three years. In most species the total PA content reached a maximum at the early flowering stage [224]. Similarly large qualitative and quantitative differences were found to exist in the PA patterns between populations of either *Senecio vulgaris* and *S. vernalis* [98]). As shown for *S. jacobaea* the large phenotypic variations in PA content and composition are mainly due to genetic differences [225]. In addition to phenotypic variations, distinctive chemotypes have been described for *S. jacobaea* and *S. erucifolius* (Appendix I) [64].

Senecio species synthesize PAs in roots but store them preferentially in the inflorescences. Translocation studies with ^{14}C-labeled senecionine *N*-oxide revealed that in *S. vulgaris* the alkaloid is exported from the roots exclusively via the phloem path and efficiently channeled to the sites of storage (i.e. inflorescences and peripheric stem tissue) [213]. This process is specific; PAs are not able to enter the phloem path in non-PA-producing plants. Phloem mobiltity of PAs is further indicated by phloem-sucking insects. The honeydew of the green peach aphid (*Myzus persica*) feeding on *S. vulgaris* was found to contain senecionine (**1**) and its *N*-oxide [226], and a specialist aphid (*Aphis jacobaeae*) was shown to sequester large quantities of PAs from three *Senecio* species [227].

The subcellular storage of PAs was studied using cell suspension cultures of PA producing plants. These cells, although lacking the capacity to synthesize PAs, retained the ability to accumulate PA *N*-oxides [228]. In experiments with isolated protoplasts and vacuoles the vacuolar storage of PA *N*-oxide and its translocation into the vacuole by specific carrier-mediated transport was demonstrated [229, 230]. Cells of non-PA-producing plants were unable to sequester PA *N*-oxides.

5.3 Metabolism, Turnover, Degradation

Relatively little is known about the metabolic fate of PAs within the plant. Birecka and co-workers thoroughly analyzed the tissue-specific PA concentrations during the ontogenesis of *Heliotropium* species and performed pulse-chase tracer experiments. They found that PAs undergo very slow, if any, metabolic changes [190, 221]. *S. vulgaris* root cultures incorporate efficiently labeled putrescine into PAs (20-30% total tracer incorporation); the senecionine *N*-oxide formed did not undergo any significant degradation or turnover [195, 196]. In *S. erucifolius* root cultures a "population" of labeled PA molecules remained constant over the whole culture period (15 to 19 days), except a slow transformation of the *N*-oxides of senecionine (**1**) into the *N*-oxides of seneciphylline (**8**) and acetylseneciphylline (**9**) [215]. *Senecio vulgaris* plants allowed to take up ^{14}C-labeled putrescine or senecionine *N*-oxide were analyzed up to five weeks after tracer application. No significant differences were found in the PA levels between the first and the fifth week following tracer application (Hartmann unpublished results). There is not a single confirmed reference indicating PA turnover or PA degradation in living vegative plant organs.

The situation may be different in germinating seeds. Seeds of *Crotalaria scassellatii* contain ca 2% PAs as tertiary alkaloids. During germination the PAs are rapidly *N*-oxidized and gradually degraded; five-week-old seedlings are almost depleted of PAs [231]. The "seed PAs" may be compared with the toxic nonprotein amino acids, such as cannavanine, which not only represents an effective chemical barrier to predation, but also a readily available source of stored nitrogen for the germinating seed [232].

5.4 Plant-to-Plant-Transfer

Previous reviews on the occurrence of PAs in the plant kingdom [1, 2, 3, 8] record *Castilleja rhexifolia* as the only example so far of a PA containing species of the family Scrophulariaceae. Stermitz and co-workers who detected PAs in the North American hemiparasitic genus *Castilleja* [233-235] first assumed that the alkaloids were produced by the plant itself, but later recognized that *Castilleja* obtains PAs solely via root parasitism from PA containing hosts such as *Senecio atratus* and *S. triangularis* [236] and *Liatris punctata* [47]. More recently a PA transfer from *S. triangularis* to its root parasites *Pedicularis groenlandica* and *P. bracteosa* (Scrophulariaceae) was observed [237]. It is reasonable to assume that this plant-to-plant-transfer occurs via phloem connections. Interestingly, a similar alkaloid transfer has been observed between quinolizidine alkaloid producing species (Fabaceae) and parasitic plants [236, 237]; quinolizidine alkaloids are also phloem-mobile compounds [22].

6. ROLE OF THE PA *N*-OXIDES IN PLANTS

The abundant occurrence of PAs in the *N*-oxide form is well documented since the beginning of PA research [see for review 3, 238, 239]. Areshkina [240] was the first who established that the PA *N*-oxides present in *S. platyphyllus* were genuine constituents and not artifacts. Reports concerning the relative proportions of *N*-oxides and tertiary PAs in plant sources are often contradictory. The major reason for these inconsistencies is that the *N*-oxides are less stable than the respective tertiary bases. Thus, PA *N*-oxides genuinely present in the plant are easily reduced during the extraction procedure [195, 222]. PAs are routinely isolated and separated from plant sources after reduction (H^+/Zn dust) as free bases (tertiary alkaloids), which are much easier to purify than the respective *N*-oxides [8]. However, in all studies where serious attention was paid to the identity of the genuine alkaloid form, *N*-oxides were identified as the dominating if not exclusive alkaloid form, at least in vegetative plant parts, e. g. *Senecio* [64, 76, 98], *Adenostyles* [31], *Chromolaena* [37], *Hackelia* [118], *Heliotropium* [216, 221], *Crotalaria* [231].

In *Senecio* species PAs are exclusively synthesized as *N*-oxides [12, 95, 195] which are the specific molecular form for long-distance translocation [213], carrier-mediated transport into the cell vacuole and storage within the vacuole [229, 230] (see section 5.2). On the basis of these results it has been suggested that the polar, salt-like *N*-oxides are molecules which are better suited for selective alkaloid transport and safe vacuolar storage than the tertiary form, which in contrast to the *N*-oxide, in unprotonated form may passively permeate biomembranes. This idea is further supported by the fact that certain insects adapted to PA containing plants and capable of sequestering plant PAs, store the alkaloids as *N*-oxides (see section 8.2). Both plants and PA storing insects are capable of the specific *N*-oxidation of tertiary PAs.

For *Crotalaria* species, however, it has frequently been reported that tertiary PAs were found to dominate in seeds [224, 241, 242]. In fact, seeds of *Crotalaria scassellatii* were shown to contain the species specific PAs exclusively in the tertiary form; not even traces of *N*-oxides were detectable. During seed germination, however, the tertiary PAs were rapidly and completely *N*-oxidized, and only PA *N*-oxides were found in the vegetative plant tissues [231]. In accord with the above-suggested view, the lipophilic free base should be better suited to be stored in desiccated seeds, whereas the polar *N*-oxide is the preferred form of storage in the aqueous cell vacuole.

It seems well documented now that many (if not all) species of the three families Asteraceae, Boraginaceae and Fabaceae synthesize, translocate and store PAs exclusively as *N*-oxides in their vegetative tissues. But there are open questions concerning the role of the *N*-oxide form: the otonecine derivatives (e.g. senkirkine **20**) which are derived from the respective retronecine *N*-oxides (see sectin 3.1) do not form *N*-oxides. In *Senecio vernalis* root cultures, in contrast to the PA *N*-oxides which are exclusively localized in the cells, senkirkine is partly released into the culture medium [95]. Furthermore it is still unknown whether alkaloids of the phalaenopsine type (Orchidaceae) exist as *N*-oxides or not.

7. ACTIVITIES OF PLANT PAS

7.1 Toxicity

The majority of PAs (i.e. the 1,2-dehydropyrrolizidines) cause serious diseases in domestic animals and humans [see for review 3, 14-16]. It is well known that the PAs per se are nontoxic to mammals. PA toxicity has been attributed to the production of toxic pyrrolic metabolites, formed through the action of the mixed function oxidase enzymes which are found in the liver microsomes [see for review 3, 18, 243]. Cytochrome P-450 monooxygenases are assumed to play the major role in pyrrole formation [244]. The instable PA pyrroles are highly reactive alkylating agents. Their toxicity is presumed to occur through covalent binding and subsequent inactivation of essential biological nucleophiles such as proteins or nucleic acids which could alter cell function and lead to cell damage and cell death and even may initiate cancer formation [3, 18, 243]. Besides pyrrole formation *N*-oxidation and hydrolysis of PAs represent further pathways for PA metabolism in the liver. Both pathways may serve as detoxification mechanisms. Unlike the highly reactive tertiary dehydroalkaloids, the corresponding *N*-oxides are relatively unreactive and are primarily excreted in the urine due to their water solubility. However, after ingestion, PA *N*-oxides are easily reduced in the gut through the activity of the intestinal flora and are passively absorbed as the tertiary alkaloids. Thus, regarding its potential toxicity, it makes no difference whether a PA is ingested as tertiary alkaloid or its *N*-oxide; a direct absorption of the polar PA *N*-oxide from the gut is unlikely to occur. Products of the hydrolysis of potentially toxic PAs are the respective necine bases (e.g. retronecine, heliotridine) and necic acids, neither of which are hepato- or cytotoxic [3, 18, 245].

PAs which are hepatotoxic in mammals are strong mutagens in non-mammalian systems such as *Drosophila melanogaster* [246-248]. It is assumed that the metabolic activation to genotoxically reactive metabolites occurs via mechanisms which are similar to those in the liver of mammals. Thus, PAs should be cytotoxic in all organisms and tissues that possess the potential to transform 1,2-unsaturated PAs into metabolically active pyrroles.

7.2 Role in Plant Protection

The idea that PAs serve as protective chemicals for plants has generally been accepted [19-22, 249]. Direct experimental evidence, however, is sparse. The potential toxicity of most PAs against mammalian herbivores is often used as a strong argument favouring an important role of PAs in plant defense. But the noxious action of PAs in mammals is not instantaneous, but appears only after an extended prepatent-period which sometimes is up to 18 months [250, 251]. Is is difficult to see any benefit to a plant damaged by a mammalian herbivor that suffers from the defense compounds only weeks later. Poisoning of livestock, even when common, should therefore be regarded as an unfortunate accident caused by placing mammalian herbivores in juxtaposition with such plants [251]. What is more important, grazing animals usually avoid PA plants unless there is shortage of other herbal food, apparently because of their deterrent taste [19]. PAs appear to protect plants from damage not only by vertebrates but perhaps more importantly also by insects. Strong feeding deterrency appears to be a particular property of PAs. PA-contaminated food is rejected by vertebrates such as toads, frogs, lizards, birds, mice [according to ref. 19], as well as insects such as locusts [252], ants, cockroaches and various lepidopteran larvae [according to ref. 19]. Feeding deterrent activity of 14 PAs (at concentrations of 1 to 2 mM) were compared in bioassays with larvae of the spruce budworm (*Choristoneura fumiferana*) [253]. Great differences were observed; some PAs were inactive whereas those with α,ß-unsaturation in the necic acid moiety, e.g. senkirkine (**20**), lasiocarpine (**237**) and senecionine (**1**), were among the most active PAs. The complexity of the field situation often precludes a clear demonstration of effectiveness of PAs in plant defense against insects. Fortunately the acquisition of plant PAs by adapted insects supports much more conclusively a function of PAs as powerful defensive chemicals (see section 8.4).

There are only single reports on other biological activities of certain PAs, e.g. moderate antimicrobial [254] and allelopathic effects [255].

8. ROLE OF PAs IN PLANT - INSECT - RELATIONSHIPS

8.1 Sequestration of PAs by Adapted Insects

In respect to their attractiveness for specialized insects PAs are unique among the some 10,000 known plant alkaloids. A great number of insects belonging to different orders such as Lepidoptera, Coleoptera, Orthoptera and Homoptera are known to sequester PAs from plants (Table 2). These insects, that store plant-derived PAs as protective chemicals, advertise their unpalatability to potential predators by conspicuous warning coloration (aposematic signals).

Use of plant PAs for their own benefit is well known from many lepidopterans. In particular, moths of the families Arctiidae and Ctenuchidae and butterflies of the subfamilies Danainae and Ithomiinae have been shown to store, or are suspected of storing [256-258], PAs they obtain from plant sources (Table 2). PAs may be taken up either by larvae, as in most Arctiidae, or by adults as in most Danainae and Ithomiinae (Table 2). It seems certain that all sequestrating species which feed as larvae on PA-containing plants store and retain the acquired PAs through to the adult stage. Some of these species are specialists that are restricted to a certain food plant, e.g. *Tyria jacobaeae*, feeding on *Senecio jacobaea*; others are polyphagously feeding on various PA plants. These generalists are often attracted by PAs to PA-containing food sources. PAs seem to function as attractants and feeding-stimulants. *Creatonotos* larvae, for instance, even consume glass-fibre discs impregnated with PAs [19]. This feeding behaviour is called "pharmacophagy", which by definition means that species of this category search for certain secondary compounds (e.g. PAs), take them up, and utilize them for a specific purpose other than primary metabolism or (merely) food-plant

Table 2. Insects known to sequester PAs from plant sources

ORDER; Family *Genus*	PA-source or feeding behaviour	PA-uptake as larva	PA-uptake as adult	PA-derived pheromone	References
LEPIDOPTERA					
Arctiidae					
Amerila phaedra	pharmacophagous	-	+	-	258,**
Arctia caja	polyphagous	+	-	-	262, 263, 277
Cisseps fulvicollis	pharmacophagous	-	+	+	257, 267
Creatonotos transiens	polyphagous*	+	-	+	268, 269, 266
Gnophaela latipennis	*Hackelia*	+	-	?	118
Hyalurga syma	*Heliotropium*	+	?	-	134
Nyctemera coleta	polyphagous*	+	+	?	258, **
Phragmatobia fuliginosa		+	-	+	270, 271, 277
Pyrrharctia isabella	polyphagous*	+	-	+	270, 271
Spilosoma spp.	polyphagous*	+	-	-	277, **
Utetheisa lotrix	*Crotalaria*	+	-	+	272, **
ornatrix	*Crotalaria*	+	-	+	273, 274
pulchelloides	*Heliotropium*	+	-	+	275,
Tyria jacobaeae	*Senecio*	+	-	-	262-264, 276, 277
Ctenuchiidae					
Euchromia spp.	pharmacophagous	-	+	-	258, **
Danainae		-			
Amauris	pharmacophagous	-	+	+	278

Danaus plexippus	pharmacophagous	-	+	-	275, 279, 280
Danaus spp.	pharmacophagous	-	+	+	275, 278, 281
Euploea spp.	pharmacophagous	-	+	+	275, 278
Euploea spp.	*Parsonsia*	+	?	+	282
Idea spp.	*Parsonsia*	+	-	+	282, 283
Ithomiinae					
Tellervo zoilus zoilus	*Parsonsia*	+	?	+	282
Tithorea spp.	*Prestonia*	+	?	+	282, 284
Mechanitis spp.					
and 44 further genera	pharmacophagous	-	+	+	71, 72, 261
ORTHOPTERA					
Pyrgomorphidae					
Zonocerus variegatus	polyphagous*	+	+	-	37, 285, 286
COLEOPTERA					
Chrysomelidae					
Oreina cacaliae	*Adenostyles*	+	+	-	31, 287-289
speciosissima	*Adenostyles*	+	+	-	31, 288
Coccinella spp.	*Aphis jacobaeae*	+	+	-	227
HOMOPTERA					
Aphididae					
Aphis jacobaeae	*Senecio*	+	+	-	227

* Attraction by PAs or pharmacophagous feeding behaviour has been demonstrated.
** A Biller, L Witte, M Boppré, T Hartmann, unpublished results.

recognition [259]. Pharmacophagous feeding is most impressive in species acquiring PAs as adults either from PA-containing nectar [71, 72, 257, 260] or most frequently from withered twigs or dead parts of PA plants. They wet the plant material with fluid from their proboscis and then reimbibe it with the extracted PAs [19, 20]. The average amounts of PAs sequestered pharmacophagously by adult Ithomiinae is in the range of approximately 2 - 7% of dry weight but may reach extremely high values, e.g. 20% in *Scada* spp. [71, 261]. Lepidopterans which sequester PAs as larvae from their food plants such as *Tyria jacobaeae* contain lower concentrations which are in the range of about 1 to 2% of dry weight [262-264]. The PA concentrations found in the insects are generally well above those of the respective food-plant tissues.

The number of Lepidoptera that are attracted by PA baits are considerably greater than the number of species listed in Table 2 [257, 256, 258]; but, although likely, it is still unknown whether they are able to sequester PAs. This is also true for other insects such as certain Chrysomelidae and Diptera [19, 21].

PA sequestration in insects outside the lepidopterans has been demonstrated for the grasshopper *Zonocerus*, the aphid *Aphis jacobaeae* and leaf beetles of the genus *Oreina* (Table 2). The phloem-feeder *A. jacobaeae* already mentioned (see section 5.2) as indicator for the phloem mobility of PAs in its host-plant, was shown not only to sequester plant PAs but also to pass over these PAs to ladybirds feeding on them. PA levels up to 5 mg per g fresh weight were measured in the ladybirds, which is almost half of the level of endogenous alkaloids (coccinellines) produced by ladybirds [227].

Numerous leaf beetles (Chrysomelidae) are well known for their chemical defense. They synthesize or sequester a great variety of compounds that are released from specialized exocrine glands [306]. The species *Oreina cacaliae* and *O. speciosissima* were found to sequester PAs (mainly seneciphylline *N*-oxide) from their food plant *Adenostyles alliariae* [31, 287-289]. Extraordinarily high concentrations of PA *N*-oxides (i.e, 0.1 - 0.2 $mol \cdot l^{-1}$) were found in the defensive secretions of the two species.

8.2 Uptake and the Role of PA *N*-Oxides in Insects

Why are insects from various unrelated taxa able to sequester PAs? Preliminary evidence has been presented that a specific carrier in the midgut of *Creatonotos* is responsible for resorption of the polar PA *N*-oxides [265]. By contrast, recent evidence from feeding experiments with senecionine *N*-oxide in which the *N*-oxide oxygen was selectively ^{18}O-labeled showed that in different arctiid larvae [290], as well as the grasshopper *Zonocerus* [291], senecionine *N*-oxide is reduced in the gut. The lipophilic tertiary senecionine is passively taken up into the hemolymph where it is rapidly re-*N*-oxidized. If [^{18}O]*N*-oxide was fed orally, the PA *N*-oxide recovered from the hemolymph was almost completely ^{16}O-labeled. This uptake behaviour explains why dietary tertiary PAs and PA *N*-oxides are taken up and stored with the same efficiency [264]. Hemolymph of various arctiids contains a very active soluble mixed-function *N*-oxygenase which converts any tertiary PA into the respective *N*-oxide [290]. As in plants, the occurrence of PA *N*-oxides is also long known from storing insects; as in plants usually mixtures of *N*-oxides and tertiary PAs have been reported to occur [72, 292]. Recent studies revealed that, so far tested, all PA sequestering lepidopterans, as well as *Zonocerus* and *Oreina*, store PAs exclusively as *N*-oxides [31, 37, 134, 264, 266, 277, 287-289, 293].

In arctiid larvae the PA *N*-oxides seem to be rather equally distributed between hemolymph and integument [264, 269]. The latter, due to its size, contains the largest total amount [264, 266, 277, 294].

8.3 Transformation of Plant Acquired PAs by Insects

It is well known that a number of species of the Arctiidae, Danainae and Ithomiinae which are able to sequester PAs, require these plant alkaloids as precursors for the biosynthesis of male courtship pheromones (Table 2) [see for review 19-21]. The structures so far isolated from the androconial organs of Lepidoptera are shown in Fig. 6. (7*R*)-hydroxydanaidal (**i-1**) and danaidal (**i-2**) are known pheromones of the arctiids, danaidone (**i-3**) and rarely **i-1** and **i-2** are found in the Danainae, whereas the Ithomiinae contain the nitrogen-free ithomiinae lactone (**i-5**) [295] and rarely nordanaidone (**i-4**). Although the formation of the respective pheromones from ingested PAs of plant origin is well documented, until recently the biosynthetic pathway was obscure. The formation of **i-1** from heliotrine (**201**) in *Creatonotos transiens* has now been elucidated [296]. The biosynthetic sequence illustrated in Fig. 7 has been established. By application of specifically

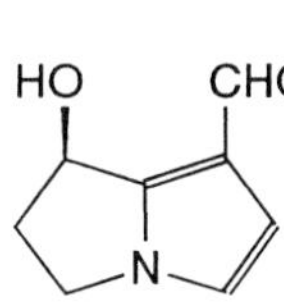

i-1

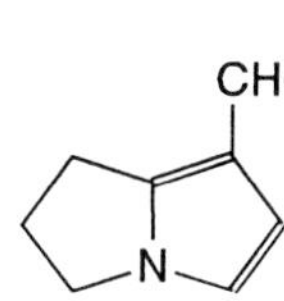

i-2

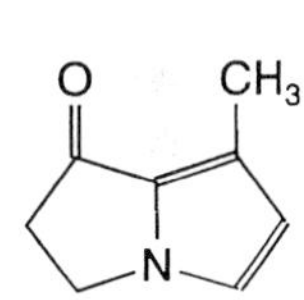

i-3

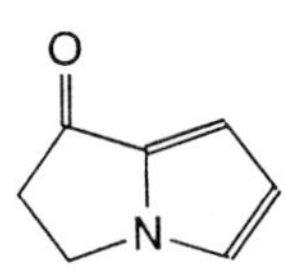

i-4

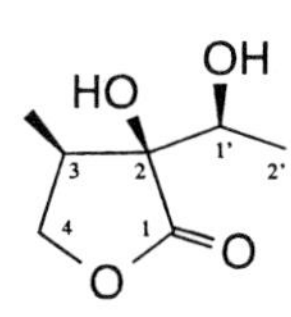

i-5

Fig. 6. PA derived pheromones of Lepidoptera. **i-1** =(7*R*)-hydroxydanaidal; **i-2** = danaidal; **i-3** = danaidone; **i-4** =nordanaidone; **i-5** = ithominae lactone.

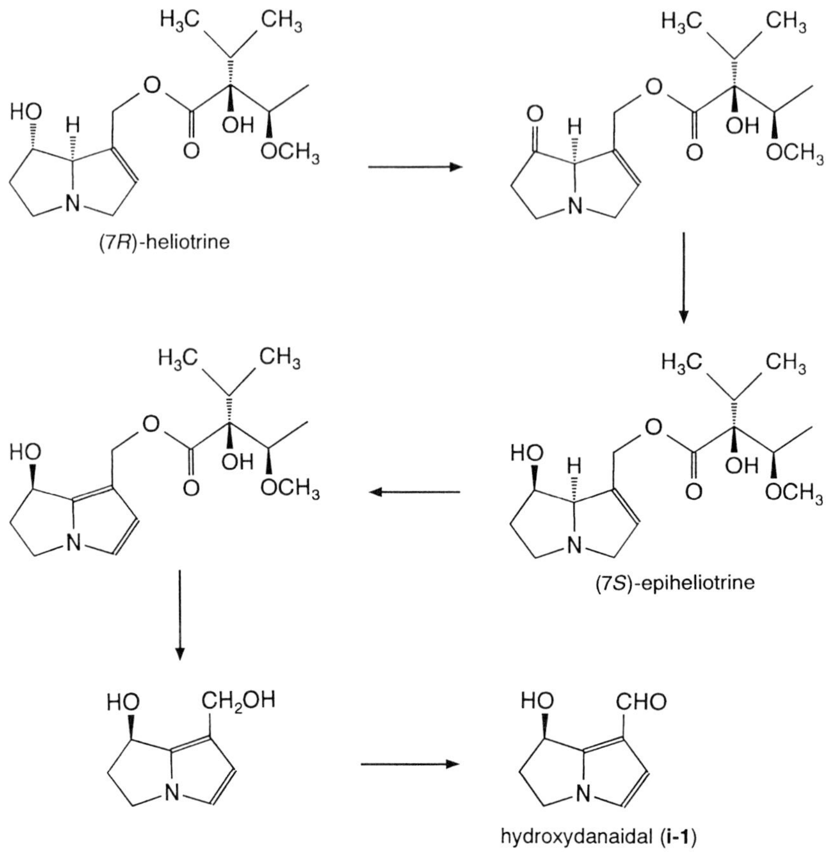

Fig. 7. Transformation of heliotrine (**201**) into hydroxydanaidal (**i-1**) by *Creatonotos transiens* .

C7-deuterated samples of heliotrine (7*S*-configurated) and epiheliotrine (7*R*-cofigurated) it could be demonstrated that the epimerization at C7 already known to take place in *Creatonotos* larvae [297] proceeds via the corresponding ketone followed by stereospecific reduction. The subsequent steps comprise aromatization of the dihydropyrrole ring, ester hydrolysis and oxidation of the resulting primary alcohol into the aldehyde. *Utetheisa ornatrix* which produces **i-1** easily from macrocyclic PAs with the 7*R* configuration [298] is unable to catalyze the C7 epimerization, and thus to produce **i-1** from heliotrine [296]. *Arctia caja* (not known to synthesize pheromones) and *Phragmatobia* (known to produce **i-1** and **i-2**) feeding as larvae on heliotrine were found to contain in the adult stage 20-30% and 40-50%, respectively, of epiheliotrine [277].

Interestingly, epimerization at the C7 (again 7*S* into 7*R*) and in addition at the 3'-carbon (3'*R* into 3'*S*) of the necic acid has been observed with PAs of the lycopsamine type in different insects. Adults of *Mechanitis polymnia* (Ithomiinae) transformed rinderine (**200**), echinatine (**223**) and intermedine (**187**) efficiently into lycopsamine (**215**) (Fig. 8); males were more efficient than females [299]. Indicine (**212**) with the same stereochemistry at C7 and C3' as lycopsamine was not transformed. Comparable transformations have been suggested for the arctiid *Hyalurga syma* which mainly sequesters intermedine (**187**) and lycopsamine (**215**) but feeds on a host plant that contains 3'-acetylrinderine (**203**) as the major PA [134]. Even the taxonomically unrelated grasshopper *Zonocerus* was found to transform about 20% of ingested intermedine/rinderine into lycopsamine/echinatine; a direct inversion of configuration at C7 was demonstrated by injection of rinderine into the hemolymph and identification of intermedine [37].

Aplin and co-workers [276] isolated a "PA metabolite" from pupae of *Tyria jacobaeae* which later [300] was identified as callimorphine (**i-6**) (Fig. 9). Callimorphine is not known from plants but has been detected in a number of arctiid moths, e.g. *Arctia, Callimorpha* [300], *Creatonotos* [269, 297], *Hyalurga* [134], and *Gnophaela* [118]. The last-quoted authors found callimorphine as the sole PA in larvae, pupae and adults of insects that had been fed on *Hackelia californica* (Boraginaceae) as laval food plant. For chemical reasons they regarded a conversion of the plant acquired PAs into callimorphine as most unlikely (compare **i-6** with the *H. californica* PAs **261**, **263**) and instead suggested hydrolysis of the PAs and reesterification of the resulting necine base (i.e. retronecine) as a possible mechanism [118]. In fact, this suggestion has recently been confirmed [264, 269]. From pupae of *Tyria jacobaeae* which as larvae had ingested ^{14}C-labeled retronecine, up to 40% of total radioactivity was recovered as radioactively labeled callimorphine. The radioactivity was restricted to the retronecine moiety of callimorphine. Feeding of [^{14}C]isoleucine to *Tyria* larvae led also to the formation of labeled callimorphine, which after isolation and hydrolysis was shown to contain the radioactivity exclusively in the necic acid moiety. In *Tyria* the esterification of retronecine is restricted to the very early stages of pupation. Neither larvae nor mature pupae are able to synthesize callimorphine. Two novel PAs, creatonotine (**i-8**) and isocreatonotine (**i-9**) (Fig. 9), were isolated from adults of *Creatonotos transiens* which as larvae received retronecine or ester alkaloids as the sole PAs [269]. Again the synthesis of the two alkaloids is restricted to the early stages of pupation. The creatonotines are accompanied by small amounts of callimorphine and the other related monoesters shown in Fig. 9. None of these structures is known from plant sources. The two arctiids are able to hydrolyze various PAs from their plant food sources and produce their own PAs by "partial biosynthesis", i.e. esterification of retronecine which is of plant origin with a necic acid synthesized by the insect. In some specimens of field-caught adults of *C. transiens* (Bali, Indonesia) the creatonotines were found to be the major PAs [290]. The "insect PAs" are found exclusively as *N*-oxides.

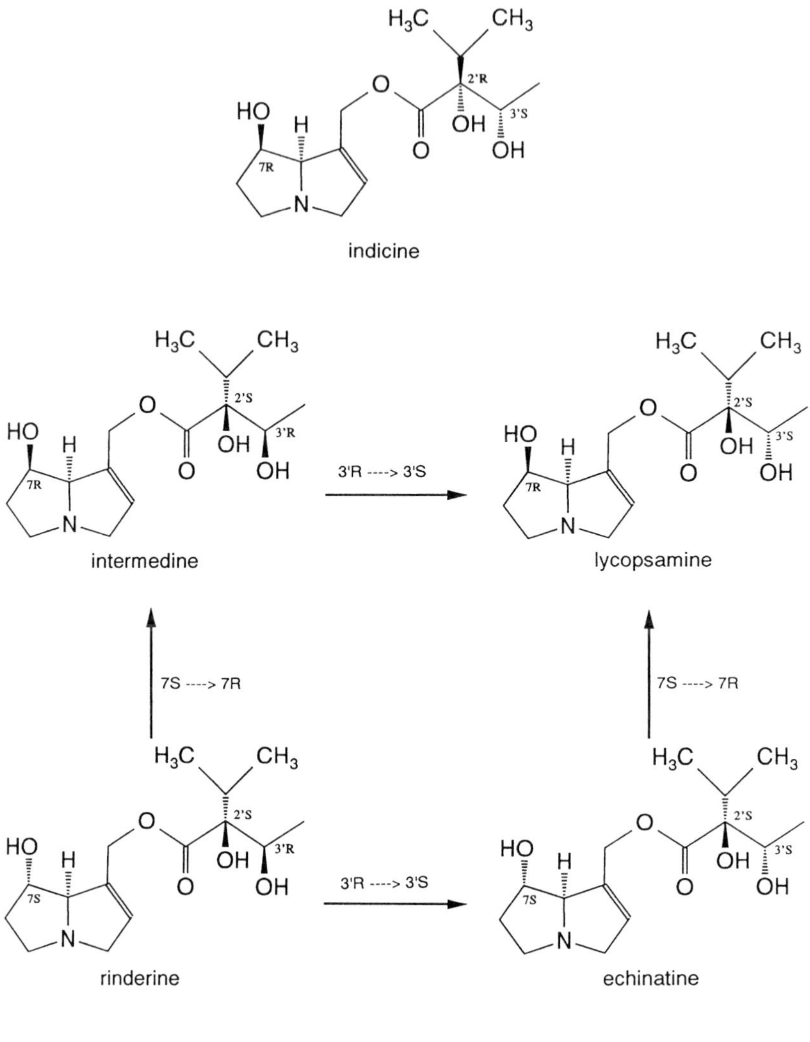

Fig. 8. Transformation of lycopsamine type PAs by Ithomiinae butterflies.

8.4 Functions of PAs in Insects

Storage for defense seems to be the primary function in most, if not all, insect-PA-relationships. As already pointed out the great majority of lepidopterans that sequester PAs are aposematically coloured and behave conspicuously [21]. In addition to the general phenomenon of aposematism there are exciting further examples favouring protective functions of sequestered PAs.

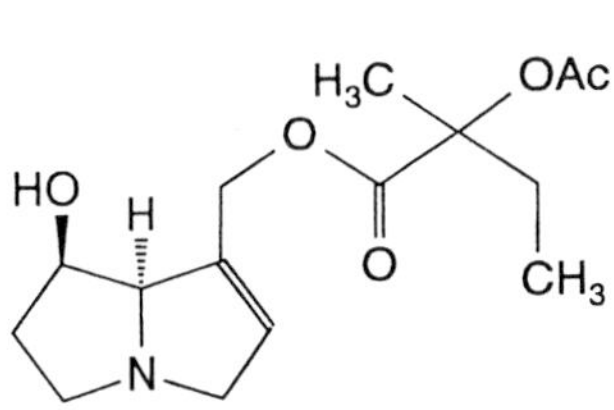

i-6

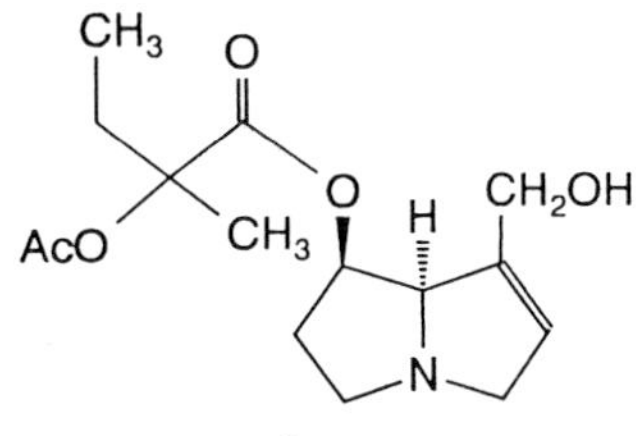

i-7

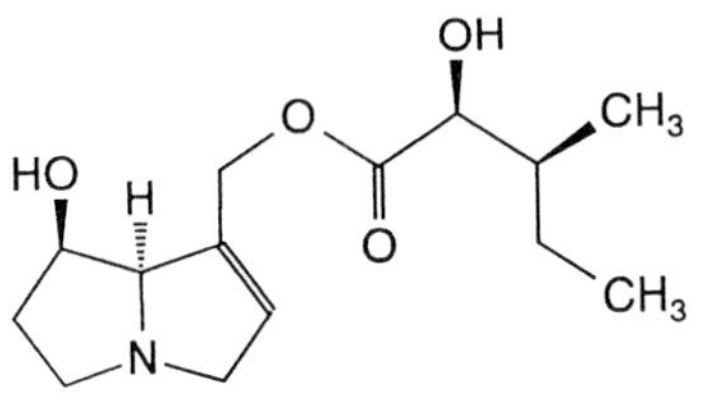

i-8

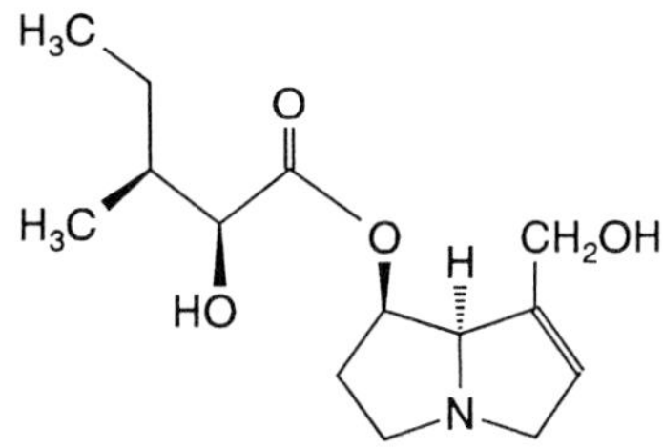

i-9

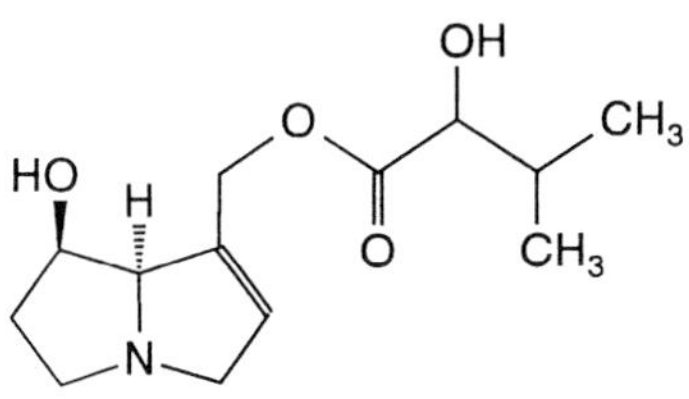

i-10

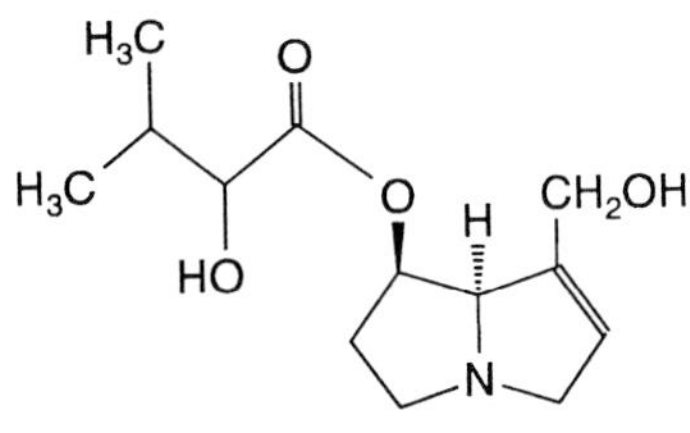

i-11

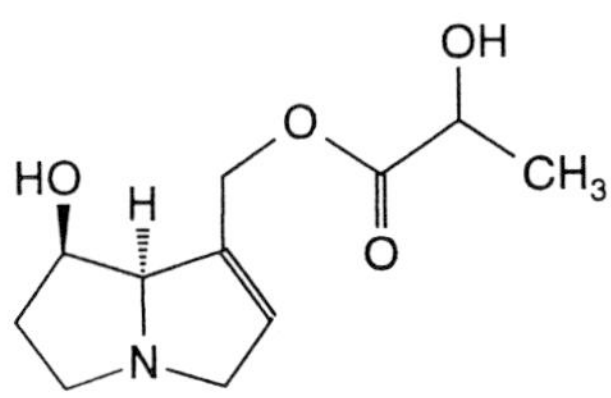

i-12

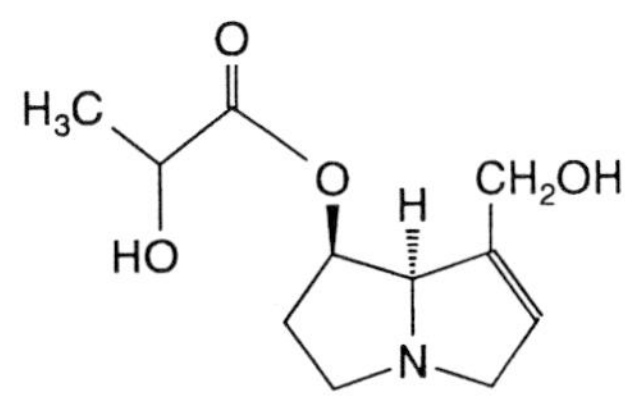

i-13

Fig. 9. "Insect PAs" synthesized in arctiids by esterification of retronecine derived from plant acquired PAs and necic acids produced by the insect. **i-6** = callimorphine, **i-7** = isocallimorphine, **i-8** = creatonotine, **i-9** = isocreatonotine, **i-10** = O^9-(2-hydroxyisovaleryl)-retronecine, **i-11** = O^7-(2-hydroxyisovaleryl)-retronecine, **i-12** = O^9-(2-hydroxypropionyl)-retronecine, **i-13** = O^7-(2-hydroxypropionyl)-retronecine.

The eggs are perhaps the most endangered stage in the life cycle of an insect. Motionless and often conspicuous, they are highly vulnerable to predators and parasites. Many insects defend their eggs by endowing them with deterrent chemicals. The eggs of several PA sequestering lepidopterans were shown to contain PAs. In the arctiid *Utetheisa ornatrix*, which sequester PAs in larvae from their food plants, both parents provide PAs for egg defense [273]. Females receive PAs from males during copulation and transmit these PAs together with PAs of their own to the eggs. This biparental contribution to egg defense was shown in laboratory tests with males raised on a diet containing monocrotaline (**284**) exclusively and females raised on an usaramine (**15**) containing diet. After mating and oviposition the eggs were shown to contain the two PAs; the male's nuptial gift accounted for about 30% of total PAs. There is good evidence that the level of the male courtship pheromone (i.e. hydroxydanaidal) which is related to the male's systemic load of PAs, signals the female the amount of protective chemicals she can expect to gain during copulation [300]. The eggs are well protected against arthropodan predators such as coccinellid beetles [273], but not against entomopathogenic fungi [301]. A similar biparental contribution to egg defense has been described for *Creatonotos transiens* [266]. Males of *Danaus gilippus* [302] and a number of Ithomiinae [72, 261] that sequester PAs pharmocophagously as adults, store large amounts of PAs in the spermatophores which they transfer to the females during mating. Again the eggs are endowed with PAs; the concentration may be as high as 1% of the fresh weight [71].

The giant tropical orb spider, *Nephila clavipes*, is an important potential predator of butterflies. PA-protected butterflies such as *Utetheisa* [303], *Hyalurga* [134] or various Ithomiinae [72, 304] are rejected by the spider; they are immediately cut out of the web and liberated unharmed. *Utetheisa* adults, that as larvae were raised on PA-free diet or freshly emerged Ithomiinae butterflies that are still free of PAs are readily eaten. The rejection of a potential PA-containing or PA-treated prey by *Nephila* seems to be very specific; most other unpalatable aposematic butterflies are freely eaten by the spider. Thus, a *Nephila* bioassay has been established to identify PA-protected butterflies or trace PA containing material [72, 134, 261]. The sequestration of PAs protect larvae of *Utetheisa* against predation by wolf spiders [305].

Chemical defense of leaf beetles of the genus *Oreina* (Chrysomelidae) is diverse and very striking [306]. Some species (e.g. *O. gloriosa*) synthesize cardenolides *de novo* and store these toxins exclusively in the pronotal and elytral glands [289]. When the beetles are disturbed, they liberate the content of their glands and spread them out over the cuticle. In other species (e.g. *O. cacaliae*), the beetles sequester PAs from their host plant and liberate PAs instead of cardenolides in their defensive secretion when disturbed [288, 289]. Recently the palatability of the two species to wild caught red-winged blackbirds (*Agelaius phoeniceus*) were tested. *O. gloriosa* were eaten in 55% of the trials and in 95% when the secretion had been removed physically; *O. cacaliae* were eaten in only 21% of the cases and 36% if the secretions had been removed [307]. The result suggests that PAs provide the beetle with better protection from the avian predator than do cardenolides.

Cardenolide-mediated protection of herbivorous insects is well known in many species, including the famous monarch butterflies (*Danaus plexippus*), that sequester cardenolides from Asclepiadaceae [308]. All danaine butterflies except the monarch, use sequestered PAs as pheromone precursors (Table 2). The monarch pharmacophagously searches for PAs and stores them in addition to the cardenolides which it acquired from its larval food plant. The fact that the monarch does not produce PA-derived pheromones strongly indicates that the monarch utilizes PAs as defensive chemicals. Recently sequestration of PAs has been shown for the North American overwintering population of the monarch. Apparently the monarchs utilize PA plants both en route to their overwintering site and at the site [279, 280]. This dual cardenolide-PA based defense may again indicate the great defensive potential of PAs.

9. THE LOLINES

The 1-aminopyrrolizidines (lolines) are biogenetically unrelated to the 1-hydroxymethylpyrrolizidines discussed so far. They represent a small distinctive group of alkaloids that are traditionally discussed together with the other PAs. All naturally occurring lolines are simple *N*-substituted derivatives of norloline (**363**). They are found in certain grasses (Poaceae) and the genus *Adenocarpus* (Fabaceae, Genisteae) (see Appendix I and II). In *Adenocarpus* the lolines (mainly decorticasine, **370**) are found preferably in the inflorescences and seeds [335, 336]. The occurrence of lolines in grasses received considerable interest since it became clear that their abundance is obligatorily linked to an infection with a fungal endophyte. The biosynthesis of the lolines has not been investigated. It may be presumed that the ring structure is formed from spermidine in a similar manner as the necine base from homospermidine [313].

Although loline (**364**) was first isolated from *Lolium cuneatum* 1955 [314] and the structure elucidated in 1965 [315, 116], it was not until 1981 that Buckner and co-workers reported that the concentrations of the lolines were greater in populations of *Lolium-Festuca* hybrids that contained a higher percentage of specimens infected with endophytic fungi [317]. Subsequently this fungus has been identified as *Acremonium coenophialum* Morgan-Jones & Gams [318], the imperfect stage of *Epichloe typhina* (Fr.)Tul. Recent field studies [319, 320] with tall fescue (*Festuca arundinacea*) plants as well as *in vitro* infection experiments [313, 321] with different grass species have convincingly demonstrated that lolines are only produced in endophyte infected plants and are not present in noninfected grasses. The systemic endophyte is tranmitted only by seed [322]. In a flowering *Festuca arundinacea* the highest tissue concentrations of the major alkaloids *N*-acetylloline (**369**) and *N*-formylloline (**368**) were found in the spikelets (4.6 mg/g dry weight), the lowest in the leaf blades (0.07 mg/g) (see [313] and ref. therein).

Endophyte infected tall fescue produces at least three different kinds of alkaloidal toxins: (i) lolines which are not produced by either the plant or the fungus alone; (ii) ergopeptide alkaloids produced by the fungus [323] and (iii) diazaphenanthrene alkaloids (e.g.

perloline, perlolidine) which are present in both non-infected as well as fungus infected plants. It is well documented that the endophyte infection is the cause of the fescue toxicosis in cattle primarily observed in the transition zone of the U.S.A., whereas uninfected tall fescue is nontoxic [324]. The specific syndroms or disorders associated with endophyte infected *F. arundinacea* include fescue foot, summer syndrome, fat necrosis, poor reproductive performance, and agalactia. The causative agents for fescue toxicosis have not been demonstrated, but current evidence implicates the ergopeptide alkaloids [325]. The lolines are related to the amount of fungus and the severity of the toxic syndroms, but have not been shown to be the toxic factors [324].

The toxic properties of infected grasses are not restricted to livestock; enhanced resistence to insect herbivory is also a characteristic property of endophyte-infected grasses. The relationship between host grass and fungal endophyte is a mutualistic one, with both partners gaining substantial benefits from the other [326]. Endophyte-infected grasses (e.g. *Lolium, Festuca*) always exhibit a better resistance to a variety of insect herbivores than uninfected plants [323, 326-328]. The lolines are one important component in the chemical defense of this mutualistic association. Lolines were found to be feeding deterrents and toxic factors to aphids [329, 330], the greenbug (*Schizaphis graminum*)[329], the large milkweed bug (*Oncopeltus fasciatus*) [328, 329], fall armyworm larvae (*Sphodoptera frugiperda*)[331, 332] and to root-feeding scarabaeid grubs (*Popillia* spec.) [333].

10. CONCLUSIONS

The general picture of the pyrrolizidine alkaloids as a complex, highly specific and functionally important system of plant secondary constituents, is becoming more and more complete:

> The biosynthesis of the necine base has been elucidated by sophisticated tracer techniques and the enzymatic characterization of the biosynthetic pathway is on the way. By contrast, however, the biosynthesis of the necic acids, with partly unique structures, has largely been neglected.
>
> In the Asteraceae PAs are synthesized as *N*-oxides in the roots. From the roots they are translocated into the shoots via the phloem path and channeled to the preferred sites of accumulation where they are stored in the cell vacuole. In the vegetative plant, PAs are metabolically stable products which, except transformation into species specific derivatives, show no significant turnover or degradation. As expected for a constitutive plant chemical defense agent, PAs are synthesized and translocated to the strategic sites of defense (e.g. inflorescences) where they are stored at appropriate concentrations.

TERTIARY ALKALOID | ALKALOID *N*-OXIDE

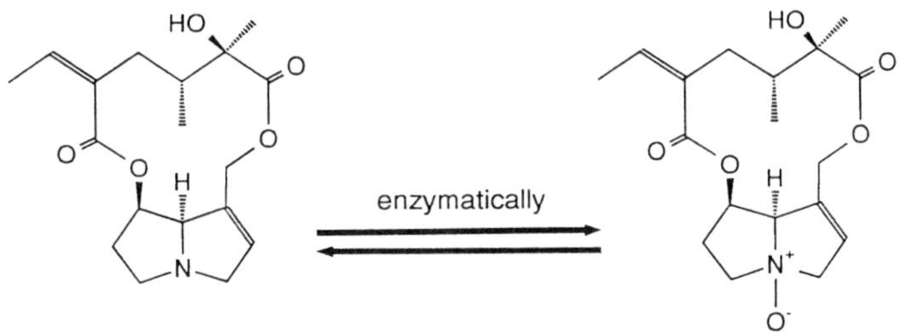

Senecionine | Senecionine *N*-oxide

TERTIARY ALKALOID	ALKALOID *N*-OXIDE
- unpolar when unprotonated	- very polar, salt-like
- able to unspecifically permeate biomembranes (unprotonated form)	- not able to unspecifically permeate biomembranes
- potentially toxic	- nontoxic

Fig. 10. Tertiary amine and amine *N*-oxide, the two molecular forms of pyrrolizidine alkaloids and their functional differences.

By contrast to the Asteraceae, in the Boraginaceae and presumably also Fabaceae the shoots are the preferred but not exclusive sites of PA biosynthesis. This implies that the site of storage and not the site of synthesis is important.

PAs are regarded as highly efficient defense molecules. Direct evidence favouring this role in plants is weak, but adapted insects which use plant acquired PAs in their own fascinating defense strategies offer strong experimental evidence in support of this hypothesis.

In both PA producing plants and PA acquiring insects the alkaloids are genuinely present and kept as *N*-oxides. The *N*-oxides are not only the more selective molecules which are incapable of passively permeating biomembranes unless a specific carrier system exists, but are also the non-toxic alkaloid forms (Fig. 10). Thus, the PA *N*-oxides may be regarded as "safe molecules" which can easily and without hazard be handled by both plants and adapted insects. On the other hand, the *N*-oxides are easily converted into the potentially toxic form in the gut of a non-adapted herbivore (Fig. 10).

REFERENCES

1. IW Sauthon J Buckingham, Eds., Dictionary of Alkaloids, Chapman and Hall, London (1989).
2. DJ Robins, Prog Chem Org Nat Prod 41, 115 (1982).
3. AR Mattocks, Chemistry and Toxicology of Pyrrolizidine Alkaloids, Academic Press, London 1986.
4. NJ Leonard, in: The Alkaloids, vol 1, pp. 108, RHF Manske and HL Holmes, Eds., Academic Press, New York (1950).
5. FL Warren, Fortschr Chem Org Naturstoffe 24, 129 (1966).
6. LB Bull, CCJ Culvenor, AT Dick, The Pyrrolizidine Alkaloids, North-Holland Publ, Amsterdam (1968).
7. LW Smith and CCJ Culvenor, J Nat Prod 44, 129-152 (1981).
8. AFM Rizk, Ed., Naturally Occurring Pyrrolizidine Alkaloids, CRC Press, Boca Raton 1990.
9. DJ Robins, Chem Soc Rev 18, 375 (1989).
10. DJ Robins, Experientia 47, 1118 (1991).
11. T Hartmann, in: Cell Cultures and Somatic Cell Genetics of Plants, Vol. 5, pp. 277, F Constabel and IK Vasil, Ed., Academic Press, New York, 1988.
12. T Hartmann, in: Biotechnology in Agriculture and Forestry, Vol. 26, Medicinal and Aromatic Plants VI, YPS Bajaj, Ed., Springer, Berlin 1994.
13. T Hartmann, in: Frontiers and New Horizons in Amino Acid Research, pp. 65, K Takai, Ed., Elsevier, Amsterdam 1992.
14. RJ Huxtable, Perspect Biol Med 24, 1 (1980).
15. JN Roitman, in: Xenobiotics in Foods and Feeds, ASC Symp Ser 234, 345 (1983).
16. R Schoental, Tox Let 10, 323 (1982).
17. PR Cheeke, in: Toxicants of Plant Origin, vol 1, Alkaloids pp. 1, PR Cheeke, Ed., CRC Press, Boca Raton 1989.
18. CK Winter and HJ Segall, in: Toxicants of Plant Origin, vol 1, Alkaloids pp. 43, PR Cheeke, Ed., CRC Press, Boca Raton 1989.
19. M Boppré, Naturwissenschaften 73, 17 (1986).
20. D Schneider, in: Perspectives in Chemoreception and Behavior, pp. 123, Eds., RF Chapman, EA Bernays, JG Stoffolano, Springer, Heidelberg 1987.
21. M Boppré, J Chem. Ecol 16, 165 (1990).
22. T Hartmann, in: Herbivores: their Interactions with Secondary Plant Metabolites, 2 ed, Vol. I pp. 79, GA Rosenthal and MR Berenbaum Ed., Academic Press, San Diego 1991.
23. DJ Robins, Nat Prod Rep 10, 487 (1993).
24. JB Harborne, Nat Prod Rep 10, 327 (1993).
25. CCJ Culvenor, Bot Notiser 131, 473 (1978).
26. F Abe and T Yamauchi, Chem Pharm Bul 35, 4661 (1987).
27. F Abe, T Nagao, H Okabe, T Yamauchi, Chem Pharm Bul 38, 2127 (1990).
28. F Abe, T Nagao, H Okabe, T Yamauchi, Phytochemistry 30, 1737 (1991).
29. F Abe, T Yamauchi, S Yaga, K Minato, Chem Pharm Bul 39, 1576 (1991).
30. P Schmid, J Lüthy, U Zweifel, A Bettschart, C Schlatter, Mitt Gebiete Lebensm Hyg 78, 208 (1987).
31. M Rowell-Rahier, L Witte, A Ehmke, T Hartmann, JM Pasteels, Chemoecology 2, 41 (1991).
32. E Röder and C Plassmeier, Sci Pharm 59, 301 (1991).
33. H Wiedenfeld and E Röder, Planta Med 57, 578 (1991).

34. CM Passreiter, G Willuhn, E Röder, Planta Med 57, A101 (1991).
35. RKS Negi, TM Fakhir, TR Rajagopalan, Indian J Chem 28B, 524 (1989).
36. Y Asada, M Shiraishi, T Takeuchi, Y Osawa, T Furuya, Planta Med 51, 539 (1985).
37. A Biller, M Boppré, L Witte, T Hartmann, Phytochemistry 35, 615 (1994).
38. TR Rajagopalan and RKS Negi, Indian J Chem 24B, 882 (1985).
39. RH Barbour and DJ Robins, Phytochemistry 26, 2430 (1987).
40. D Cheng and E Röder, Planta Med 52, 484 (1986).
41. H Hendriks, W Balraadjsing, HJ Huizing, AP Bruins, Planta Med 53, 456 (1987).
42. K Liu, E Röder, HL Chen, XJ Xiu, Phytochemistry 31, 2573 (1992).
43. H Hendriks, HJ Huizing, AP Bruins, J Chromatogr 428, 352 (1988).
44. H Niwa, H Ishiwata, K Yamada, J Nat Prod 48, 1003 (1985).
45. JR Matheson and DJ Robins, Fitoterapia LXIII, 557 (1992).
46. E Röder and C Plassmeier, Planta Med 59, 91 (1993).
47. EW Mead, M Looker, DR Gardner, FR Stermitz, Phytochemistry 31, 3255 (1992).
48. Y Asada, T Furuya, N Murakami, Planta Med 42, 202 (1981).
49. E Röder and A Abdel Ghani, Sci Pharm 58, 403 (1990).
50. E Röder, H Wiedenfeld, P Knoezinger-Fischer, Planta Med 50, 203 (1984).
51. J Gonzales Urones, P Basabe Barcala, I Sanchez Marcos, R Fernandesz Moro, M Lopez Esteban, A Fernandez Rodiguez, Phytochemistry 27, 1507 (1988).
52. L Witte, L Ernst, V Wray, T Hartmann, Phytochemistry 31, 1027 (1992).
53. LH Zalkow, CF Asibal, JA Glinski, SJ Bonetti, LT Gelbaum, D van Derveer, G Powis, J Nat Prod 51, 690 (1988).
54. E Röder, H Wiedenfeld, R Kersten, Sci Pharm 58, 1 (1990).
55. MJ Pestchanker and OS Giordano, J Nat Prod 49, 722 (1986).
56. K Liu and E Röder, Phytochemistry 30, 1303 (1991).
57. GS Hirschmann, EA Ferro, L Franco, L Recalde, C Theoduloz, J Nat Prod 50, 770 (1987).
58. F Bohlmann, C Zdero, J Jakupovic, M Grenz, V Castro, RM King, H Robinson, LPD Vincent, Phytochemistry 25, 1151 (1986).
59. N Guner, Acta Pharm Turc 30, 79 (1988).
60. C Bicchi, P Rubiolo, C Frattini, J Nat Prod 54, 941 (1991).
61. GS Hirschmann and J Jacupovic, Planta Med 54, 360 (1988).
62. ME Stelljes, RB Kelley, RJ Molyneux, JN Seiber, J Nat Prod 54, 759 (1991).
63. E Röder, H Wiedenfeld, A Pfitzer, Phytochemistry 27, 4000 (1988).
64. L Witte, L Ernst, H Adam, T Hartmann, Phytochemistry 31, 559 (1992).
65. LH Villarroel, R Torres, V Fajardo, Bol Soc Chil Quim 32, 5 (1987).
66. H Wiedenfeld, H Hendriks, AP Bruins, E Röder, Sci Pharm 57, 97 (1989).
67. FH Guidugli, MJ Pestchanker, MSA DeSalmeron, OS Giordano, Phytochemistry 25, 1923 (1986).
68. AC Ray, HJ Williams, JC Reagor, Phytochemistry 26, 2431 (1987).
69. GJ Kapadia, A Ramdass, F Bada, Int J Crude Drug Res 28, 67 (1990).
70. MJ Pestchanker, MS Ascheri, OS Giordano, Planta Med 51, 165 (1985).
71. KS Brown Jr, Ann Missouri Bot Gard 74, 359 (1987).
72. KS Brown Jr, Nature 309, 707 (1984).
73. GS Hirschmann, G Bannerjee, J Jakupovic, Rev Latinoam Quim 16, 109 (1985).
74. GS Hirschmann, BEA Ferro, C Cespedes, L Recalde, C Theoduloz, Fitoterapia 58, 263 (1987).

75. O Were, M Benn, RM Munavu, J Nat Prod 54 491 (1991).
76. O Were, M Benn, RM Munavu, Phytochemistry 32, 1602 (1993).
77. AG Gonzalez, G De la Fuente, M Reina, LA Loyola, Planta Med 52, 160 (1986).
78. C Bicchi, A D'Amato, E Cappelletti, J Chomatog 349, 23 (1985).
79. C Bicchi, R Caniato, R Tabacchi, G Tsoupras, J Nat Prod 52, 32 (1989),
80. B Sener, F Ergun, S Kusmenoglu, AE Karakaya, Gazi Univ Eczacilik Fak Derg 5, 157 (1988).
81. E Röder and K Liu, Phytochemistry 30, 1734 (1991).
82. DL Cheng, JK Niu, E Röder, Phytochemistry 31, 3671 (1992).
83. MW Bredenkamp, A Wiechers, PH van Rooyen, Tetrahedron Lett 26, 5721 (1985).
84. MW Bredenkamp, A Wiechers, PH van Rooyen, Tetrahedron Lett 26, 929 (1985).
85. C Zdero, F Bohlmann, RM King, L Haegi, Phytochemistry 29, 509 (1990).
86. LH Villarroel, P Varea, R Caceres, Bol Soc Chil Quim 33, 107 (1988).
87. C Zdero, F Bohlmann, JR Liddell, Phytochemistry 28, 3532 (1989).
88. LH Villaroel, R Torres, J Navarro, V Fajardo, Fitoterapia LVI, 250 (1985).
89. E Röder, T Bourauel, R Kersten, Phytochemistry 32, 1053 (1993).
90. W Ahmed, AQ Khan, A Malik, J Nat Prod 55, 1764 (1992),
91. W Ahmed, AQ Khan, A Malik, F Ergun, B Sener, Phytochemistry 32, 224 (1993).
92. W Ahmed, AQ Khan, A Malik, F Ergun, B Sener, Fitoterapia LXIV, 361 (1993).
93. M Benn and O Were, Phytochemistry 31 3295 (1992).
94. SP Betancourt, A Proliac, J Raynaud, Pharm Acta Helv 65, 175 (1990).
95. G Toppel, L Witte, B Riebesehl, K v Borstel, T Hartmann, Plant Cell Rep 6, 466 (1987).
96. E Röder, T Hille, H Wiedenfeld, Sci Pharm 54, 347 (1986).
97. MJ Pestchanker, MS Ascheri, OS Giordano, Phytochemistry 24, 1622 (1985).
98. K von Borstel, L Witte, T Hartmann, Phytochemistry 28, 1635 (1989).
99. E Jares and AB Pomilio, J Nat Prod 50, 514 (1987).
100. LA Pieters and AJ Vlietinck, Planta Med 54, 178 (1988).
101. O Lock de Ugaz, J Franco, G Seminario, F Delle Monache, B Millan, RP Ubillas Sanches, EO Schlemper, MS Tempesta, Phytochemistry 29, 2373 (1990).
102. E Röder, H Wiedenfeld, R Schraut, Phytochemistry 23, 2125 (1984).
103. E Röder, T Sarg, S El-Dahmy, A Abdel Ghani, Fitoterapia LXIII, 405 (1992).
104. RB Kelley and JN Seiber, Phytochemistry 31, 2369 (1992).
105. RB Kelley and JN Seiber, Phytochemistry 31, 2513 (1992).
106. E Röder and B Rengel-Mayer, Planta Med 59, 192 (1993).
107. G Wassel, B El-Menshawi, A Saeed, G Mahran, Acta Pharm Suec 24, 199 (1987).
108. KM Larson, MR Roby, FR Stermitz, J Nat Prod 47, 747 (1984).
109. CD Dodson and FR Stermiz, J Nat Prod 49, 727 (1986).
110. E Röder, H Wiedenfeld, KJ Kabus, Sci Pharm 58, 9 (1990).
111. G Wassel, B El-Menshawi, A Saeed, G Mahran, J Reisch, Sci Pharm 55, 163 (1987).
112. FR Stermitz, MA Pass, RB Kelley, JR Liddell, Phytochemistry 33, 383 (1993).
113. AR Mattocks and CD Pigott, Phytochemistry 29, 2871 (1990).
114. HA Kelly and DJ Robins, Fitoterapia LXIII, 91 (1992).
115. N Guner, Marmara Univ Eczacilik Derg 3, 57 (1987).
116. T Sarg, S El-Dahmy, E Abdel Aziz, A Abdel Ghani, E Röder, Fitoterapia LXIII, 466 (1992).
117. E Röder, K Liu, T Bourauel, Phytochemistry 30, 3107 (1991).

118. KM L'Empereur, Y Li, FR Stermitz, J Nat Prod 52, 360 (1989).
119. JN Roitman, Aust J Chem 41, 1827 (1988).
120. AM Rizk, FM Hammouda, E Röder, H Wiedenfeld, SI Ismail, NM Hassan, HA Hosseiny, Sci Pharm 56, 105 (1988).
121. N Guner, Acta Pharm Turc 30, 53 (1988).
122. S Mohanraj, PS Subramanian, W Herz, Phytochemistry 21, 1775 (1982).
123. JG Davicino, MJ Pestchanker, OS Giordano, Phytochemistry 27, 960 (1988).
124. FM Hammouda, AM Rizk, SI Ismail, SZ Atteya, HA Ghaleb, KM Madkour, AE Pohland, G Wood, Pharmazie 39, 703 (1984).
125. N Guner, Plant Med Phytother 20, 287 (1986).
126. T Constantinidis, C Harvala, AL Skaltsounis, Phytochemistry 32, 1335 (1993).
127. S Ravi, AJ Lakshmanan, W Herz, Phytochemistry 29, 361 (1990).
128. SC Jain and M Purohit, Chem Pharm Bull 34, 5154 (1986).
129. CF Asibal, LT Gelbaum, LH Zalkow, J Nat Prod 52, 726 (1989).
130. E Röder, E Breitmaier, H Birecka, MW Frohlich, A Badzies-Crombach, Phytochemistry 30, 1703 (1991).
131. A Malik and K Rahman, Heterocycles 27, 707 (1988).
132. T Winkler, R Heckendorn, Heterocycles 27, 2331 (1988).
133. FR Stermitz and KM L'Empereur, Tetrahedron Lett. 29, 4943 (1988).
134. JR Trigo, L Witte, KS Brown Jr, T Hartmann, LES Barata, J Chem Ecol 19, 669 (1993).
135. HA Kelly and DJ Robins, Fitoterapia 61, 89 (1990).
136. E Röder and B Rengel, Phytochemistry 29, 690 (1990).
137. Y Li and FR Stermitz, J Nat Prod 51 1289 (1988).
138. AM Rizk, FM Hammouda, SI Ismail, NM Hassan, HA Hosseiny, E Röder, H Wiedenfeld, HA Ghaleb, KM Madkour, Int J Crude Drug Res 26, 112 (1988).
139. E Röder and T Bourauel, Phytochemistry 31 3613 (1992).
140. AS Mellidis and VP Papageorgiou, Chem Chron 17, 67 (1988).
141. HJ Huizing, Pharm Weekblad Sci Ed 9, 185 (1987).
142. TA Jaarsma, E Lohmanns, TWJ Gadella, TM Malingre, Plant Syst Evol 167, 113 (1989).
143. TA Jaarsma, E Lohmanns, H Hendriks, TWJ Gadella, TM Malingre, Plant Syst Evol 169, 31 (1990).
144. JJ Vollmer, NC Steiner, GY Larsen, KM Muirhead, RJ Molyneux, J Chem Educ 64, 1027 (1987).
145. P Stengl, H Wiedenfeld, E Röder, Dtsch Apoth Ztg 122, 851 (1982).
146. K Jenett-Siems, M Kaloga, E Eich, Phytochemistry 34, 437 (1993).
147. BE van Wyk and GH Verdoorn, Biochem Syst Ecol 17, 385 (1989).
148. MC Williams and RJ Molyneux, Weed Sci 35, 476 (1987).
149. R Rastogi and TR Rajagopalan, Pharmazie 37, 75 (1982).
150. D Cheng, Y Liu, TT Chu, Y Cui, J Cheng, E Röder, J Nat Prod 52, 1153 (1989).
151. OP Suri, RS Jamwal, RK Khajuria, CK Atal, CN Haksar, Planta Med 44, 181 (1982).
152. CN Haksar, OP Suri, RS Jamwal, CK Atal, Indian J Chem 21B, 492 (1982).
153. DL Cheng, SB Tu. AA Enti, E Röder, Sci Pharm 54, 351 (1986).
154. AR Mattocks and N Nwude, Phytochemistry 27, 3289 (1988).
155. B Abegaz, G Atnafu, H Duddeck, G Snatzke, Tetrahedron 43, 3263 (1987).
156. H Wiedenfeld, E Röder, E Anders, Phytochemistry 24, 376 (1985).
157. G Toppel, L Witte, T Hartmann, Phytochemistry 27, 3757 (1988).

158. E Röder, XT Liang, KJ Kabus, Planta Med 58, 283 (1992).
159. CCJ Culvenor, GM O'Donovan, LW Smith, Aust J Chem 20, 757 (1967).
160. GH Verdoorn and BE van Wyk, Phytochemistry 31, 1029 (1992).
161. K Mothes, HR Schütte, M Luckner Eds, Biochemistry of Alkaloids, VCH Verlagsgesellschaft, Weinheim 1985.
162. E Nowacki and RU Byerrum, Life Sci 5:157 (1962).
163. CA Hughes, R Letcher, FL Warren, J Chem Soc 4974 (1964).
164. W Bottomley and TA Geissman, Phytochemistry 3:357 (1964).
165. NM Bale and DHG Crout, Phytochemistry 14:2617 (1975).
166. DJ Robins and JR Sweeney, J Chem Soc, Chem Commun 120 (1979).
167. DJ Robins and JR Sweeney, J Chem Soc, Perkin Trans I 3083 (1981).
168. DJ Robins and JR Sweeney, Phytochemistry 22:457 (1983).
169. HA Khan and DJ Robins, J Chem Soc, Chem Commun 146 (1981).
170. HA Khan and DJ Robins, J Chem Soc, Perkin Trans I 101 (1985).
171. G Grue-Sorensen and ID Spenser, J Am Chem Soc 103:3208 (1981).
172. HA Khan and DJ Robins, J Chem Soc, Chem Commun 554 (1981)
173. G Grue-Sorensen and ID Spenser, Can J Chem 60:643 (1982).
174. ID Spenser, Pure & Appl Chem 57:453 (1985).
175. HA Khan and DJ Robins, J Chem Soc, Perkin Trans I 819 (1985).
176. J Rana and DJ Robins, J Chem Res (S), 146 (1983).
177. DJ Robins, J Chem Soc, Chem Commun 1289 (1982).
178. Kelly HA and Robins DJ, J Chem Soc, Chem Commun 329 (1988).
179. G Grue-Sorensen and ID Spenser, J Am Chem Soc 105:7401 (1983).
180. J Rana and DJ Robins, J Chem Soc, Perkin Trans I 983 (1986).
181. AR Battersby, J Staunton, and MC Summers, J Chem Soc, Perkin Trans I 1052 (1976).
182. RC Richards and ID Spenser, Tetrahedron 39:3549 (1983).
183. J Rana and E Leete, J Chem Soc, Chem Commun, 1742 (1985).
184. EK Kunec and DJ Robins, J Chem Soc, Chem Commun 250 (1986).
185. EK Kunec and DJ Robins, J Chem Soc, Perkin Trans I, 1437 (1989).
186. HA Kelly and DJ Robins, J Chem Soc, Perkin I 2195 (1987).
187. HA Kelly and DJ Robins, J Chem Soc, Perkin Trans I 177 (1987).
188. DB Hagan and DJ Robins, J Chem Res (S) 292 (1990).
189. AA Denholm, HA Kelly, DJ Robins, J Chem Soc, Perkin Trans I 2003 (1991).
190. H Birecka and JL Catalfamo, Phytochemistry 21, 2645 (1982).
191. P Hülsmeyer, L Witte, T Hartmann, Planta Med 58, A597 (1992).
192. EK Kunec and DJ Robins, J Chem Soc, Chem Commun 1450 (1985).
193. EK Kunec and DJ Robins, J Chem Soc, Perkin Trans I 1089 (1987).
194. HA Kelly, EK Kunec, M Rodgers, DJ Robins, J Chem Res (S) 358 (1989).
195. T Hartmann and G Toppel, Phytochemistry 26, 1639 (1987).
196. T Hartmann, H Sander RD Adolph, G Toppel, Planta 175, 82 (1988).
197. H Birecka, M Birecki, MW Frohlich, Plant Physiol 84, 42 (1987).
198. H Sander and T Hartmann, Planta (in press)
199. F Böttcher, RD Adolph, T Hartmann, Phytochemistry 32, 679 (1993).
200. GH Tait, Biochem. Soc Trans 7, 199 (1979).

201. S Yamamoto, S Nagata, K Kusaba, J Biochem 114, 45 (1993).
202. KS Srivenugopal and PR Adiga, Biochem J 189, 553 (1980).
203. F Böttcher, D Ober, T Hartmann, Canad J Chem (in press).
204. DHG Crout, J Chem Soc (C) 1968 (1966).
205. DHG Crout, J Chem Soc (C) 1233 (1967).
206. DHG Crout, MH Benn, H Imaseki, TA Geismann, Phytochemistry 5, 1 (1966).
207. DHG Crout, NM Davies, EH Smith, D Whitehouse, J Chem Soc, Perkin Trans I 671 (1972).
208. NM Davies and DHG Crout, J Chem Soc, Perkin Trans I 2079 (1974).
209. NM Bale, RH Cahill, NM Davies, MB Mitchell, DHG Crout, J.Chem Soc, Perkin Trans I 101 (1978).
210. RH Cahill, DHG Crout, MB Mitchell, US Muller, J Chem Soc, Perkin Trans I 173 (1983).
211. DJ Robins, NM Bale, DHG Crout, J Chem Soc, Perkin Trans I 2082 (1974).
212. JA Devlin and DJ Robins, J Chem Soc, Perkin Trans I 1329 (1984).
213. T Hartmann, A Ehmke, U Eilert, K vBorstel, C Theuring, Planta 177, 98 (1989).
214. H Birecka, M Birecki, EJ Cohen, AJ Bitonti, PP McCann, Plant Physiol 86, 224 (1988).
215. H Sander and T Hartmann, Plant Cell, Tissue Organ Cult 18, 19 (1989).
216. H Birecka, T DiNolfo, WB Martin, MW Frolich, Phytochemistry 23, 991 (1984).
217. C Wirz, L Witte, T Hartmann, Planta Med 59, A646 (1993).
218. NM van Dam and T Hartmann, Planta Med 59, A646 (1993).
219. HJ Huizing, EC Pfauth, TM Malingre, JH Sietsma, Plant Cell, Organ Cult 2, 227 (1983.
220. L Heide, B Rengel, E Röder, M Tabata, Planta Med 55, 684 (1989).
221. JL Catalfamo, WB Martin, H. Birecka, Phytochemistry 21, 2669 (1982).
222. T Hartmann and M Zimmer, J Plant Physiol 122, 67 (1986).
223. RJ Molyneux and AE Johnson, J Nat Prod 47, 1030 (1984)
224. AE Johnson, RJ Molyneux, GB Merill, J Agric Food Chem 33, 50 (1985)
225. K Vrieling, H de Vos, CAM van Wijk, Phytochemistry 32, 1141 (1993).
226. RJ Molyneux, BC Campbell, DL Dreyer, J Chem Ecol 16, 1899 (1990).
227. L Witte, A Ehmke, T Hartmann, Naturwissenschaften 77, 342 (1990).
228. K vBorstel and T Hartmann, Plant Cell Rep 5, 39 (1986).
229. A Ehmke, K v Borstel, T Hartmann, in: Plant Vacuoles, their Importance in Solute Compartmentation in Cells and their Application in Plant Biotechnology, pp. 301, B Martin Ed., Plenum Press, New York 1987.
230. A Ehmke, K vBorstel, T Hartmann, Planta 176, 83 (1988).
231. G Toppel, L Witte, T Hartmann, Phytochemistry 27, 3757 (1988).
232. GA Rosenthal, in: Herbivores: their Interactions with Secondary Plant Metabolites, 2 ed, Vol. I p. 1, GA Rosenthal and MR Berenbaum Ed., Academic Press, San Diego 1991.
233. FR Stermitz and TR Suess, Phytochemistry 17, 2142 (1978).
234. MR Roby and FR Stermitz, J Nat Prod 47, 846 (1984).
235. FR Stermitz, GH Harris and W Jing, Biochem Syst Ecol 14, 499 (1986).
236. FR Stermitz and GH Harris, J Chem Ecol 13, 1917 (1987).
237. MJ Schneider and FR Stermitz, Phytochemistry 29, 1811 (1990).
238. JD Phillipson, Xenobiotica 1, 419 (1971)
239. JD Phillipson and SS Handa, J Nat Prod 41, 385 (1978).
240. LJ Areshkina, Biochemistry (USSR) 22, 487 (1957).
241. CCJ Culvenor and LW Smith, Austr J Chem 10, 464 (1957).

242. AR Mattocks, Xenobiotica 1, 451 (1971).
243. AR Mattocks, in: Pharmacology and future of man, vol. 2, pp. 114, GE Acheson, Ed., S Karger, New York 1973.
244. DR Buhler and B Kedzierski, in: Biological reactive intermediates, pp. 611, Eds., JJ Kocsis, DJ Jollow, CM Witmer, JO Nelson, R Snyder,Plenum Press, New York 1986.
245. H Mori, S Sugie, N Yoshimi, Y Asada, T Furuya, GM Williams, Cancer Res 45, 3125 (1985).
246. H Frei, J Lüthy, J Brauchli, U Zweifel, FE Würgler, C Schlatter, Chem Biol Interact 83, 1 (1992).
247. JA Zijlstra and EW Vogel, Mutation Res 201, 27 (1988).
248. JA Zijlstra and EW Vogel, Mutation Res 202, 251 (1988).
249. EA Bernays, in: Nitrogen as an Ecological Factor, pp. 321, Eds. JA Lee, S McNeill, JH Rorison, Beackwell, Oxford 1983.
250. RJ Molyneux, AE Johnson, LD Stuart, Human Toxicol 30, 201 (1988).
251. RJ Molyneux and MH Ralphs, J Range Manage 45, 13 (1992).
252. EA Bernays and RF Chapman, Ecol Entomol 2, 1 (1977).
253. MD Bentley, DE Leonard, WF Stoddard, LH Zalkow, Ann Entomol Soc Am, 77, 393 (1984).
254. G Marquina, A Laguna, P Franco, L Fernandez, R Perez, O Valiente, Pharmazie 44, 870 (1989).
255. K Fujida, Y Shoyama, H Matsunaka, I Nishioka, Phytochemistry 27, 1564 (1988).
256. GJ Goss, Environ Entomol 8, 487 (1979).
257. TE Pliske, Environm Entomol 4, 453 (1975).
258. M Boppré, Ecol Entomol 6, 445 (1981).
259. M. Boppré, J Chem Ecol 10, 1151 (1984).
260. CCJ Culvenor, JA Edgar, LW Smith, J Agric Food Chem 29, 958 (1981).
261. KS Brown, Rev Brasil Bio 44, 435 (1984).
262. RT Aplin and M Rothschild, in: Toxins of Animal and Plant Origin, Vol. 2, pp. 579, A de Vries and K Kochva Ed., Gordon & Breach, London 1972.
263. M Rothschild, RT Aplin, PA Cockrum, JA Edgar, P Fairweather, R Lees, Biol J Linn Soc 12, 305 (1979).
264. A Ehmke, L Witte, A Biller, T Hartmann, Z Naturforsch 45c, 1185 (1990).
265. M Wink and D Schneider, Naturwissenschaften 75, 524 (1988).
266. E von Nickisch-Rosenegk, D Schneider, M Wink, Z. Naturforsch 45c, 881 (1990).
267. SB Krasnoff and DE Dussourd, J Chem Ecol 15, 47 (1989).
268. M Boppré and D Schneider, Zool J Linn Soc 96, 339 (1989).
269. T Hartmann, A Biller, L Witte, L Ernst, M Boppré, Biochem Syst Ecol 18, 549 (1990).
270. SB Krasnoff and WL Roelofs, J Chem Ecol 15, 1077 (1989).
271. SB Krasnoff and WL Roelofs, Zool J Linn Soc 99, 319 (1990).
272. CCJ Culvenor and JA Edgar, Experientia 28, 627 (1972).
273. DE Dussourd, K Ubik, C Harvis, J Resch, J Meinwald, T Eisner, Proc Natl Acad Sci USA 85, 5992 (1988).
274. F Bogner and T Eisner, J Chem Ecol 17, 2063 (1991).
275. JA Edgar, J Zool Lond 196, 385 (1982).
276. RT Aplin, MH Benn, M Rothschild, Nature 219, 747 (1968).
277. E von Nickisch-Rosenegk and M Wink, J Chem Ecol 19, 1889 (1993).
278. JA Edgar, M Boppré, D Schneider, Experientia 35, 1447 (1979).
279. RB Kelley, JN Seiber, AD Jones, HJ Segall, LP Brower, Experientia 43, 943 (1987).

280. ME Stelljes and JN Seiber, J Chem Ecol 16, 1459 (1990).
281. JA Edgar, PA Cockrum, JL Frahn, Experientia 32, 1535 (1976).
282. JA Edgar, in: The Biology of Butterflies, p. 91, RI Vane-Wright and PR Ackery Ed., Academic Press, New York 1984.
283. R Nishida, CS Kim, H Fukami, R Irie, Agric Biol Chem 55, 1787 (1991).
284. JR Trigo and KS Brown, Chemoecology 1, 22 (1990).
285. E Bernays, JA Edgar, M Rothschild, J Zool Lond 182, 85 (1977).
286. M Boppré, U Seibt, W Wickler, Entomol Exp Appl 35, 115 (1984).
287. JM Pasteels, M Rowell-Rahier, T Randoux, JC Braekman, D Daloze, Entomol Exp Appl 49, 55 (1988).
288. A Ehmke, M Rowell-Rahier, JM Pasteels, T Hartmann, J Chem Ecol 17, 2367 (1991).
289. JM Pasteels, F Eggenberger, M Rowell-Rahier, A Ehmke, T Hartmann, Naturwissenschaften 79, 521 (1992).
290. A Biller, R Lindigkeit, M Boppré, T Hartmann, J Chem Ecol (in preparation)
291. R Lindigkeit, A Biller, HM Schiebel, T Hartmann, J Chem Ecol (in preparation).
292. AR Mattocks, Xenobiotica 1, 451 (1971).
293. A Biller and T Hartmann, in: Proc 8th Int Symp Insect-Plant Relationships, p.83, SBJ Menken, JH Visser, P Harrewijn Ed., Acad Publ, Dordrecht 1992.
294. A Egelhaaf, K Cölln, B Schmitz, M Buck, M Wink, D Schneider, Z Naturforsch 45c, 115 (1990).
295. S Schulz, Liebigs Ann Chem 829 (1992).
296. S Schulz, W Francke, M Boppré, T Eisner, J Meinwald, Proc Natl Acad Sci USA 90, 6834 (1993).
297. M Wink, D Schneider, L Witte, Z Naturforsch 43c, 737 (1988).
298. WE Conner, B Boach, E Benedict, J Meinwald, T Eisner, J Chem Ecol 16, 543 (1990).
299. JR Trigo, LES Barata, KS Brown Jr, (in preparation).
300. WE Conner, B Roach, E Benedict, J Meinwald, T Eisner, J Chem Ecol 16, 543 (1990).
301. GK Storey, DJ Aneshansley, T Eisner, J Chem Ecol 17, 687 (1991).
302. DE Dussourd, CA Harvis, J Meinwald, T Eisner, Experientia 45. 896 (1989).
303. T Eisner, BioScience 32, 321 (1982).
304. AR Masters, Biotropica 22, 298 (1990).
305. T Eisner and M Eisner, Psyche 98, 111 (1991).
306. JM Pasteels, M Rowell-Rahier, JC Braekman, D Daloze, in: Novel Aspects of the Biology of Chrysomelidae, p. 289, PH Jolivet, ML Cox, E Petitpierre Ed., Kluwer Academic Publishers (1994).
307. M Rowell-Rahier, JM Pasteels, A Alonso-Mejia, LP Brower, Animal Behavior (in press).
308. SB Malcolm, in: Herbivores: their Interactions with Secondary Plant Metabolites, 2 ed, Vol. I p. 251, GA Rosenthal and RU Berenbaum Ed., Academic Press, San Diego 1991.
309. MR Grue and JR Liddell, Phytochemistry 33, 1517 (1993).
310. JR Liddell and CG Logie, Phytochemistry 34, 1198 (1993).
311. Y Li, Huaxue Xuebao 48, 415 (1990).
312. E Röder, T Sarg, S El-Dahmy, A Abdel Ghani, Phytochemistry 34, 1421 (1993).
313. LP Bush, FF Fannin, MR Siegel, DL Dahlman, HR Burton, Agic Ecosystems Environ 44, 81 (1993).
314. SY Yunosov and ST Akramov, J Gen Chem USSR (Engl Transl) 25, 1765 (1955).
315. ST Akramov and SY Yunusiv, Chem Nat Comp (Engl Transl) 1, 203 (1965).
316. SG Yatesan and HL Tookey, Austr J Chem 18, 53 (1965).
317. RC Buckner, LP Bush, PB Burrus, in: Proc 14th Int Grassland Congress, JA Smith and VW Hays, Eds., pp 157, Westview Press, Boulder, CO 1981.

318. G Morgan-Jones and W Gams, Mycotaxon 15, 311 (1982).
319. LP Bush, PL Cornelius, RC Buckner, SR Varney, RA Chapman, PB Burrus, CW Kennedy, TA Jones, MJ Saunders, Crop Sci 22, 941 (1982).
320. TA Jones, RC Buckner, PB Burrus, LP Bush, Crop Sci 23, 1136 (1983).
321. MR Siegel, GCM Latch, LP Bush, FF Fannin, DD Rowan, BA Tapper, CW Bacon, MC Johnson, J Chem Ecol 16, 3301 (1990).
322. TA Jones, RC Buckner, PB Burrus, Can J Plant Sci, 65, 317 (1985).
323. GB Garner, GE Rottinghaus, CN Cornell, H Testereci, Agric Ecosystems Environ 44, 65 (1993).
324. RW Hemken and LP Bush, in: Toxicants of Plant Origin, vol 1, Alkaloids pp. 281, PR Cheeke, Ed., CRC Press, Boca Raton 1989.
325. FN Thompson and JK Porter, Vet Hum Toxicol 32 (Supplement), 51 (1990).
326. K Clay, in: Microbial Mediation of Plant-Herbivore Interactions, pp. 199, VA Krischik and CG Jones, Ed., John Wiley & Sons, Inc, 1991.
327. CR Funk, PM Halisky, MC Johnson, MR Siegel, AV Stewart, S Ahmad, RH Hurley, IC Harvey, Biotechnology 1, 189 (1983).
328. SG Yates, JC Fenster, RJ Bartelt, J Agric Food Chem 37, 354 (1989).
329. MC Johnson, DL Dahlman, MR Siegel, LP Bush, GCM Latch, DA Potter, DR Varney, Appl Environ Microbiol 49, 568 (1985).
330. MR Siegel, GCM Latch, LP Bush, FF Fannin, DD Rowan, BA Tapper, CW Bacon, MC Johnson, J Chem Ecol 16, 3301 (1990).
331. WE Riedell, RE Kieckhefer, RJ Petroski, RG Powell, J Entomol Sci 26, 122 (1991).
332. TN Hardy, K Clay, AM Hammond, Environ Entomol 15, 1083 (1986).
333. CG Patterson, DA Potter, FF Fannin, Entomol exp appl 61, 285 (1991).
334. JR Liddell, CG Logie, Phytochemistry 34, 1629 (1993).
335. G Veen, R Greinwald, P Canto, L Witte, FC Czygan, Z Naturforsch 47c, 341 (1992).
336. R Greinwald, P Bachmann, L Witte, JR Acebes-Grinoves, FC Czygan, Biochem Syst Ecol 20, 69 (1992).
337. GH Verdoorn and BE van Wyk, Phytochemistry 31, 369 (1992).
338. FR Stermitz, GH Harris, KM Hagglund, LA Wright, Lindleyana 5, 158 (1990).

Appendix I. Alkaloid types and alkaloid composition of plant species which have been investigated

	Type	Alkaloids	References
Apocynaceae			
Alafia multiflora Stapf	E	**311**	2
Anodendron affine Druce	M2	**356, 357**	1, 2
Parsonsia eucalytophylla (F. Muel.)	C1	**187, 215**	2, 3
heterophylla A. Cunn.	C4, M1	**269, 271, 337**	1-3
laevigata Alston	C2, C4	**252, 267-274, 277**	26-29
spiralis Wall.	C4	**269, 271, 272, 275, 276**	1, 2, 3
straminea [(R. Br.) F. Muel.]	C1	**87, 215**	2,3
Urechtites karwinsky Muel	M2	**350**	1, 2
Asteraceae			
Eupatorieae			
Adenostemma involucratum King.&H.Robinson	C1	**200, 215**	71
brasilianum Cass.	C1	**200, 215**	71
Ageratum conyzoides L.	C1	**2, 5, 223**	33
Conoclinium coelestinum (L.) DC	C1	**187, 223**	3
Chromolaena odorata King & Robinson	B1, C1	**106, 109, 181, 187, 200, 203**	37
Eupatorium altissimum L.	B1, C1	**135, 200**	3
betonicaeforme (DC)Baker	C1	**212**	71
cannabinum L.	C1	**174, 175, 181, 183-185, 223, 224, 227, 253, 254**	2, 3, 30, 41
*coelestinum**	C1	**215**	1
compositifolium Walt.	C1	**187, 215**	3
*fortunei**	C1	**181, 200, 204**	42
itatiayense Hieron	C1	**187**	71
laevigatum Lam	C1	**223**	71
macrocephalum Less	C1	**212**	71
macrophyllum L	C1	**200**	71
maculatum L.	C1	**176, 223**	1-3
maximiliani Schrad ex DC	C1	**200**	71, 72
rotundifolium L. var. *ovatum*	C1	**175, 223, 225, 227, 228**	43
serotinum Michx.	C1	**181, 200**	1-3
stoechadosmum Hance	C1	**168, 181**	2, 3
vauthierianum DC	C1	**200**	72
Liatris punctata Hook	A1, C1	**1, 4, 33, 187, 188, 208, 215, 245**	47
Trichogonia gardneri A. Gray	C1	**187, 215**	71, 72
Senecioneae			
Adenostyles alliariae (Gouen) Kerner	A1	**1, 4, 8, 9, 10, 27, 28**	1, 3, 30-32
glabra (Mill.) DC.	A1	**1, 8, 9, 10**	3, 32
leucophylla (Willd.)Rchb.	A1	**1, 8, 10**	32
rhombifolia (Willd.) Pim.	A1, B1	**8, 27, 126**	3
Arnica amplexicaulis Nutt.	M2	**354, 355**	34
chamissonis ssp. *foliosa* (Nutt.) Maguire	M2	**354. 355**	34
montana L.	M2	**354, 355**	34
sachalinensis (Regel.) A. Gray	M2	**354, 355**	34
Brachyglottis repanda Forst.&Forst.	A1	**1, 20**	1, 3
Cacalia floridana	A1	**33, 34, 40, 42**	1, 3
hastata L.	A1	**59**	1
hastata L. ssp. *orientalis* Kitamura	A1	**4**	3
robusta Tolmatch	A1	**59**	1, 2

Appendix I (continued)

	Type	Alkaloids	References
Cacalia yatabei Maxim	A1	**53**	1-3
Cirsium wallichii DC	A1	**36**	35
*Crassocephalum crepidioides**	A1	**31, 35**	36
*Doronicum macrophyllum**	A1	**33, 40, 42**	1-3
pardalianches L.	A1	**33**	38
Emilia flammea Cass.	A1	**33, 51**	1-3, 39
sonchifolia DC.	A1	**1, 20, 42**	1-3, 40
Erechtites heiracifolia (L.) Raf. ex DC.	A1	**1, 8**	1-3
Farfugium japonicum Kit.	A1, B1	**20, 43, 137**	1-3, 44
Gynura scandens O. Hoffm.	A1	**17, 18**	1, 3
segetum (Lour.) Merr.	A1	**1**	3
sarmentosa D.C.	A1	**1, 20, 33**	45
Homogyne alpina (L.) Cass.	A1	**20, 21, 43, 44**	30, 46
discolor (Jacq.) Cass.	A1	**20**	46
sylvestris Cass.	A1	**43, 44**	46
Kleinia kleinioides (Sch. Bip.) M.R.F. Taylor	A1	**78, 79**	1, 2
Ligularia brachyphylla Hand.-Mazz.	A1	**49, 52**	1-3
clivorum Maxim.	A1	**52**	1-3
dentata (A. Gray) Hara	A1	**45-48, 52**	1-3
elegans Cass. (syn. *L. macrophylla* (Ledeb)DC)	A1	**49, 52**	1-3
*japonica**	A1	**1, 27, 44**	48
Petasites albus L.	A1	**20**	3
hybridus (L.) P. Gaertn. et al.	A1	**1, 4, 20**	1-3
japonicus Maxim.	A1	**20, 43, 44**	1-3
laevigatus (Willd.) Reichenb.	A1	**1, 20, 27**	1-3
paradoxus Retz.	A1	**8, 20, 22, 43, 44**	49
Senecio abrotanifolius ssp. *abrotanifolius* L.	A1, A3	**42, 94, 95**	1, 50
abrotanif. ssp. *abrotanif.* var. *tiroliensis* Gams	A1, A3	**42, 94, 95**	50
adnatus DC.	A1	**27**	1, 2
adonidifolius Loisel.	A1	**1, 34, 50**	3, 51, 52
aegypticus L.	A1	**8, 16, 33**	1-3
aetnensis Jan.	A1	**81**	2
alpinus (L.) Scop.(syn. *S. cordatus* Koch).	A1	**1, 4, 8, 10, 30, 31, 37, 84**	1-3, 60
ambrosioides (syn. *S. brasiliensis* Less.)	A1	**1, 8, 14, 16**	3
amphilobus C. Koch	B2	**146**	2
ampullaceus Hook	A1	**1, 8, 14**	2, 3
angulatus L.	A4	**96, 99**	1, 2
anonymus Wood.	A1	**1, 8, 14, 15, 20, 22-25, 33**	53
antieuphorbium (L.) Sch. Bip.	A1	**4, 20, 78, 81**	2, 3
aquaticus Hill.	A1	**1, 8, 33**	1-3, 54
argentinus Baker	A1	**1, 72**	55
argunensis Debeaux	A1	**1, 4, 6, 8, 12, 33**	56
*asperum**	B1	**136**	1
aucheri DC	A1, A2	**81, 89**	1, 2
aureus L.	A1	**1, 33, 34, 40**	1-3
auricola Bourg.	A1	**22**	1-3
barbertonicus Klatt.	A1	**83**	1, 2
bipinnatisectus Belcher (syn. *Erechtites atkinsoniae*)	A1	**14**	2, 3
brachypodus DC.	A4	**96**	1, 2
brasiliensis DC	A1	**1, 4, 8, 14, 15, 31**	2, 3, 57

Appendix I (continued)

	Type	Alkaloids	References
Senecio bupleuroides DC.	A1	**14**	2, 3
cacaliaster Lam.	A3, B1	**95, 107, 114, 117**	1, 3
campestris (Retz.) DC.var. *maritimus*	A1	**1**	1, 2
cannabifolius Less.	A1	**8, 30, 32**	3
carthamoides Greene	A1	**1, 8**	2, 3
caudatus DC.	B1, B2	**105, 107, 110, 114, 130, 131, 133, 149-155**	1, 58
chrysanthemoides Schrank & Baier	A1	**8**	2, 3
chrysocoma Meerb.	B1, M1	**120, 123, 126, 127, 339a, 339b**	309, 310
cilicius Boiss.	A1	**1, 4, 27**	59
cineraria DC.	A1	**1, 8, 14, 31, 33**	1-3
cissampelinus DC	B3	**157-160**	1, 2
clevelandi Greene	A1	**42**	1
colaminus Cuatr.	A2	**89**	2
congestus (R.Br.) DC.	A1	**1, 27, 28**	3
cruentus DC.	A1	**1, 8, 14, 16**	1-3
cymbalarioides	A1	**1, 8, 14, 16**	3
deferens Griseb.	A1, B2	**1, 14, 15, 20, 21, 34 or 44, 148**	61
desfontainei Druce (S. *coronopifolius*)	A1	**1, 8, 16, 33**	2, 3
dimorphophyllus Greene	A1	**1, 4**	62
discolor DC.	A1	**1, 14**	2, 3
dolichodoryius Cuatr.	A1	**50**	1, 58
doria L. (S. *paucifolius*)	B1	**118**	1-3, 63
doronicum L.	A3, B2	**94, 95, 146**	1-3, 30
douglasii DC. (syn. *S. longilobus* Benth.)	A1	**1, 8, 14, 16**	3
durieui Gay	A1	**4**	3
eremophilus Richards	A1	**1, 8, 14, 16**	1-3
erraticus Bertol.	A1	**1, 4, 8, 12, 33**	1-3
erucifolius L.	A1, A2	**1, 8, 12, 14**	1-3
erucifoline chemotype	A1	**4, 6, 8, 10, 12**	64
eruciflorine chemotype	A1,A2	**1, 4, 6, 8, 10, 84**	64
faberi Hemsl.	A1	**4**	3
filaginoides (H.et A.)DC	A1	**1, 14**	55
fistulosus Poepp. ex DC	A1	**1, 8, 72**	65
floridanus Sch. Bip. (syn. *Brachyglottis floridiana*)	A1	**33, 34, 39, 40**	1-3
fluviatilis Wallr.	A1	**8, 33, 34**	1-3
formosus H.B.&K.	A1	**4, 14**	3
francheti C. Winkl	B1	**126 etc.**	1, 2
fremonti Torr. & A. Gray	A1	**1, 8**	3
fuchsii Gmel.	B1	**106, 109, 112, 120, 121, 123, 124, 126**	1-3, 30, 31, 66
gallicus Chaix.	A1	**1, 20, 45**	51
gillesianus Hieron.	A1	**1, 14**	55, 67
glabellus (Turcz.) DC.	A1	**1, 4, 20, 33, 34**	2, 3, 68, 69
glaberrimus DC.	A1	**14**	2, 3
glandulosus Don ex Hook et Arn	A1	**4, 14, 15**	55, 70
graminifolius Jacq.	A1	**14**	1-3
grandifolius Less.	A1	**8, 20, 22, 27**	3, 58
griesbachii Baker	A1	**1, 14, 77**	1-3, 73, 74

Appendix I (continued)

	Type	Alkaloids	References
Senecio hadiensis Forsk (syn. *S.petianus*)	A4	**96-98, 101-104**	75, 76
halimifolius L.	A4	**96**	1, 2
hydrophyllus Nutt.	A1, B1	**1, 4, 27, 28, 106-109, 111, 115,**	
		116, 120, 122, 123, 125-129	62
hygrophyllus R.A. Dyer et C.A. Smith	A1, A4	**27, 58, 96**	1, 2
ilicifolius Thunb.	A1	**1, 4, 14**	2, 3
illinitus Phill.	A1	**1, 20, 21**	77
inaequidens DC	A1, A2	**1, 4, 8, 14, 15, 20-22, 33, 34,**	
		39-42, 84	1, 2, 60, 78, 79
incanus L. ssp. *carniolicus* (Willd.) Br.	A1	**4, 8, 30**	1-3
integerrimus Nutt.	A1	**1, 4, 27**	2, 3
integrifolius (L.) Clairv.	A1, B1, D	**1, 4, 14, 20, 105, 297**	80
integrifolius L. var. *fauriri* Levl. et Vant.	A1, B1	**61, 135, 138, 142**	81
isatideus DC.	A1	**14**	2, 3
jacobaea L.	A1,	**1, 4, 8, 30, 31, 35, 37**	1-3
jacobine chemotype	A1, A2	**1, 4, 6, 8, 10, 16, 30, 31, 33, 35,**	
		37, 84	64
erucifoline chemotype	A1, A2	**1, 4, 6, 8, 10, 12-14, 16, (30),**	
		(31), 84	64
kaschkarovii C. Winkl.	B1	**112-114, 127**	82
kaempferi DC.	B1	**126**	2
kirkii Hoo f. ex Kirk	A1	**20, 21**	1-3
kleinia Sch. Bip.	A1	**4, 20**	2, 3
krylovii Schischk.	A1	**8**	3
kubensis Grossh.	A1	**8**	2, 3
laricifolius H.B.K.	A1	**1, 8, 20, 26**	1
latifolius DC.	A1, A2	**14, 87, 88**	1, 3, 83, 84
lautus Forst. ex Willd.	A1	**1**	2, 3
leucostachys Baker	A1	**1, 4, 72**	55
longiflorus Sch. Bip.	A1	**1, 8, 81**	1-3
longilobus Benth..	A1	**1, 4, 8, 14**	2, 68
macrophyllus Bieb.	B2	**146**	1, 2
mandralicae Jacobs.	A1, A2	**81, 82, 90**	58
magnificus F. Muell.	A1, A2	**1, 2, 4, 78, 80-82, 90**	2, 3, 85
megaphyllus Green.	A1	**3, 7**	1, 58
minimus Poir.	A1	**8**	3
mikanioides Otto	B1, B3	**111, 115, 120, 122, 123, 125-129,**	
		159-163	1, 2, 62
morrisonensis Hayata	A1	**4**	2, 3
multivenius Benth. in Oerst.	A1	**1, 8**	58
murorum Remy	A1	**4, 15**	86
nebrodensis L. var. *sicula*	A1	**1, 4**	2, 3
nemorensis L.	A3	**92, 93, 95**	1, 2
othonnae Bieb.	A1	**8, 33, 38, 39, 42**	1-3
othonnaeflorus DC.	A1	**14**	87
ovirensis DC	B1	**135**	3
paludosus L.	A1	**8, 31**	2, 3
pampaenus Cabrera	A1	**1**	2, 3
pancicii Degen.	A1	**1, 8**	3
paniculatus Berg. (syn. *S. grandiflorus*)	A1	**1, 27**	2, 3
patagonicus Hook and Arn.	A1	**1, 8, 72**	65, 88
paucicalyculatus Klatt	A1	**14**	2, 3

Appendix I (continued)

	Type	Alkaloids	References
Senecio paucifolius S. G. Gmel.	A1	**8**	3
pauciligulatus R.A. Dyer et C.A. Smith	A4	**96**	1, 2
pellucidus (syn. *S. ruderalis*)	A1	**14**	2, 3
persoonii De Not.	A1	**8, 12**	89
*petasis**	A1	**54**	1
petasitis DC.	A1	**1**	2, 3
phillipicus Rogel et Köern	A1	**8, 14**	77
pierotii Miq.	A1	**20, 22**	3
pimpinellifolius H.B.K.	A1	**1**	58
platyphylloides Somm. & Lev.	A1	**8, 27**	2, 3
pojarkovae Schischk.	A1, A2	**8, 88**	3
praealtus Bertol. (syn. *S. borysthenicus*)	A1	**8**	1-3
propinquus Schischk.	A1	**8**	2, 3
procerus L.	A1	**20**	1-3
pseudo-arnica Less.	A1	**1**	2, 3
pterophorus DC.	A1, A2, A4	**1, 8, 9, 10, 14, 89, 96**	1-3, 334
pubigerus L.	A2	**89**	2
pulviniformis Hieron.	A2	**89, 90**	1, 2
quebradensis Greenm.	A1	**20,34**	58
quadridentatus Labill. (syn. *Erechtites quadritentata*)	A1	**1, 8, 14**	1, 2, 3
racemosus DC.	A1, B1	**8, 29, 120, 123, 126, 139, 140, 143**	2, 3, 90-92
ragonesei Cabr.	A1	**1, 4, 14**	55
renardii Winkl.	A1	**8, 20, 33**	1-3
retrorsus DC.	A1	**14**	3
rhombifolius (Willd.) Sch. Bip. (syn. *S. platyphyllus*)	A1, B1	**8, 27, 126**	1-3
richii A. Gray	A1	**81, 82**	
riddellii Torr. & A. Gray	A1	**14, 16**	1-3
rivularis DC.	B1	**135**	1-3
rosmarinifolius L.	A4	**96**	1, 2
rudbeckiaefolius Meyer et Walp.	A2	**89**	2
ruderalis Harvey	A1	**14**	3
ruwenzoriensis S. Moore (syn. *S. othonniformis*)	A1	**14, 54, 55**	1-3, 93
salignus DC.	B1	**135**	58
sanguisorbae DC.	A1	**1, 4, 33**	94
sarracenicus L.	B1	**126**	1, 2
scandens Wall.	A1	**1, 8**	2, 3
sceleratus Schweickerdt	A1, A2	**14, 87, 88**	1-3
schvetzovii Korsh	B2	**146**	2
seratophiloides Griseb	A1, A2	**1, 14, 15, 72, 84**	55
serra Hook.	B1, M2	**111, 120, 122, 123, 125, 128, 129, 34**	62
spartioides Torr. & A. Gray	A1	**1, 8, 10, 14**	1, 3
spathulatus A. Rich.	A1	**1, 4, 8**	2, 3
squalidus L.	A1	**1, 4, 84**	2, 3, 95
stapeliaeformis Phill.	B3	**156**	1, 58
stenocephalus Maxim	A1	**8**	2, 3
subalpinus Koch	A1	**1, 4, 8, 30**	2, 3
subulatus Don ex Hook et Arn.	A1	**1, 14, 29**	1, 55, 70
swaziensis Compton	A1, A2	**14, 86**	1-3

Appendix I (continued)

	Type	Alkaloids	References
Senecio sylvaticus L.	B1	**106, 112, 120, 123, 126**	1, 2, 96
syringifolius O. Hoffm.	A4	**96-100**	76
taiwanensis Hayata	A4	**96**	1, 2
tenuifolius Burm.	A1	**1, 4, 20, 21**	3
tomentosus[*]	A1	**1, 33**	2, 3
tournefortia Lapeyr.	A1	**27**	2
triangularis Hook	A1, A4, B1	**1, 4, 14, 27, 96, 106, 107, 112-114**	1, 2, 3
uintahensis A. Nelson	A1	**120**	3
umgeniensis Thell	A1	**15**	1, 58
usaramoensis[*]	A1	**11**	1
usgorensis Cuatr.	A1	**3**	1, 58
uspallatensis Hook. & Arn	A1	**14, 72**	1, 55, 97
variablilis Sch. Bip	B1	**107, 110, 114**	1, 58
venosus Harvey	A1	**14 oder 20**	2, 3
vernalis Waldst. & Kit.	A1, A2	**1, 4, 8, 14, 20, 84**	1-3, 95, 98
viminalis Bremek	A1	**1, 14**	2, 3
vira-vira G. Hieron	A1	**28, 67, 72**	99
viscosus L.	A1	**1, 4**	1-3
vulgaris L.	A1, A2	**1, 4, 8, 10, 14-16, 84**	2, 95, 98, 100
warszewiczii A. Br. et Bouche	A2	**89**	2
werneriaefolius A. Gray	A1	**1, 14**	3
Syneilesis palmata Maxim	A1	**1, 65, 66**	1-3
Tussilago farfara L.	A1, M2	**1, 20, 354**	1-3
Werneria decora Blake	A1, B1	**14, 105**	101
Boraginaceae			
Alkanna tinctoria Tausch	B1	**106, 112, 132**	1, 3, 102
orientalis (L.) Boiss.	B1	**106, 109, 112, 132**	103
Amsinckia carinata Nels. & Macbr.	C1	**181, 187, 197, 215**	104
douglasiana A. DC.	C1, C2	**170, 187, 208, 215, 244, 257**	104, 105
eastwoodiae Macbride	C1	**168, 181, 187, 189, 192**	104
furcata Suksdorf	C1, C2	**181, 187, 197, 198, 215, 257**	104
grandiflora Kleeb. ex Gray	C1, C2	**187, 208, 214-217, 257, 258**	104
hispida (Ruis & Pav.)	C1	**187, 192, 215**	1-3
intermedia Fisch. & Mey.	C1	**187, 192, 215, 229**	1-3
lunaris Macbride	C1, C2	**181, 187, 215, 216, 257, 258**	104
lycopsoides Lehm.	C1, C2	**181, 187, 188, 192, 208, 215, 257**	1, 2, 3, 104
menziesii (Lehm.) Nels. & Macbr.	C1	**170, 181, 187-190, 192, 193, 195, 208, 215-218, 220, 250**	1, 3, 104, 105
retrorsa Suksdorf	C1	**181, 187, 188, 195, 196, 208, 215, 250**	104
spectabilis F. & M. var. *microcarpa* (Greene) Jeps. & Hoov.	C1, C2	**168, 181, 187, 188, 195, 215, 257**	104
spectabilis F. & M. var. *spectabilis*	C1	**168, 181, 187, 188**	104
spectabilis F. & M. var. *nicolai* (Jepson) Munz	C1, C2	**168, 174, 181, 187, 215, 257**	104
tesselata Gray var. *gloriosa* (Suksdorf) Hoover	C1, C2	**174, 178, 179, 181, 187, 208, 215, 257**	104, 105
tesselata Gray var. *tesselata*	C1, C2	**187, 208, 215, 216, 244, 257-260**	104
vernicosa Hook. & Arn.	C1, C2	**181, 187, 189, 197, 215, 255**	104
Anchusa arvensis (L.) Bieb.	C1	**223** or isomer	2, 3
officinalis L.	C1, E	**215, 217, 171**, and **314** or isomer	1-3
Arnebia euchroma (Royle) Johnst.	B1	**106, 109**	106

Appendix I (continued)

	Type	Alkaloids	References
Arnebia hispidissima (Lehm.) DC.	C1, D	**234, 284**	107
Asperugo procumbens L.	C1	**181** or **208, 223**	2, 3
Borago officinalis L.	C1, E	**181, 187, 189, 208, 215, 217, 313**	3, 108, 109
Caccinia glauca Savi.	M1	**335**	1-3
Cerinthe minor L.	C1	**187**	110
Cordia myxa L.	B2	**146**	111
sinensis Lam.	A1	**39**	111
Cryptantha cana (A. Nels.) Payson	C1	**187-189, 215-217**	112
clevelandii Greene	C1	**187, 188, 192, 193, 199, 210**	112
confertiflora Greene	C1, C2	**187, 208, 215, 257, 258**	112
fendleri (Gray) Greene	C1, C3	**106, 109, 262, 264**	112
flava (A. Nels.) Payson	C1	**187-189, 215-217**	112
leiocarpa (F. & M.) Greene	C1	**106, 109, 187, 188, 192, 193,**	
		119, 210	112
thyrsiflora (Greene) Payson	C1	**187-189, 215-217**	112
virgata (Porter) Greene	C1	**187-189, 215, 217**	112
virginiensis (M.E. Jones) Payson	C1	**187, 208, 215, 257**	112
Cynoglossum amabile Stapf. & Drummond	C1	**208, 223**	1-3
australe R. Br.	C1	**170, 209**	1-3
germanicum Jacq.	C1	**178, 223** or isomer	113
glochidiatum Wll. ex Lindl.	C1	**208**	2, 3
lanceolatum Forsk.	C1	**170, 209**	2, 3
latifolium R. Br.	C1, C3	**106, 262**	1-3
macrostylum Bunge	C1	**223, 239**	114
montanum L.	C1	**170, 209, 223, 239**	115
officinale L.	B1, C1	**135, 223, 239, 240**	1-3
pictum Ait (syn. *C. creticum* Mill.)	C1	**223, 239**	1-3
viridiflorum Pallas ex Lehm.	C1	**178, 239**	1, 2
Echium angustifolium Mill.	C1	**234**	116
italicum L.	C1	**234**	2, 3
lycopsis L. (syn. *E. plantagineum* L.)	C1	**192, 234**	1-3
pininana Webb. & Berth.	C1	**189, 194, 195, 232, 234**	117
sericeum Vahl.	C1	**220** or **221, 234**	107
vulgare L.	C1	**136, 223, 239, 240**	1-3
Ehretia aspera Willd.	M1	**339**	1, 2
Hackelia californica (Gray) Johnston	C3	**261, 263, 265, 266**	118, 311
floribunda (Lehm.) Johnston	C3	**262**	1
longituba I. M. Johnston	B1, C3	**106, 109, 261, 262, 264**	119
Heliotropium acutiflorum Kar.	C1	**201**	3
amplexicaule Vahl.	C1	**212**	2, 3
angiospermum Murray	M2	**347**	1
arbainense Fresen	C1	**201, 235, 237**	1-3
arguzioides Kar. & Kir.C1,	D	**201, 294**	1-3
bacciferum Forssk.	C1	**201, 235**	120
circinatum Griseb.	C1	**182, 201, 223, 235, 237, 248**	121
curassavicum L.	B1, C1, M2	**135, 165, 166, 172, 177, 178, 201, 206,**	
		207, 230, 231, 237, 247-249, 351	1-3, 122, 123
dasycarpum Ledeb.	C1	**201**	2, 3
digynum Forssk. (*H. luteum* Poir)	C1	**201, 202, 235, 237**	124
eichwaldii Steud.	B1, C1	**135, 201, 202, 237**	1-3
europaeum L.	C1	**181, 182, 201, 235, 237, 238**	1-3

Appendix I (continued)

	Type	Alkaloids	References
Heliotropium hirsutissimum Grauer	C1	**181, 182, 201, 235, 237, 238, 240**	125, 126
indicum L.	C1	**181, 182, 201, 212, 213, 223, 237**	1-3
keralense Sivaraj et Manilal	B1, C1, C2	**105, 187, 256**	127
lasiocarpum Fisch & Mey.	C1	**201, 237**	1-3
marifolium Retz.	C1	**201, 212, 235, 237**	128
maris-mortui Zohary	C1	**235, 237**	1-3
molle (Torr) J.M. Johnston.	M2	**347**	1
olgae Bunge	C1, D	**201, 293**	1-3
ovalifolium Forsk.	B1, B2	**105, 145**	1
peruvianum L.	C1	**200**	71
popovii subsp. *gillianum* H. Riedl.	C1	**201**	3
ramosissimum (syn. *H. persicum*)	C1	**201**	2, 3
rotundifolium Sieber ex Lehm	C1	**235-238**	1-3, 129
spathulatum Rydb.	C1, C2	**176, 177, 207, 208, 215, 248-250, 257**	3, 130
steudneri Vatke	C1	**215**	3
strigosum Willd.	E	**310**	1
subulatum Hochst ex DC	M2	**347**	131-133
supinum L.	C1	**135, 181, 201, 205, 223, 226, 237, 239**	1-3
ternatum Vahl	M2	**347**	1
transalpinum Vell.	C1	**181, 187, 200, 203, 212, 215**	134
transoxanum Bunge	C1	**201**	3
Lappula glochidiata Brand	C1	**223**	2, 3
intermedia (Ledeb.)Popov	C1	**237**	2, 3
Lindelofia anchusioides Lehm.	C1	**168**	1, 2
angustifolia (Schrenk) Brand	C1	**208, 223**	2, 3
*longiflora**	C1	**223**	135
macrostyla (Bunge) M. Pop.	C1	**167, 168**	1, 2
olgae (Regel et Smirnow) Brand	C1	**178**	1, 2
pterocarpa (Rupr.) M. Pop.	C1	**178**	1, 2
spectabilis Lehm.	C1, D	**223, 225, 284**	1-3
stylosa (Kar. & Kir.) Brand	C1	**168, 178, 223**	1-3
taschimganica (Lipsky) Popov	C1	**178, 223, etc**	2, 3
Lithospermum erythrorhizon Sieb.+ Zuccharini	C1	**187, 195, 232**	136
Macrotomia echioides Boiss	C1	**251**	1, 2
Mertensia bakeri Greene	C1	**215**	137
ciliata (James) G.Don	C1	**187, 215**	137
Messerschmidia sibirica	B1, C1	**109, 215**	3
Moltikiopsis ciliata (Forsk.)J.M.Johnston	C1	**201, 223**	138
Myosotis scorpioides L. (syn. *M. palustris* L.)	C1	**195, 219, 221, 222**	1, 3
sylvatica Hoffm..	B1, C1	**109, 176 or 178, 239. 240**	1, 3
Neatostema apulum (L.) I.M. Johnston	C1	**208, 215**	139
Onosma heterophyllum Griseb.	B1, M2	**134, 344**	140
Paracaryum himalayense (Klotzsch) C.B. Clarke	C1	**178**	1, 2
Paracynoglossum imeretinum (Kusnez.) Pop.	C1	**223, 239**	2, 3
Rindera austroechinata M. Pop.	C1	**223**	1-3
baldschuanica Kusnezov	C1	**174, 200, 223**	1-3
cyclodonta Bge.	C1	**223**	3
echinata Regel	C1	**176, 223**	1-3
oblongifolia M. Pop.	C1	**223**	1-3
Solenanthus circinatus Ledeb.	C1	**223**	1-3

Appendix I (continued)

	Type	Alkaloids	References
Solenanthus coronatus Regel	C1	**223**	2, 3
karateginus Lipsky	C1	**223**	1-3
turkestanicus (Regel & Smirnov) Kusnezov	C1	**200**	1-3
Symphytum asperum Lepech.	B1, C1	**136, 192, 195, 220, 221, 223, 225,**	
		234, 239, 240	1-3, 141
bohemicum S. Schmidt	C1	**215, 217, 221, 234**	142
caucasicum Bieb.	B1, C1	**136, 223, 234, 237, 239**	1-3
grandiflorum DC.	C1	**215, 217, 221, 234**	143
ibericum Stev	C1	**215, 217, 221, 234**	143
officinale L.	C1, C2	**187, 189, 192, 195, 215, 217,**	
		220, 221, 234,	1-3, 141-144
orientale L.	C1	**191, 221, 234**	1-3
peregrinum Ledeb.	C1	**187, 189, 215, 217, 221**	145
tanaicense Steven	C1	**215, 217, 221, 234**	142
tuberosum L.	C1	**191, 215, 217, 221, 234**	1-3, 143
x uplandicum Nyman	C1	**187, 189, 215, 220, 221, 233, 234**	1-3
Tournefortia sarmentosa Lam.	C1	**181**	2, 3
sibirica L.	B1	**144**	2
Trachelanthus hissaricus Lipsky	C1	**174, 178**	1, 2
korolkovii Lipsky	C1	**174**	1, 2
Trichodesma africanum L.	C1, D	**187, 235, 294**	3
ehrenbergii Schweinf.	A1, C1	**20, 181**	107
incanum Alph. DC.	D	**293, 294**	1-3
zeylanicum (Burm.f.) R. Br.	C1	**181**	2, 3
Ulugbekia tschimganica (B. Fedtsh.) Zak.	C1	**241**	1, 2
Celastraceae			
Bhesa archboldiana (Merr. et Perry) Ding Hou	B1	**109**	1, 2
Convolvulaceae			
Ipomoea angulata Lamk.	M1	**340-343**	146
Fabaceae			
Adenocarpus argyrophyllus (R. Goday) RM	L	**370**	2
complicatus (L.) Gay	L	**367, 370, 371**	336
decorticans Boiss	L	**363, 367, 370, 371**	2, 336
foliolosus (Ait.) DC	L	**367, 370, 371**	336
grandiflorus Boiss	L	**370**	2
hispanicus DC.	L	**367, 370, 371**	2, 335, 336
mannii (Hook.f.)Hook.f.	L	**367, 370, 371**	336
telonensis (Lois.) DC.	L	**367, 370, 371**	336
Buchenroedera lotononoides Scott Elliot	A1	**1, 4**	147
meyeri Presl	A1	**1, 4**	147
multiflora Eckl. & Zeyh.	A1	**1, 4**	147
tenuifolia Eckl. & Zeyh. var. *tenuifolia*	A1	**1, 4**	147
tenuifolia var. *pulchella* (E. Mey.) Harv	A1	**1, 4**	147
trichodes Presl	A1	**1, 4**	147
Crotalaria aegyptiaca Benth.	D, M2	**281, 284, 294, 301, 301a,**	
		309a 345	1-3, 312
agatiflora Schweinf.	A1	**67, 69, 71, 73-75**	1-3
alata Leville	D	**284**	148

Appendix I (continued)

	Type	Alkaloids	References
Crotalaria albida Heyne ex Roth (C. montana)	D	**309**	1, 2
anagyroides H.B. & K.	A1	**1, 67**	1-3, 149
argyrolobioides Bak.	D	**286**	148
aridicola Domin.	M1, M2	**332, 358, 359**	1, 2
assamica Benth	D	**284, 309b**	3, 150, 153
astragalina Hochst. ex A. Rich. (syn. *C. impressa*)	D	**286**	148
axillaris Ait.	D	**304, 306**	1-3
balansae Micheli	D	**284**	148
barbata R. Graham	D	**282**	1-3
brachystachya Benth.	D	**286**	148
brevidens Benth.	D	**286**	148
breviflora DC	A1	**4, 15**	1-3
burhia Buch.-Ham.	D	**281, 284**	1-3
calycina Schrank	D	**284**	150
candicans W. & A.	B1, D	**141, 289, 291, 302, 303, 308**	1-3, 151, 152
cephalotes Steud. ex A. Rich.	D	**284**	3
crassipes Hook.	D	**301**	1-3
crispata F. Muell. ex Benth.	D	**283, 284, 289**	1-3
cunninghamii R. Br.	D	**284, etc**	3
cylindrocarpa DC.	M2	**345 or 346**	2
damarensis Engl.	D	**278**	1, 2
dura Wood & Evans	D	**278**	1-3
fulva Roxb.	D	**283**	1-3
globifera E. Mey	D	**278, 288, 292, 294, 296**	1-3
goreensis Guill. et Perr.	M2	**345 or 346**	1, 2
grahamiana R. Wight & Walk. Arn.	D	**284, 285, 287**	1-3
grandistipulata Harms.	M2	**344**	2
grantiana Harvey (syn. *C. virgulata*)	D, M2	**286, 288, 292, 347**	1-3, 148
incana L.	A1	**4, 15, 67**	1-3
intermedia Kotschy	A1, D	**4, 15, 284**	2, 3, 148
juncea L.	A1, D	**1, 4, 16, 294, 295**	1-3
laburnifolia L.(syn. *C. capensis* Baker)	A1	**67, 68, 73**	1-3, 337
laburnifolia subsp. *eldomae* (Bak. f.) Polhill	A1	**20, 24, 62, 67, 73**	2, 3
lachnophora A. Rich.	M2	**344**	1, 2
lachnosema Stapf.	D	**278, 279**	154
leiloba Bartl. (syn. *C. ferruginea*)	D	**284**	2, 3
leschenaultii DC.	D	**284, 289**	2, 3
leubnitziana Schinz	D	**284**	148
lukwangulensis Harms	D	**286**	148
madurensis R. Wight	A1, D	**73, 283, 289, 290, 291**	1-3
maypurensis H.B. et K.	M2	**345, 346**	1, 2
maxillaris Klotzsch	D	**286**	148
medicaginea Lam.	M2	**358, etc**	1, 2
mitchellii Benth.	D	**284, 301**	1-3
mucronata Desv. (syn. *C. strictata: C. pallida*)	A2, D	**4, 15, 19, 56, 57, 284**	1-3, 148
mysorensis Roth.	D	**284**	2, 3
nana Burm.	A2, D	**85, 280**	1-3
naragutensis Hutch.	A1	**4, 5, 15, 56**	154
natalitia Meissner	M2	**344**	1, 2
nitens Kunth.	D	**284**	3

Appendix I (continued)

	Type	Alkaloids	References
Crotalaria novae-hollandae DC.	D	**284, 301**	1-3
officinale[*]	B1	**135**	1
oxirensis Willd.	D	**284**	148
paniculata Willd.	D	**283**	1-3
paulina Schrank	D	**284, etc**	3, 148
pilosa Miller	D	**284**	148
podocarpa DC	M2	**345 or 346**	1, 2
pumila Ortega	D	**284**	148
quinquefolia L.	D	**284**	2, 3, 148
recta Steud. ex Rich.	D	**284, 294**	3
retusa L.	B1, D	**105, 284, 301, 307**	1-3, 148
rhodesiae E.G.Baker	M2	**344**	1, 2
rosenii (Pax)Milne-Readhead ex Polhill	A1	**73-75**	1, 155
rubiginosa Willd. (syn. *C. wightiana*)	D	**294, 295**	1-3
sagittalis L.	D	**284**	2, 3, 148
scassellatii Chiov.	B1, D	**105, 107, 284, 299, 304, 306**	1, 156, 157
semperflorens Vent	D	**301**	1-3
sessiliflora L.	A1, D	**4, 284, 294**	3, 158
spartioides DC.	A1	**14**	2, 3
spectabilis Roth. (syn. *C. retzii* Hichc.; *C. sericea* Retz.)	D	**284, 286, 307**	1-3, 148
stipularia Desv.	D	**284**	2, 3, 148
stolzii (Baker f.) Milne-Redh. ex Polhill	M2	**344**	1, 2
tetragona Roxb.	A1, D	**4, 294**	1-3
trifoliastrum Willd.	D, M1, M2	**284, 332, 347, 348, 358-360**	1, 2, 148, 159
usaramoensis E.G. Baker	A1, D	**1, 4, 14, 15, 284**	1-3, 148
verrucosa L.	A1, M2	**63, 64, 67, 344**	1-3
walkeri Arn.	A1	**39, 40**	1-3
Laburnum anagyroides Med	M2	**171**	1, 2
Lotononis arenicola Schlechter es De Wild.	M2	**361**	160
azurea Eckl. & Zey.	A1	**4**	147
brachyloba (E. Mey.) Benth.	A1	**27, 28**	147
brevicaulis B-E. van Wyk	A1	**4**	147
caerulescens (E. Mey.) B-E. v. Wyk	A1	**1, 4**	147
carnea[*]	M2	**361**	160
comptonii B-E. van Wyk	A1	**1**	147
divaricata (Eckl. & Zeyh.) Benth.	A1	**1, 4**	147
elongata (Thunb.) D. Dietr.	A1	**1, 4**	147
fruticoides B-E. van Wyk ined	A1	**27, 28**	147
involucrata (E. Mey.) Benth.	A1	**1**	147
lenticula (E. Mey.) Benth.	A1	**27, 28**	147
longicephala B-E. van Wyk ined.	A1	**4**	147
oxyptera Benth	M2	**361**	160
perplexa Eckl.& Zeyh.	M2	**361**	160
polycephala (E. Mey.) Benth.	A1	**1**	147
purpuresens B-E. van Wyk	A1	**1, 4**	147
rigida (E. Mey.) Benth.	A1	**4**	147
rostrata Benth.	M2	**361**	160
serpens (E. Mey.) R. Dahlgr.	A1	**4**	147
stenophylla[*]	M2	**361**	160
trisegmentata Phill. var. robusta	A1	**1, 4**	147

Appendix I (continued)

	Type	Alkaloids	References
Orchidaceae			
Catasetum maculatum Kunth.	M2	**352**	338
Chysis bractescens Lindl.	M1	**352**	1, 2
Doritis pulcherrima Lindl. (syn. *Phalaenopsis esmeralda*)	E	**320 or 321**	2
Hammarbya paludosa (L.) O. K.	E	**324, 325**	1, 2
Kingiella taenialis (Lindl.) Rolfe	E	**321**	1, 2
Liparis auriculata Rchb.f.	E	**328**	1, 2
bicallosa Schltr.	E	**328**	1, 2
hachijoensis Nakai	E	**328**	1, 2
keitaoensis Hay.	E	**327, 329**	1, 2
kumokiri F. Maekwa	E	**323**	1, 2
krameri Franch. & Sav.	E	**323**	1
loeselii (L.) L. C. Rich.	E	**330**	1, 2
nervosa Lindl.	E	**326**	2
Malaxis congesta comb. nov. (Rchb. f.)	E	**328**	1, 2
grandifolia Schltr.	E	**331**	1, 2
Phalaenopsis amabilis Bl.	E	**320**	1, 2
amboinensis J. J. Smith	E	**321**	1, 2
aphrodite Rchb.f.	E	**320**	1, 2
cornu-cervi Rchb. f.	E	**319**	1, 2
equestris Rchb. f.	E	**320, 322**	1, 2
fimbriata J.J.Smith	E	**320**	1, 2
hieroglyphica (Rchb.)H.R.Sweet	E	**320 or 321**	2
lueddemanniana Rchb.f.	E	**320 or 321**	2
mannii Rchb. f.	E	**320**	1, 2
sanderiana Rchb. f.	E	**320, 321**	1, 2
schilleriana Rchb.f.	E	**321**	1, 2
stuartiana Rchb.f.	E	**320, 321**	1, 2
sumatrana Korth & Rchb.f.	E	**321**	2
violacea Teijsm. & Binn.	E	**320 or 321**	2
Vanda cristata Lindl.	E	**314**	2
helvola Bl.	E	**312, 314**	2
hindsii Lindl.	E	**314**	2
luzonica Loher	E	**312, 314**	1, 2
Vandonopsis gigantea Pfitz.	E	**312, 314**	1, 2
lissochiloides Pfitz.	E	**312, 314**	1. 2
Poaceae			
Festuca arundinacea Schreb.	L	**364**	1, 2
Lolium cuneatum Nevski	L	**363-369**	1, 2
temulentum L.	L	**364**	1
Schismus barbatus (L.)Juel.	A1	**1**	111
Ranunculaceae			
Caltha biflora DC	A1	**1**	2, 3
leptosepala DC	A1	**1**	2, 3
Rhizophoraceae			
Cassipourea gummiflua Tulasne	M2	**362**	1, 2
gummiflua var. *verticellata* Lewis	M2	**362**	1, 2

Appendix I (continued)

	Type	Alkaloids	References
Santalaceae			
Thesium minkwitzianum B. Fedtsch.	C1, E	**167, 313, 318**	1, 2
Sapotaceae			
Minusops elengii L.	M1	**336**	2
Planchonella anteridifera Lam	E	**315-317**	1, 2
equestris[*]	C1	**164** or isomer	1
thyrsoidea White ex Walker	E	**315-317**	1, 2

[*] For a number of botanical names the authorities are missing in the original papers. As far as possible the authorities were substituted according to the Index Kewensis. In a number of cases this was impossible due to synonyms; these species need to be taxonomically confirmed. As far as possible individual species listed under different synonyms were combined; the synonymous names not primarily considered are given in parentheses.

Appendix II Senecionine-Type (A); Senecionine Group (A1)

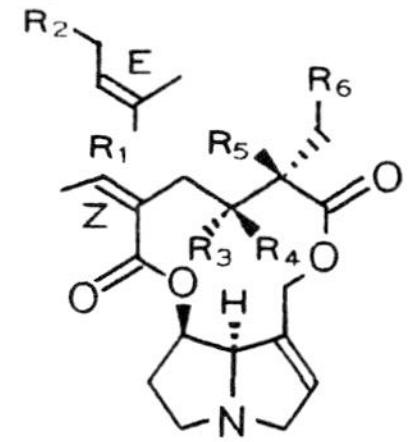

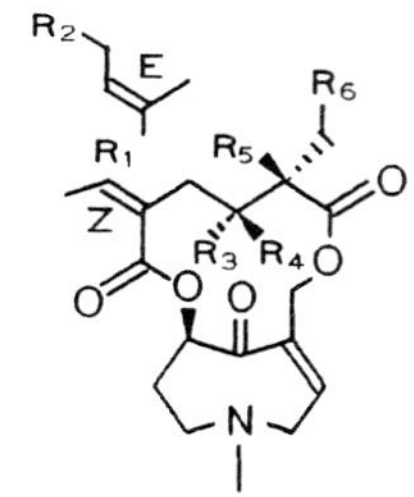

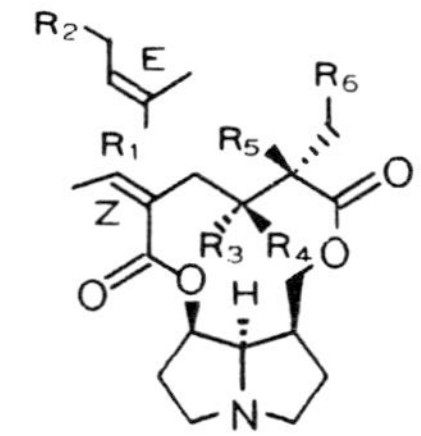

R_1	R_2	R_3	R_4	R_5	R_6						
Z		CH_3	H	OH	H	1	senecionine	20	senkirkine	27	platyphylline
Z		CH_3	H	OAc	H	2	O[12]-acetylsenecionine	21	O[12]-acetylsenkirkine		
Z		$-CH_2-O-$		OH	H	3	13,19-epoxisenecionine				
E	H	CH_3	H	OH	H	4	integerrimine	22	neosenkirkine	28	neoplatyphylline
E	H	CH_3	H	OAc	H	5	O[12]-acetylintegerrimine				
E	OH	CH_3	H	OH	H	6	eruciflorine	23	anonamine		
E	H	$-CH_2-O-$		OH	H	7	13,19-epoxyintegerrimine				
Z		$-CH_2-$		OH	H	8	seneciphylline				
Z		$-CH_2-$		OAc	H	9	O[12]-acetylseneciphylline				
E	H	$-CH_2-$		OH	H	10	spartioidine				
E	H	OH	H	CH_3	H	11	usaramoensine				
Z		CH_2OH	-O-		H	12	erucifoline				
Z		CH_2OAc	-O-		H	13	O[19]-acetylerucifoline				
Z		CH_3	H	OH	OH	14	retrorsine	24	18-hydroxysenkirkine	29	dihydroretrorsine
E	H	CH_3	H	OH	OH	15	usaramine	25	18-hydroxyneosenkirkine		
Z		$-CH_2-$		OH	OH	16	riddelliine				
Z		CH_2OH	H	OH	H	17	gynuramine	26	19-hydroxysenkirkine		
Z		CH_2OAc	H	OH	H	18	O[19]-acetylgynuramine				
E	H	H	CH_3	OH	OH	19	mucronatinine				

Appendix II Senecionine Type (A); Senecionine Group (A1)

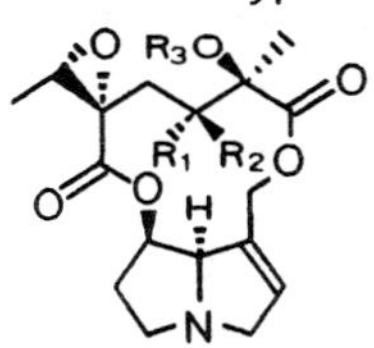

R_1	R_2	R_3		
-CH_2-		H	30	jacozine
CH_3	H	H	31	jacobine
CH_3	H	Ac		
-O-CH_2-		H	32	senecicannabine

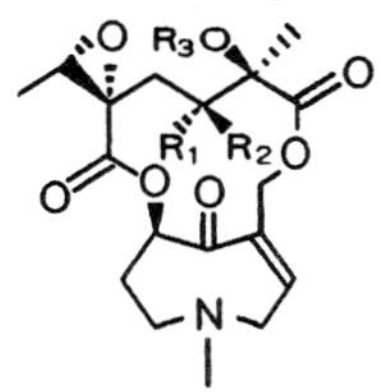

33 otosenine

34 florosenine

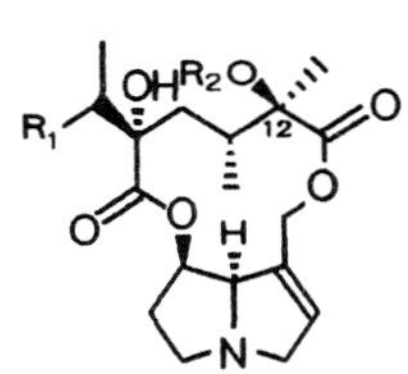

R_1	R_2		
OH	H	35	jacoline
OH	Ac	36	O[12]-acetyljacoline
OAc	Ac		
Cl	H	37	jaconine
Cl	Ac		

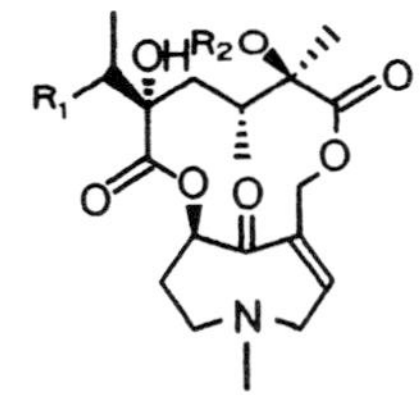

38 onetine

39 floridanine

40 floricaline

41 desacetyldoronine

42 doronine

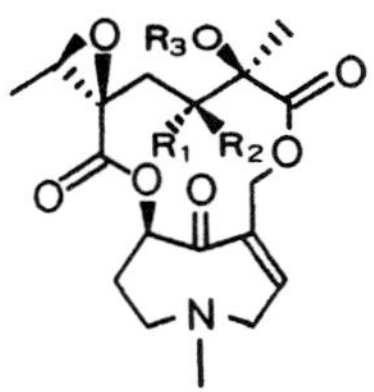

R_1	R_2	R_3		
CH_3	H	H	43	petasitenine
CH_3	H	Ac	44	neopetasitenine

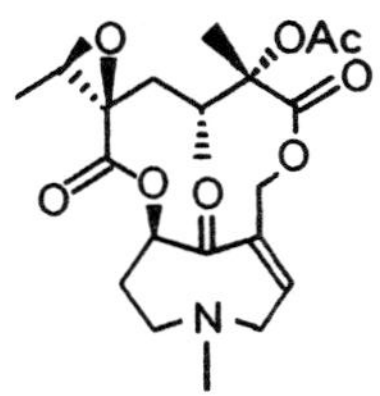

45 ligularizine

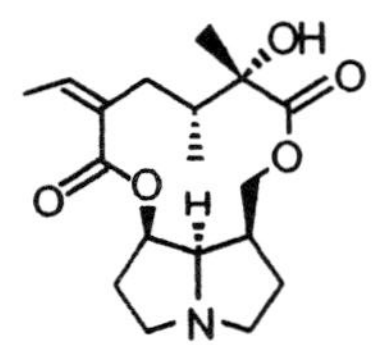

46 ligularinine

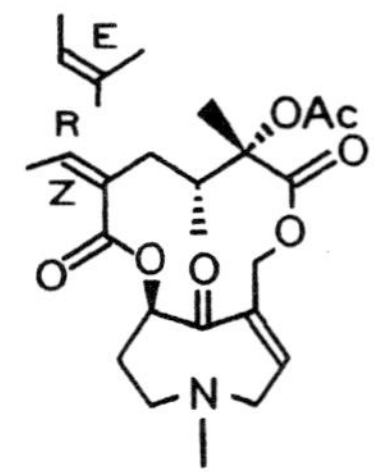

R		
Z	47	neoligularidine
E	48	ligularidine

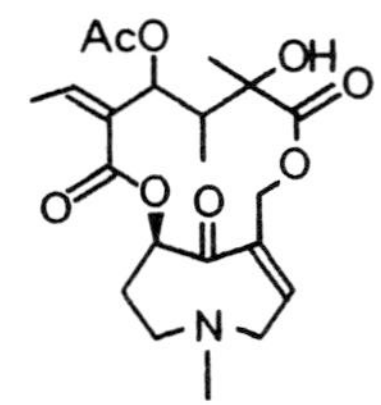

49 ligularine

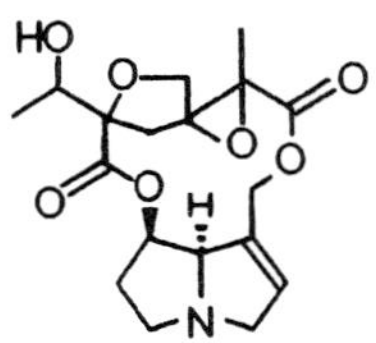

50 adonifoline

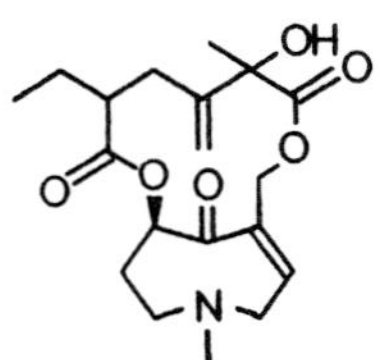

51 emiline

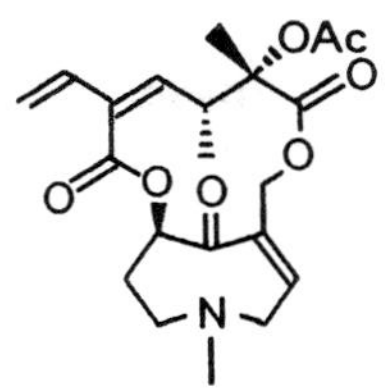

52 clivorine

Appendix II Senecionine Type (A); Senecionine Group (A1)

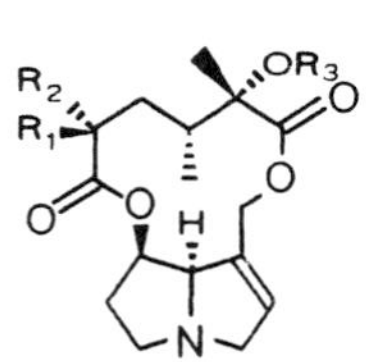

R_1	R_2	R_3		
CH_2CH_3	H	H	53	yamataimine
OH	CH_2CH_3	H	54	bisline
OH	CH_2CH_3	Ac	55	isoline

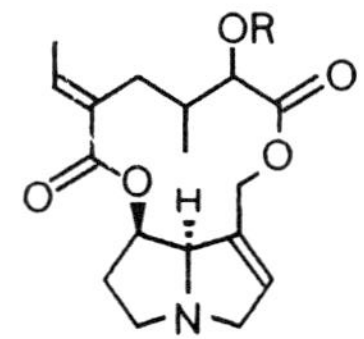

R		
H	56	nilgrine
Ac	57	crotastriatine

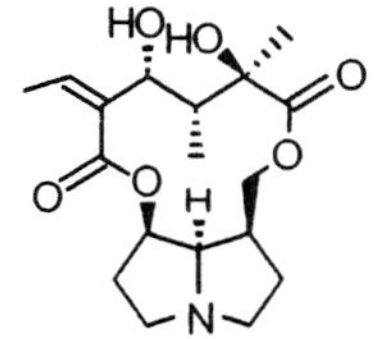

58 hygrophylline

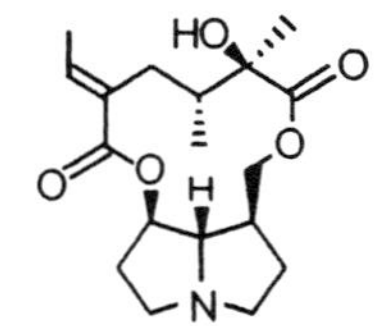

59 hastacine

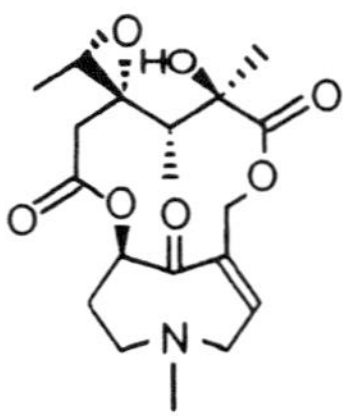

60 senecioracenine

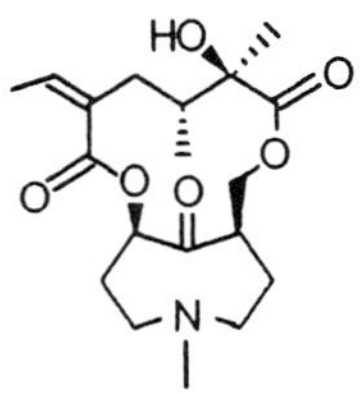

61 1,2-dihydrosenkirkine

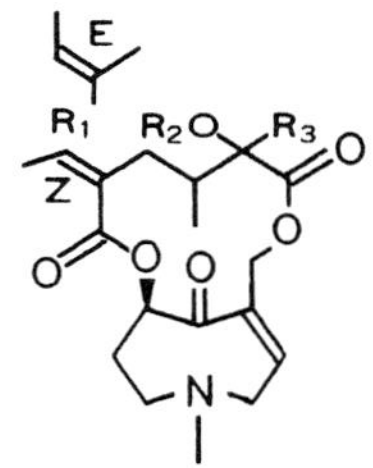

R_1	R_2	R_3		
Z	H	H	62	crotafoline
E	H	CH_3	63	crotaverrine
E	Ac	CH_3	64	O^{12}-acetylcrotaverrine

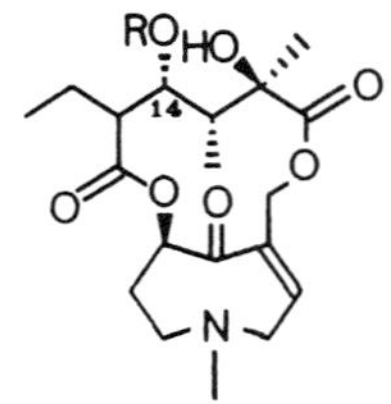

R		
H	65	syneilesine
Ac	66	O^{14}-acetylsyneilesine

Appendix II Senecionine Type (A); Senecionine Group (A1)

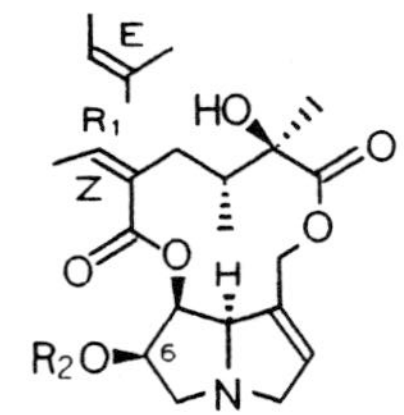

R_1	R_2		
Z	H	67	anacrotine / crotalaburnine
E	H	68	trans-anacrotine
Z	Ac	69	O^6-acetylanacrotine
E	Ac	70	O^6-acetyl-trans-anacrotine
E	Ang	71	O^6-angeloyl-trans-anacrotine

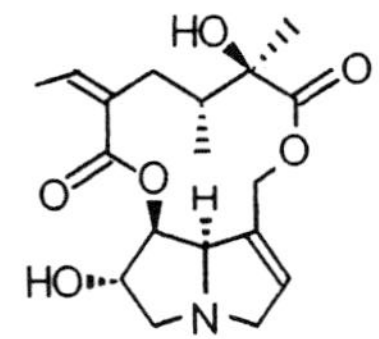

72 uspallatine

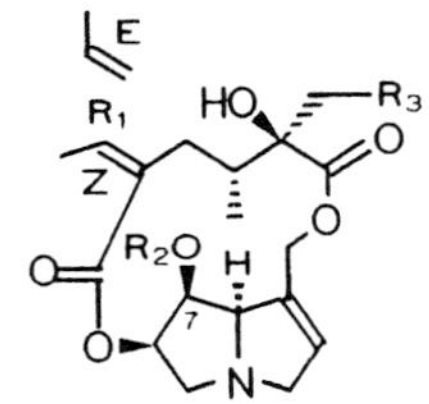

R_1	R_2	R_3		
E	H	H	73	madurensine
E	H	OH	74	crotaflorine
Z	Ac	H	75	O^7-acetyl-cis-madurensine
E	Ac	H	76	O^7-acetylmadurensine

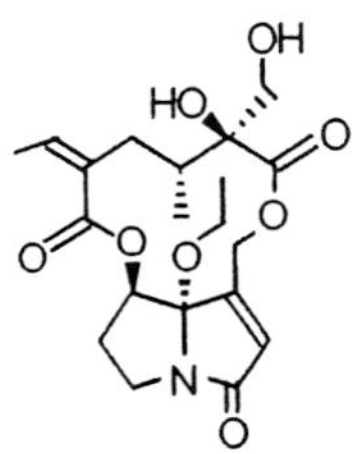

77 8-ethoxy-3-oxo-retrorsine

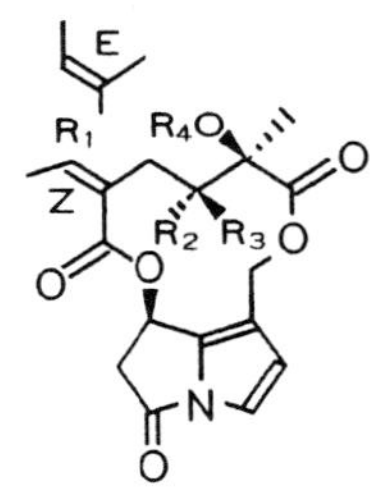

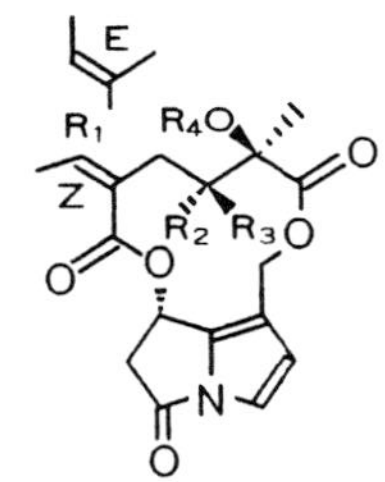

R_1	R_2	R_3	R_4				
Z	CH_3	H	OH			80	desacetylsenaetnine
Z	CH_3	H	Ac	78	isosenaetnine	81	senaetnine
E	CH_3	H	Ac			82	trans-senaetnine
Z	-CH_2-		Ac	79	dehydroisosenaetnine	83	dehydrosenaetnine

Appendix II Senecionine Type (A); Senecivernine Group (A2)

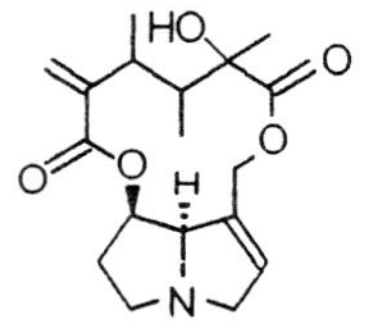

84 senecivernine

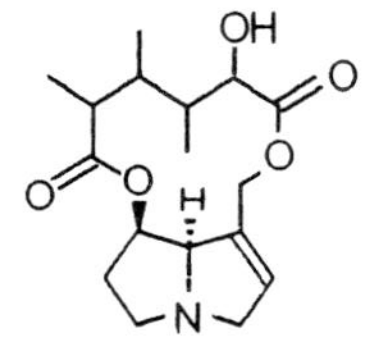

85 crotananine

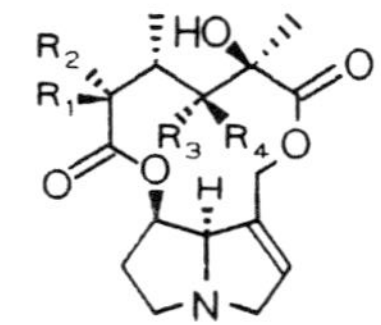

	R_1	R_2	R_3	R_4
86 swazine	$-CH_2-$		$-CH_2O-$	
87 sceleratine	CH_2Cl	OH	CH_3	OH
88 merenscine	CH_2OH	OH	CH_3	OH

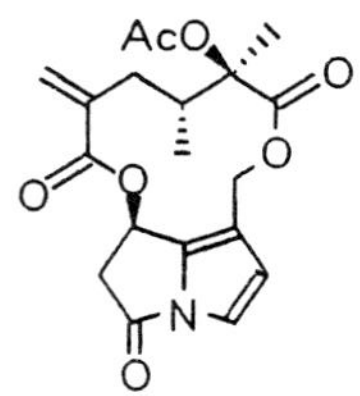

89 pterophorine

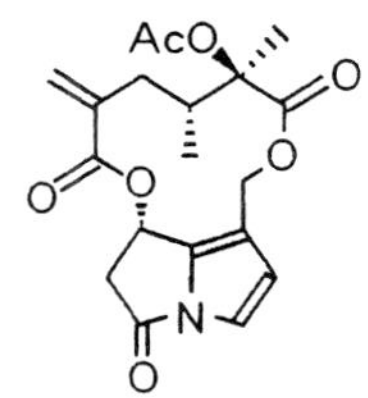

90 isopterophorine

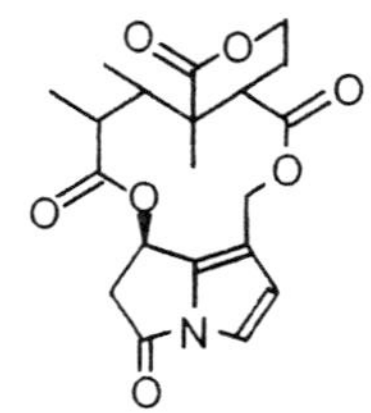

91 inaequidenine

Appendix II Senecionine Type (A); Nemorensine Group (A3)

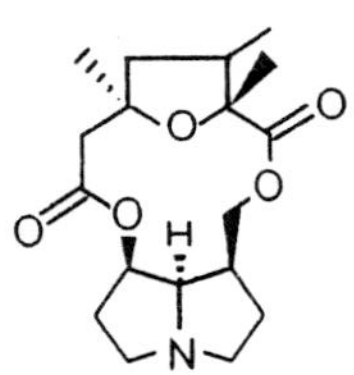

92 nemorensine

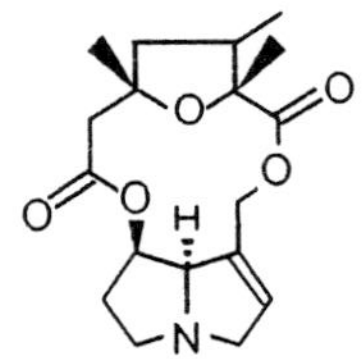

93 retroisosenine

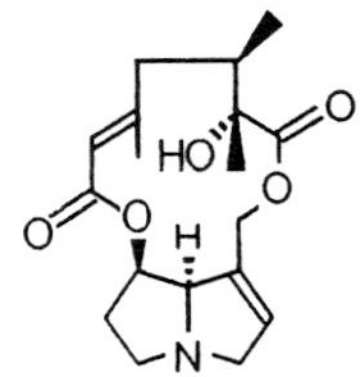

94 doronenine

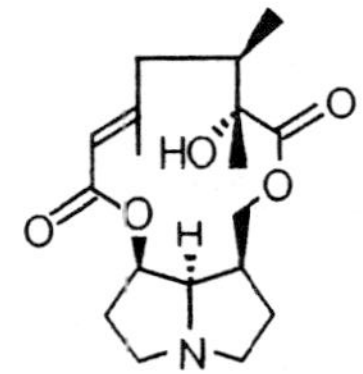

95 bulgarsenine

Rósmarinine Group (A4)

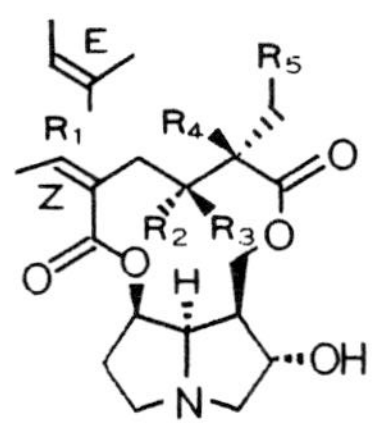

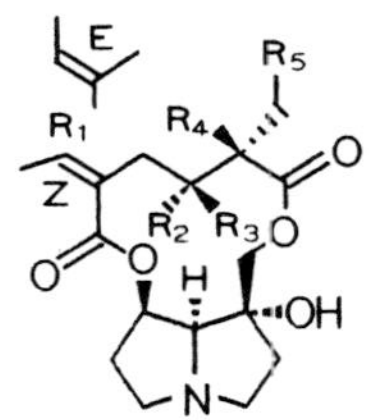

R_1	R_2	R_3	R_4	R_5		
Z	CH_3	H	OH	H	96 rosmarinine	102 hadiensine
Z	CH_3	H	OAc	H	97 O^{12}-acetylrosmarinine	103 O^{12}-acetylhadiensine
E	CH_3	H	OH	H	98 neorosmarinine	
E	CH_3	H	OAc	H		104 O^{12}-acetylneohadiensine
Z		$-CH_2-$	OH	H	99 angularine	
E		$-CH_2-$	OH	H	100 neoangularine	
Z	CH_3	H	OH	OH	101 petitianine	

Appendix II Triangularine Type (B); Triangularine Group (B1)

Ang = 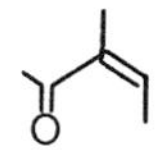 Tigl = 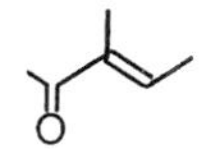 Sen = 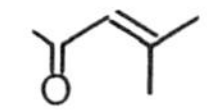 i-Val = 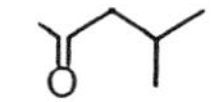

t-Sarr = 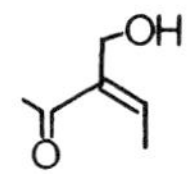 *c*-Sarr = 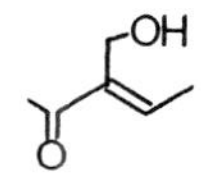 3-OH-Sen = 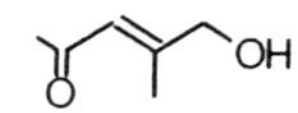

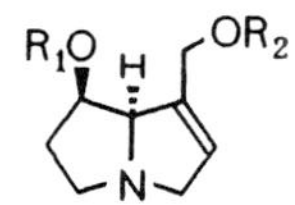

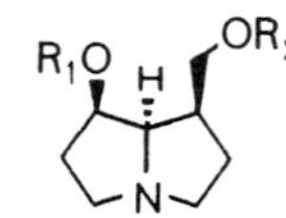

R_1	R_2				
H	H	105	retronecine	119	platynecine
Ang	H	106	7-angeloylretronecine	120	7-angeloylplatynecine
Sen	H	107	7-senecioylretronecine	121	7-senecioylplatynecine
Tigl	H	108	7-tigloylretronecine	122	7-tigloylplatynecine
H	Ang	109	9-angeloylretronecine	123	9-angeloylplatynecine
H	Sen	110	9-senecioylretronecine	124	9-senecioylplatynecine
H	Tigl	111	9-tigloylretronecine	125	9-tigloylplatynecine
Ang	*t*-Sarr	112	triangularine	126	sarracine
Ang	*c*-Sarr	113	neotriangularine	127	neosarracine
Sen	*t*-Sarr	114	7-senecioyl-9-sarracinoylretronecine		
Tigl	*t*-Sarr	115	triangularicine	128	sarranicine
Tigl	*c*-Sarr	116	neotriangularicine	129	neosarranicine
3-OH-Sen	Ang	117	sencalenine		
3-OH-Sen	*t*-Sarr	118	doriasenine		

Appendix II Triangularine Type (B); Triangularine Group (B1)

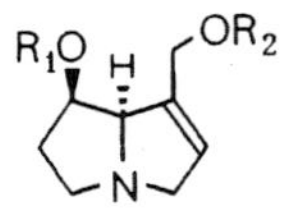

R_1	R_2		
Sen	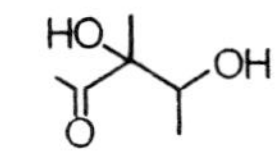	130	7-senecioyl-9-(2,3-dihydroxy-2-methyl-butanoyl)-retronecine
Sen	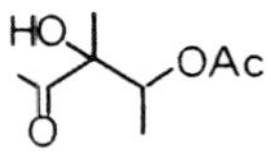	131	7-senecioyl-9-(3-acetoxy-2-hydroxy-2-methylbutanoyl)-retronecine
Ang	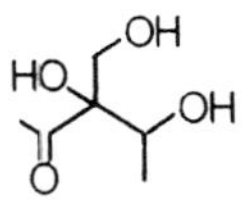	132	dihydroxytriangularine (Stereochemie unbewiesen)
Sen	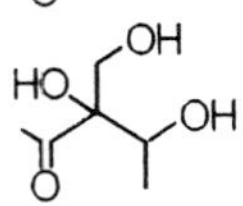	133	7-senecioyl-9-(2,3-dihydroxy-2-hydroxy-methylbutanoyl)-retronecine

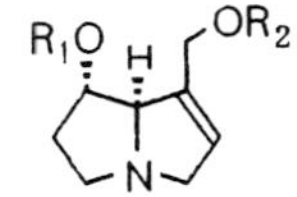

R_1	R_2		
H	H	134	heliotridine
Ang	H	135	rivularine
Ang	Ang	136	asperumine

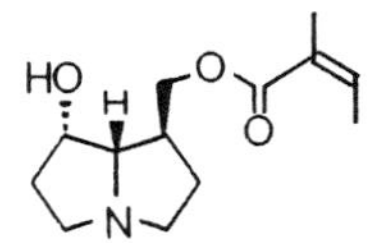

137 farfugine

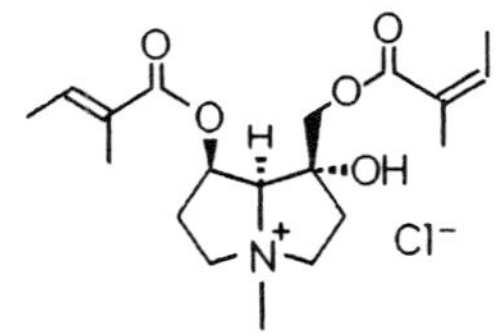

138 N-methyl-O[7],O[9]-diangeloyl-1-hydroxy-platynecinium chloride

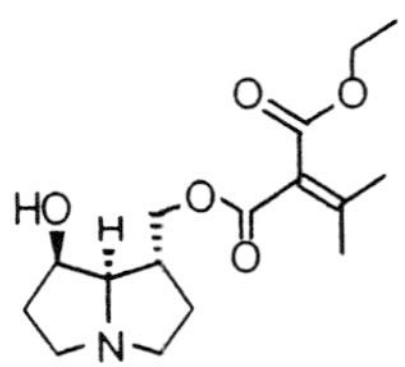

139 racemonine

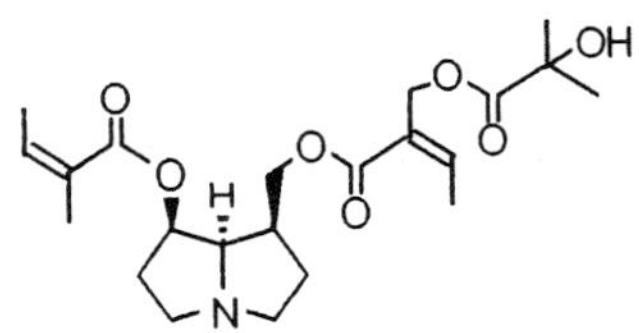

140 racemodine

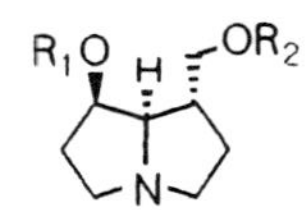

R_1	R_2		
H	H	141	(-)-turneforcidine
Ang	H	142	O[7]-angeloylturneforcidine
H	Tigl	143	razemozine
Tigl	H or inverse	144	turneforcine

Appendix II Triangularine Type (B); Macrophylline Group (B2)

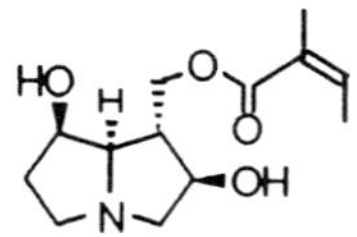

145 helifoline

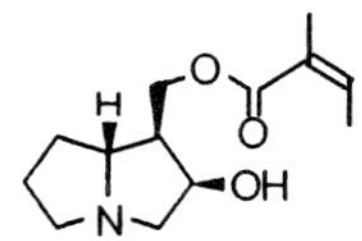

146 macrophylline

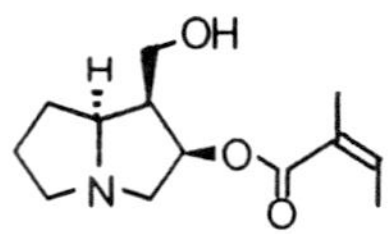

147 petasinine

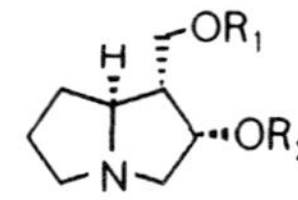

R_1	R_2		
Ang	H	148	2α-hydroxy-9-angeloyloxy-(-)-trachelanthamidine
H	Sen	149	2α-senecioyloxy-9-hydroxy-(-)-trachelanthamidine
Sen	H	150	2α-hydroxy-9-senecioyloxy-(-)-trachelanthamidine

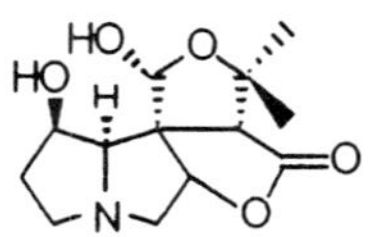

151 senecicaudatinal semiacetal A

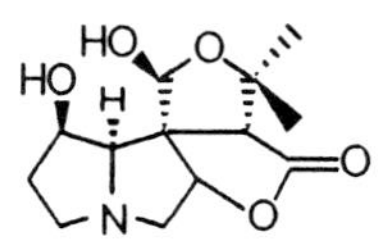

152 senecicaudatinal semiacetal B

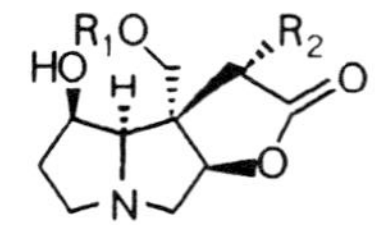

R_1	R_2		
Sen	H	153	O^9-senecioylnorsenecicaudatine
Sen	$COH(CH_3)_2$	154	O^9-senecioylsenecicaudatine
iVal	$COH(CH_3)_2$	155	O^9-isovaleroylsenecicaudatine

Appendix II Triangularine Type (B); Senampeline Group (B3)

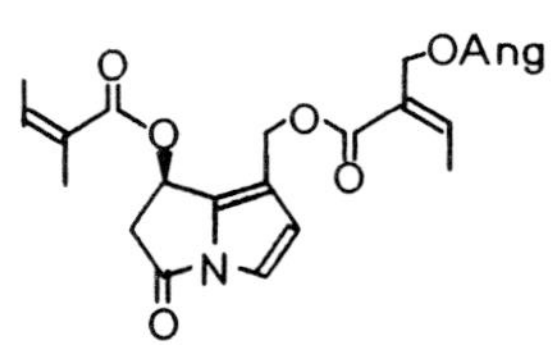

156 5-oxo-desacetyloxysenampeline G

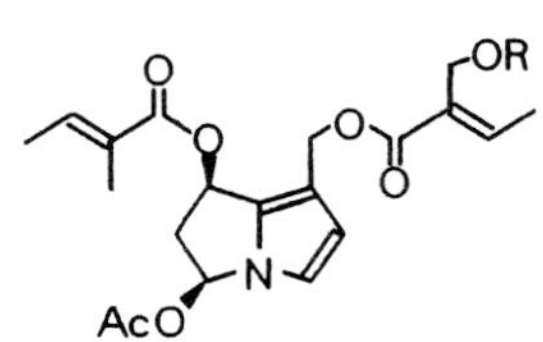

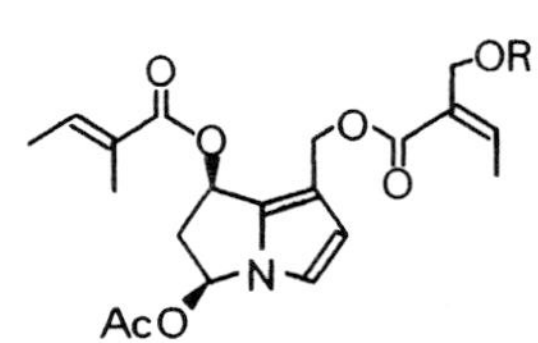

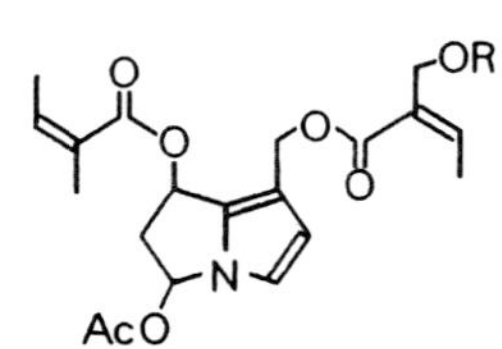

R			
Sen	157 senampeline A	159 senampeline C	161 senampeline E
Tigl	158 senampeline B	160 senampeline D	162 senampeline F
Ang			163 senampeline G

Appendix II Lycopsamine Type (C); Lycopsamine Group (C1)

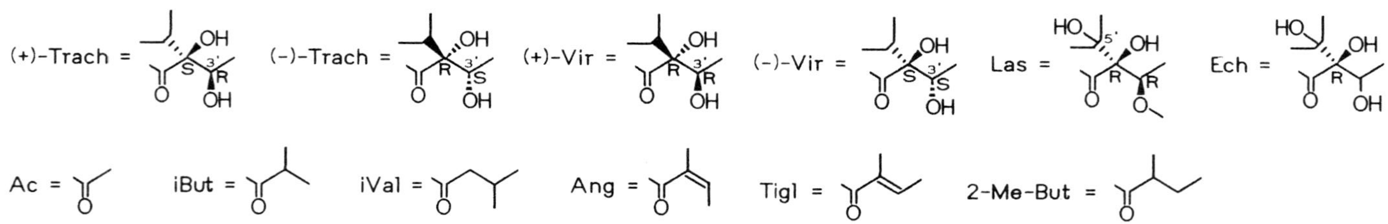

R	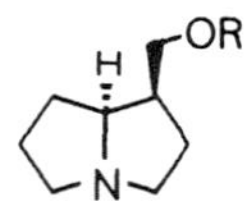	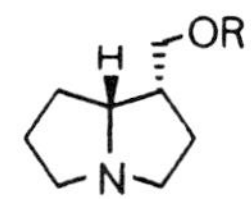	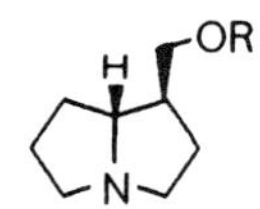	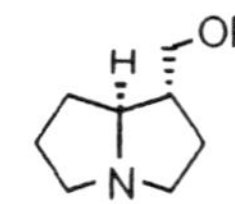
H	164 (-)-isoretronecanol	167 lindelofidine	171 laburnine	173 (-)-trachelanthamidine
(+)-Trach		168 lindelofine		174 trachelanthamine
(+)-Trach, 3'-OAng				
(+)-Trach, 3'-OTigl		169 lindelofamine		
(+)-Trach, 3'-OC_4H_7				175 3'-angeloyl- or tigloyl-trachelanthamine
(-)-Trach	165 heliocurassavicine		172 heliocurassavinine	176 heliovicine
(+)-Vir	166 heliocoromandaline			177 coromandaline
(-)-Vir		170 cynaustraline		178 viridiflorine
(-)-Vir, 3'-OAc				179 3'-acetylviridiflorine

Appendix II Lycopsamine Type (C); Lycopsamine Group (C1)

R_1	R_2	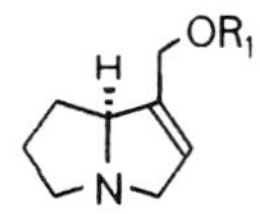	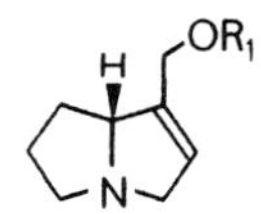	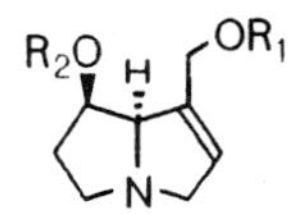	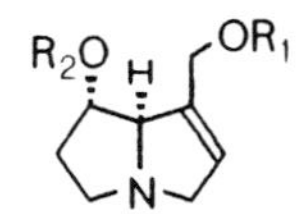
H	H	180 (-)-supinidine	186 (+)-supinidine	105 retronecine	134 heliotridine
(+)-Trach	H	181 supinine		187 intermedine	200 rinderine
(+)-Trach, 3'-OCH_3	H	182 heleurine			201 heliotrine
(+)-Trach, 3'-OCH_3	Ang				202 7-angeloylheliotrine
(+)-Trach, 3'-OAc	H			188 3'-acetylintermedine	203 3'-acetylrinderine
(+)-Trach	Ac			189 7-acetylintermedine	204 7-acetylrinderine
(+)-Trach, 3'-OAc	Ac			190 7,3'-diacetylintermedine	
(+)-Trach, 3'-OiBut	H	183 3'-isobutyrylsupinine			
(+)-Trach, 3'-OTigl	H			191 anadoline	
(+)-Trach, 3'-OC_4H_7	H	184 3'-angeloyl- or tigloylsupinine			
(+)-Trach	Ang			192 echiumine	205 7-angeloylrinderine
(+)-Trach, 3'-OAc	Ang			193 3'-acetylechiumine	
(+)-Trach	Sen			194 echiupinine	
(+)-Trach	Tigl			195 myoscorpine	
(+)-Trach, 3'-OAc	Tigl			196 3'-acetylmyoscorpine	
(+)-Trach, 3'-OC_4H_9	H	185 3'-(iso)valeroylsupinine			
(+)-Trach	2-Me-But			197 furcatine	
(+)-Trach, 3'-OAc	2-Me-But			198 3'-acetylfurcatine	
(+)-Trach	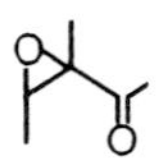			199 2",3"-epoxyechiumine	

Appendix II Lycopsamine Type (C); Lycopsamine Group (C1)

R_1	R_2	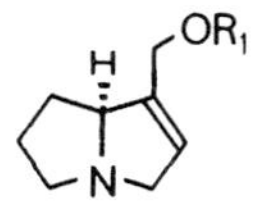	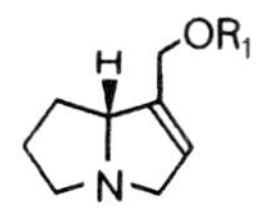	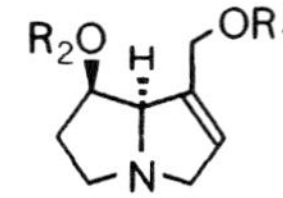	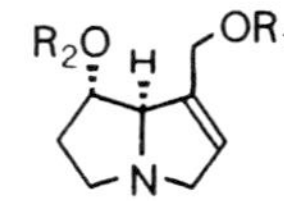
(+)-Trach	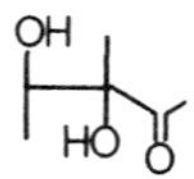			210 threo-2",3"-dihydroxy-echiumine	
(+)-Trach	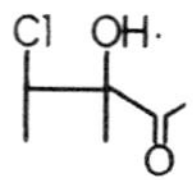			211 erythro-3"-chloro-2"-hydroxyechiumine	
(−)-Trach	H	206 heliovinine		212 indicine	
(−)-Trach, 3-OAc	H			213 3'-acetylindicine	
(−)-Trach	Ac			214 7-acetylindicine	
(+)-Vir	H	207 coromandalinine			
(−)-Vir	H	208 amabiline	209 cynaustine	215 lycopsamine	223 (+)-echinatine
(−)-Vir, 3'-OAc	H			216 3'-acetyllycopsamine	224 3'-acetylechinatine
(−)-Vir	Ac			217 7-acetyllycopsamine	225 7-acetylechinatine
(−)-Vir, 3'-OAc	Ac			218 7,3'-diacetyllycopsamine	
(−)-Vir, 3'-OTigl	H			219 scorpioidine	
(−)-Vir	Ang			220 symlandine	226 7-angeloylechinatine
(−)-Vir	Tigl			221 symphytine	
(−)-Vir, 3'-OTigl	Ac			222 7-acetylscorpioidine	
(−)-Vir, 3'-OC_4H_7	H				227 3'-angeloyl- or tigloyl-echinatine
(−)-Vir, 3'-O-iVal	H				228 3'-isovaleroylechinatine

Appendix II Lycopsamine Type (C); Lycopsamine Group (C1)

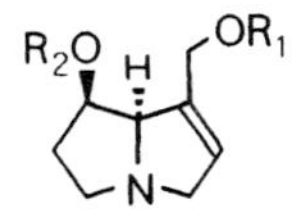

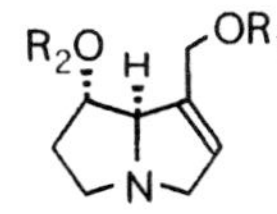

R_1	R_2				
Vir-3'-OCH_3	H	229	sincamidine		
Vir-3'-OAc	H	230	9-(3'-acetyl)-viridiflorylretronecine		
Vir-3'-OiVal	H	231	9-(3'-isovaleroyl)-viridiflorylretronecine		
Las	H			235	europine
Las-5'-OAc	H			236	5'-acetyleuropine
Las	Ang			237	lasiocarpine
Las-5'-OAc	Ang			238	5'-acetyllasiocarpine
Ech	Tigl	232	hydroxymyoscorpine		
Ech	Ac	233	uplandicine		
Ech	Ang	234	echimidine	239	(-)-heliosupine
Ech-3'-OAc	Ang			240	3'-acetylheliosupine

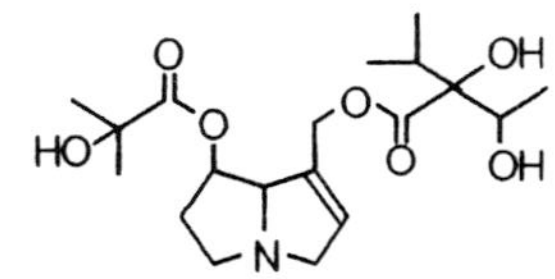

241 uluganine

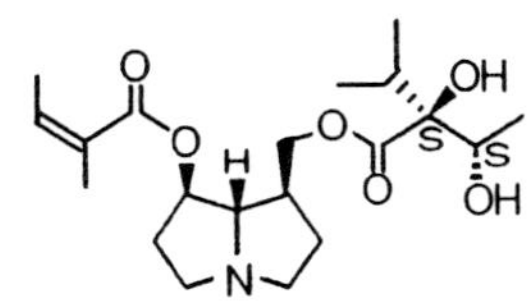

245 punctanecine
7-angeloyl-9-(-)-viridifloryl-(+)-hastanecine

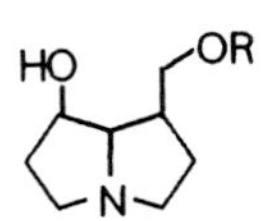

R		
H	242	turniforcidine
(-)-Vir	243	9-viridifloryl-turniforcidine
(-)-Vir, 3'OAc	244	9-(3'-acetylviridifloryl)-turniforcidine

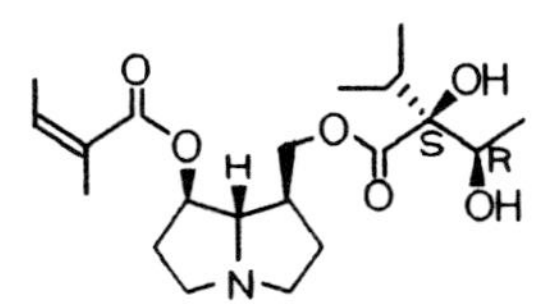

246 isomer of punctanecine
7-angeloyl-9-trachelanthylhastanecine

Appendix II Lycopsamine Type (C); Lycopsamine Group (C1)

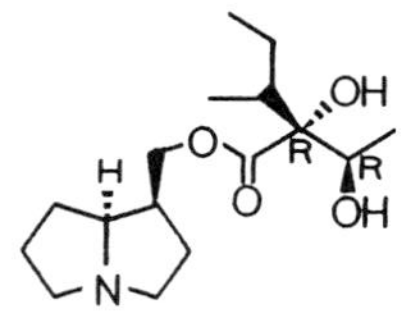

247 heliocurassavine

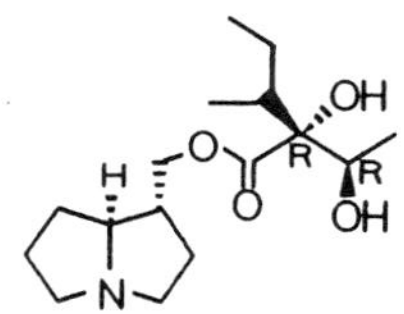

248 curassavine

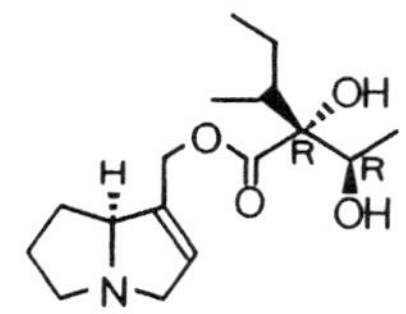

249 curassavinine

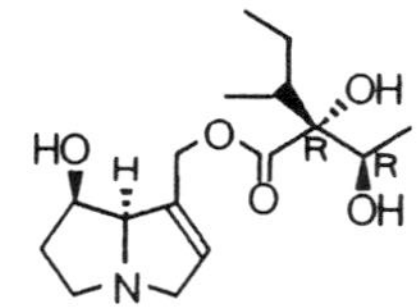

250 heliospathine

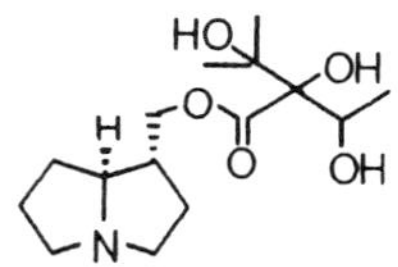

251 macrotomine

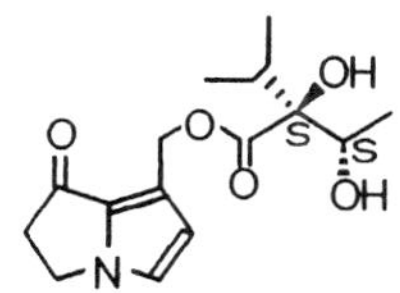

252 parsonine

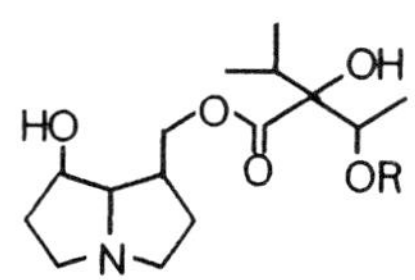

R	
H	253 isomer of dihydroechinatine
$CO-C_4H_7$	254 3'-angeloyl- or tigloylderivative of dihydroechinatine isomer

Appendix II Lycopsamine Type (C); Isolycopsamine Group (C2)

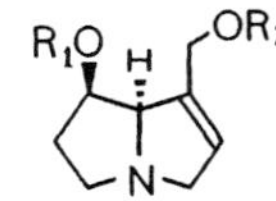

R_1	R_2		
(+)-Trach	H	255	isointermedine
(+)-Vir	H	256	isolycopsamine
(-)-Vir	H	257	tesselatine / heliospathuline
(-)-Vir	Ac	258	9-acetyltesselatine
(-)-Vir-3'-OAc	H	259	3'-acetyltesselatine
(-)-Vir-3'-OAc	Ac	260	9,3'-diacetyltesselatine

Latifoline Group (C3)

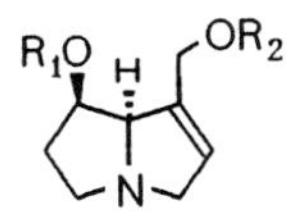

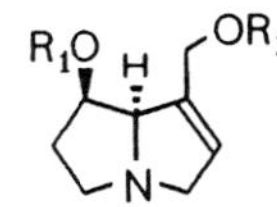

R_1	R_2		R_1	R_2	
Ac	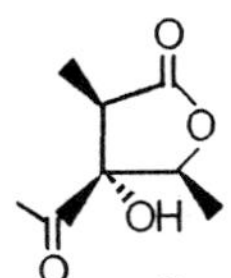	261 longitubine	Ang	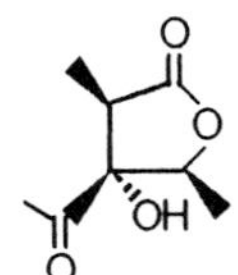	262 latifoline
H	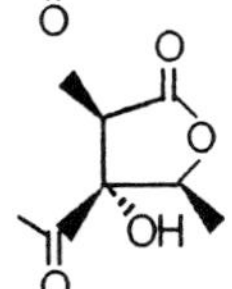	263 O^9-latifolylretronecine	Ang	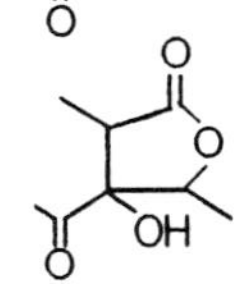	264 neolatifoline
H	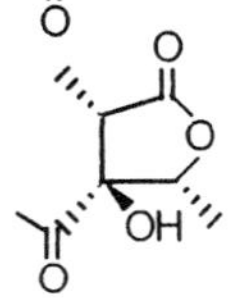	265 hackelidine	Ac	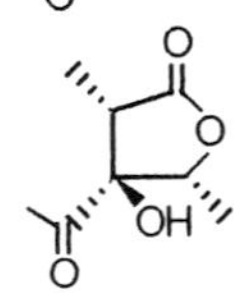	266 7-acetylhackelidine

Appendix II Lycopsamine Type (C); Parsonsine Group (C4)

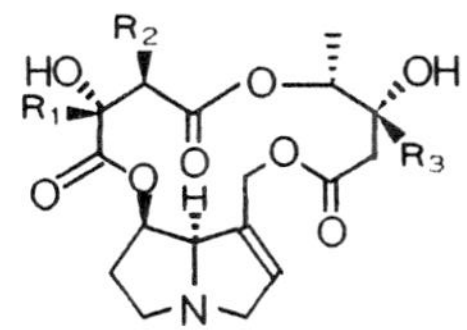

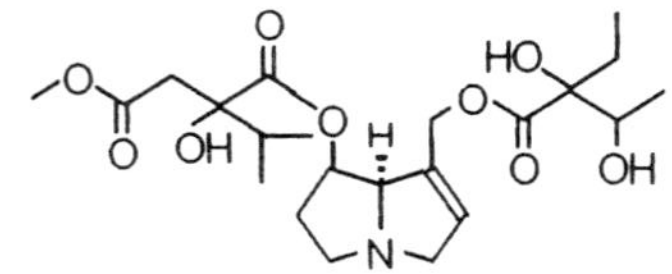

R_1	R_2	R_3		
$CH(CH_3)_2$	H	C_2H_5	267	14-deoxyparsonsianine
$CH(CH_3)C_2H_5$	H	C_2H_5	268	14-deoxyparsonsianidine
$CH(CH_3)_2$	H	$CH(CH_3)_2$	269	parsonsine
$CH(CH_3)_2$	OH	C_2H_5	270	parsonsianine
$CH(CH_3)C_2H_5$	H	$CH(CH_3)_2$	271	heterophylline
$CH(CH_3)_2$	OH	$CH(CH_3)_2$	272	spiraline
$CH(CH_3)C_2H_5$	OH	C_2H_5	273	parsonsianidine
$CH(CH_3)C_2H_5$	OH	$CH(CH_3)_2$	274	17-methylparsonsianidine
$CH(CH_3)C_2H_5$	OH	$CH(CH_3)_2$	275	spiranine
$COH(CH_3)C_2H_5$	OH	$CH(CH_3)_2$	276	spiracine

277 12-seco-14-deoxy-parsonsianine-13-methylester

Appendix II Monocrotaline Type (D)

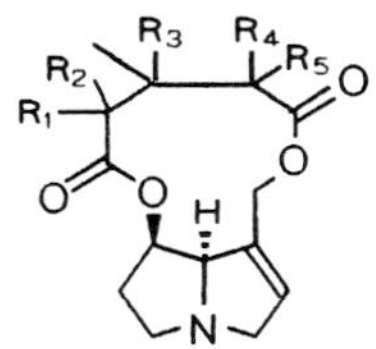

R_1	R_2	R_3	R_4	R_5	
H	H	OH	H	H	278 dicrotaline
H	H	OAc	H	H	279 acetyldicrotaline
CH_3	H	H	$CHCH_3$ (OH)	H	280 cronaburmine
CH_2CH_3	CH_3	OH	CH_3	OH	281 croburhine / crotalarine

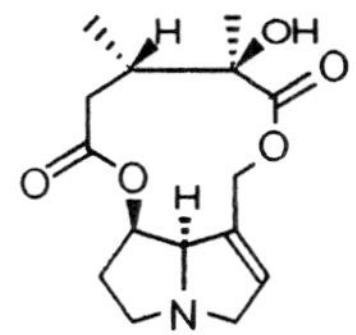

282 crobarbatine

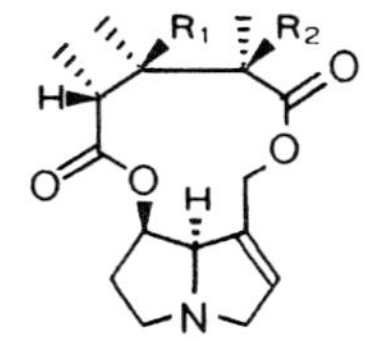

R_1	R_2		
OH	H	283	fulvine
OH	OH	284	monocrotaline
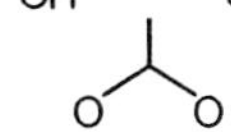		285	monocrotalinine
OAc	OH	286	spectabiline
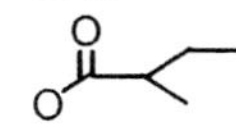	OH	287	grahamine

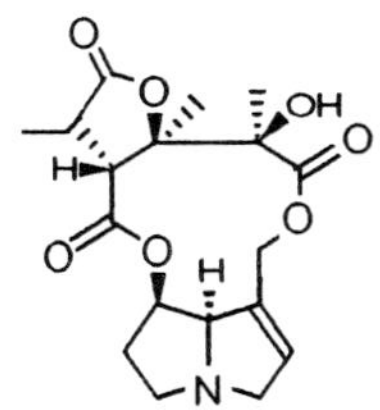

288 grantianine

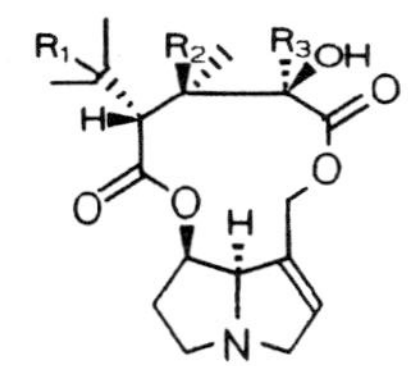

R_1	R_2	R_3		
—O—		CH_3	292	grantaline
H	H	CH_3	293	incanine
H	OH	CH_3	294	trichodesmine
H	OH	CH_2OH	295	junceine
OH	OH	CH_3	296	globiferine

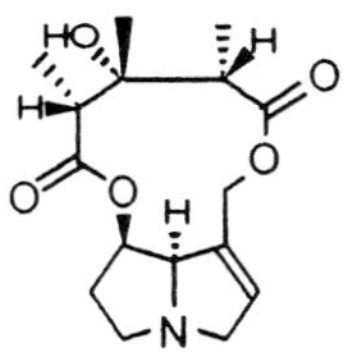

289 crispatine

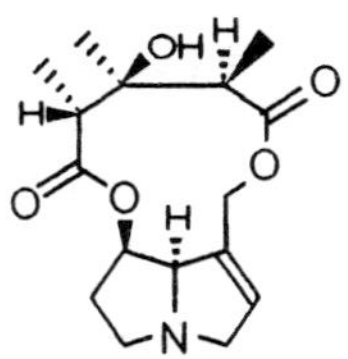

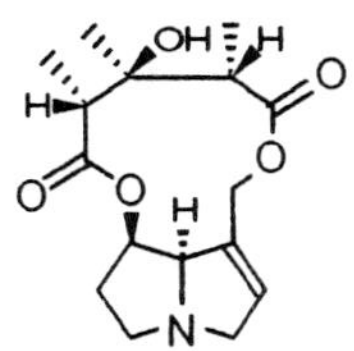

necine-acid relative configuration

290 cromadurine 291 isocromadurine

Appendix II Monocrotaline Type (D)

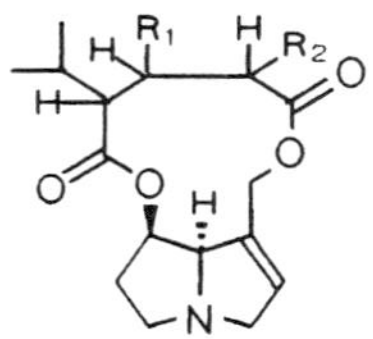

R_1	R_2		
OH	CH_3	297	aucherine
OH	CH_2CH_3	298	othonnine
H	$CHCH_3$ | OH	299	desoxyaxillarine

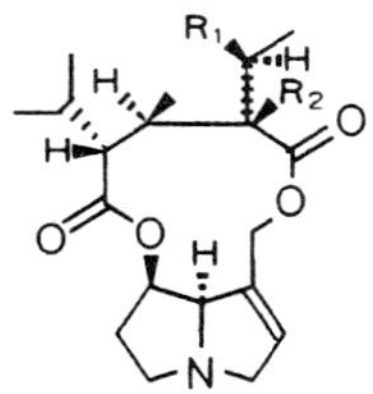

R_1	R_2		
H	OH	304	axillaridine
OH	H	305	12-desoxyaxillaridine
OH	OH	306	axillarine

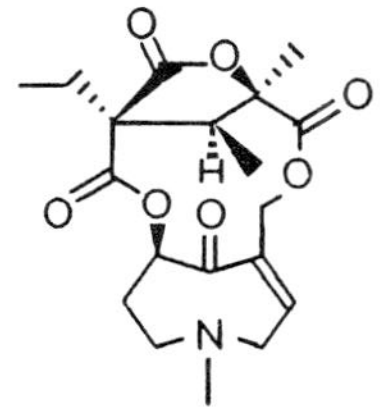

300 retusamine

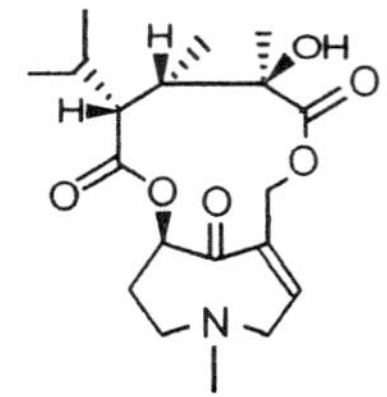

301 crosemperine

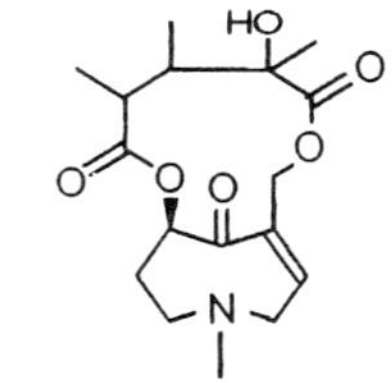

301a croaegyptine

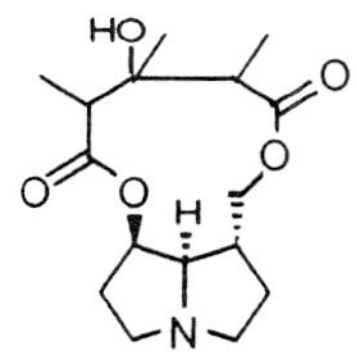

302 crocandine
303 isocrocandine

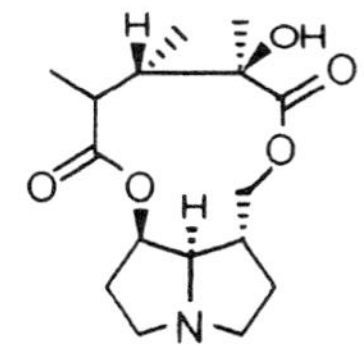

307 retusine

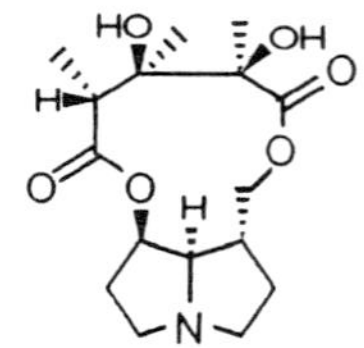

308 cropodine

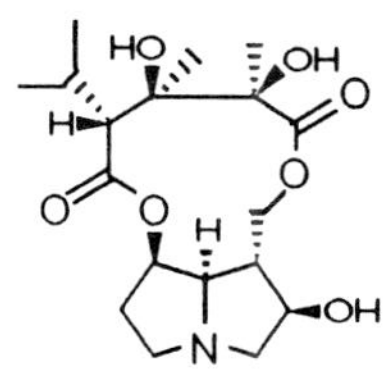

309 croalbidine

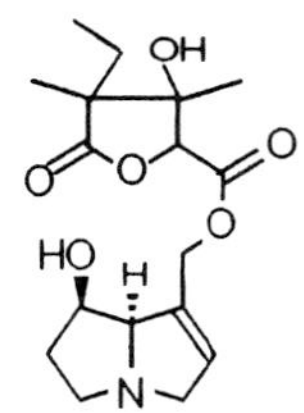

309a crotalarialactone

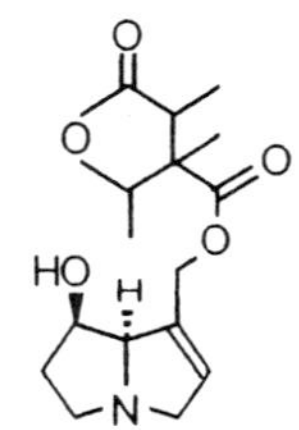

309b assamicadine

Appendix II Phalaenopsine Type (E)

R	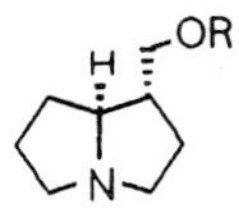	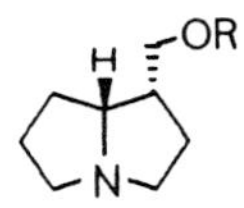	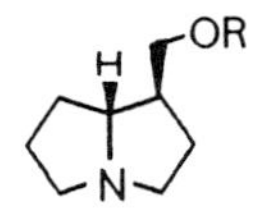
Ac		312 acetyllindelofidine	314 acetyllaburnine
Tigl			315 tigloyllaburnine
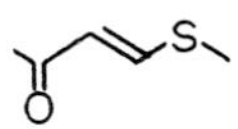			316 planchonelline
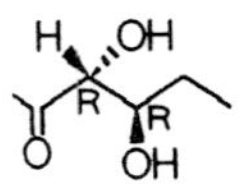	310 strigosine		
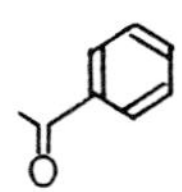			317 benzoyllaburnine
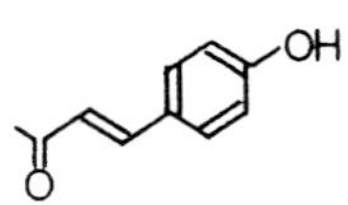		313 thesinine	
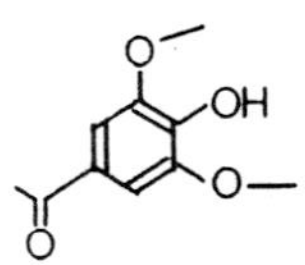	311 alafine		

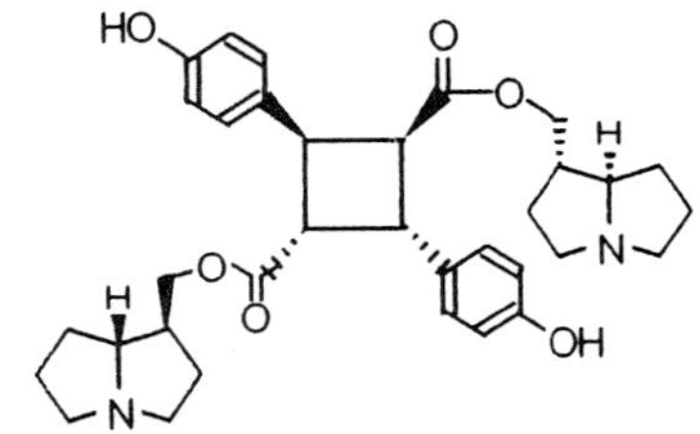

318 thesine

Appendix II Phalaenopsine Type (E)

R

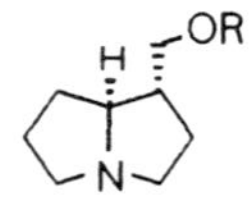

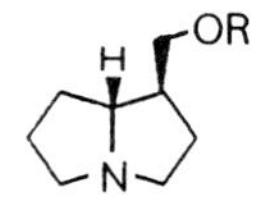

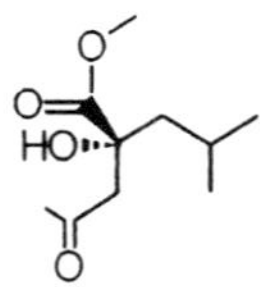

319 cornucervine

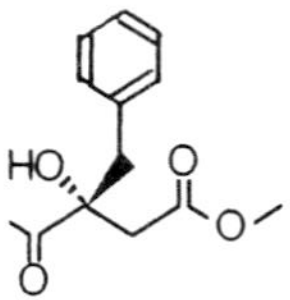

320 phalaenopsine T   321 phalaenopsine La

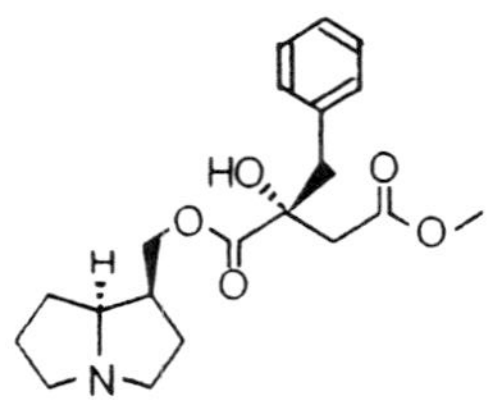

322 phalaenopsine Is

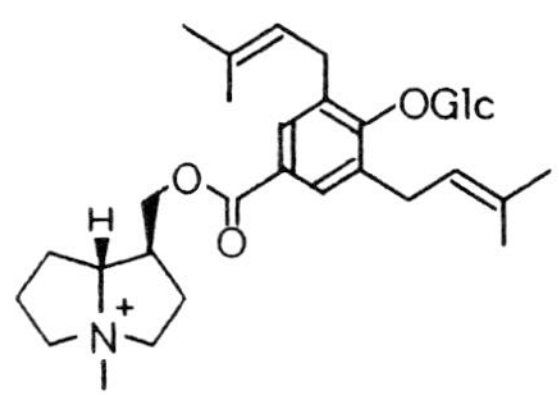

323 kumokirine

R

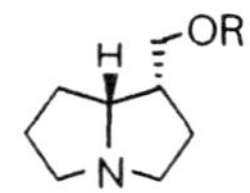

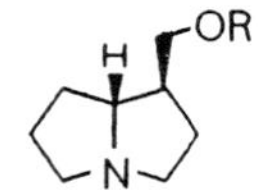

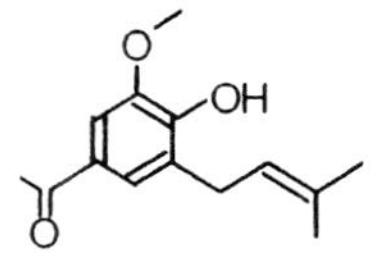

327 keitine

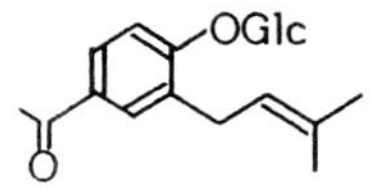

328 malaxine

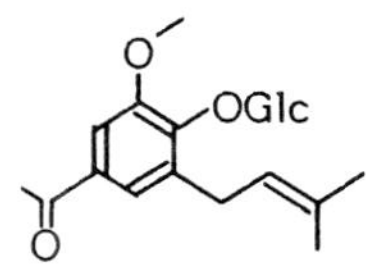

324 hammarbine   329 keitoaine

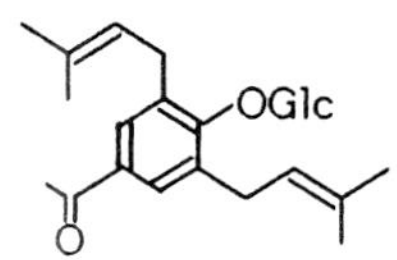

325 paludosine   330 auriculine

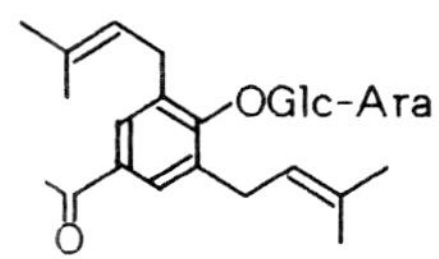

326 nervosine   331 grandifoline

Appendix II Miscellaneous PAs (M); Unusual Necine esters (M1)

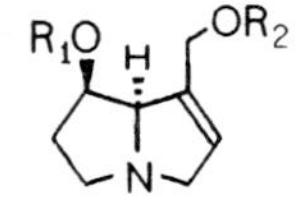

R_1	R_2	
H	Ac	332 O[9]-acetylretronecine
Ac	H	333 O[7]-acetylretronecine
Ac	CH_3	334 O[7]-acetyl-O[9]-methylretronecine
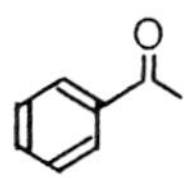	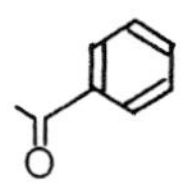	335 O[7,9]-dibenzoylretronecine

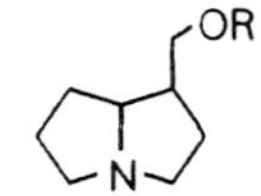

R	
Tigl	336 1-hydroxymethylpyrrolizidine tiglate
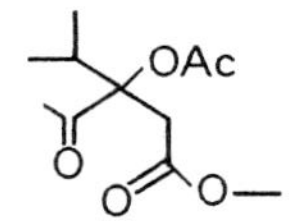	337 1-hydroxymethylpyrrolizidine-methyl-(2-isopropyl-O[2]-acetyl)-malate

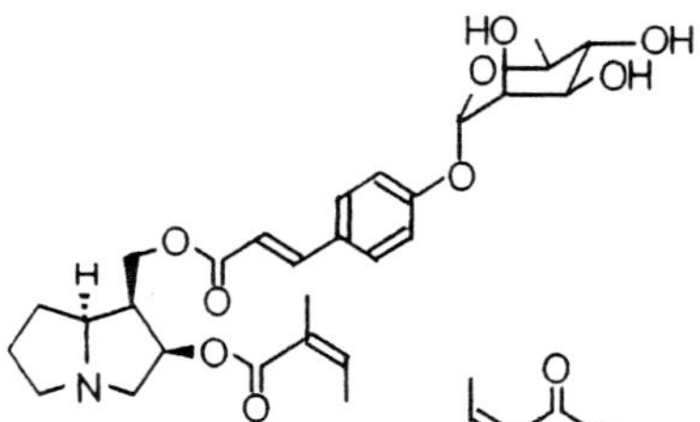

338 petasinoside

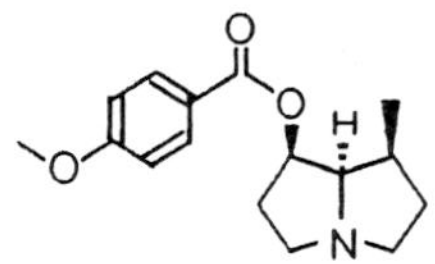

339 ehretinine

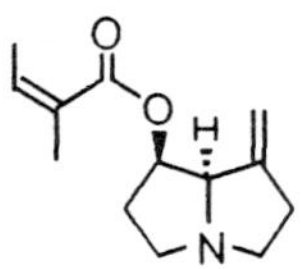

339a
7ß-angelyl-1-methylene-8α-pyrrolizidine

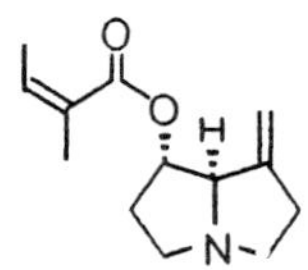

339b
7α-angelyl-1-methylene-8α-pyrrolizidine

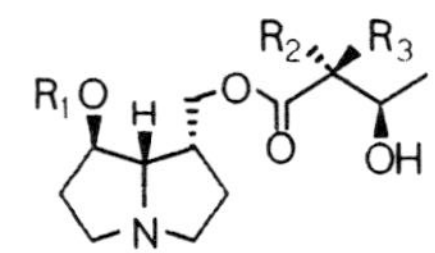

R_1	R_2	R_3	
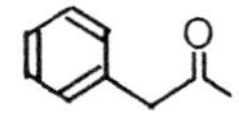	OH	CH_3	340 ipanguline A
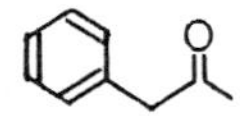	CH_3	OH	341 isoipanguline A
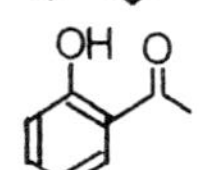	OH	CH_3	342 ipanguline B
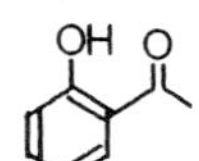	CH_3	OH	343 isoipanguline B

Appendix II Miscellaneous PAs (M); Simple Necine Derivatives (M2)

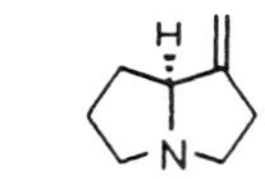

344 1-methylene-8α-pyrrolizidine

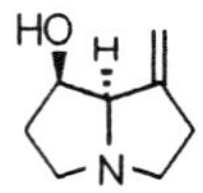

345 7β-hydroxy-1-methylene-8α-pyrrolizidine

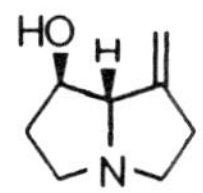

346 7β-hydroxy-1-methylene-8β-pyrrolizidine

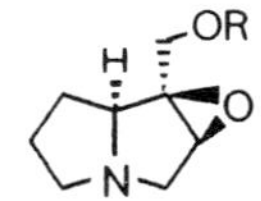

R	
H	347 1β,2β-epoxy-1α-hydroxymethyl-8α-pyrrolizidine/subulacine
CH_3	348 1β,2β-epoxy-1α-methoxymethyl-8α-pyrrolizidine

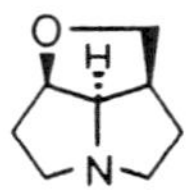

349 anhydroplatynecine

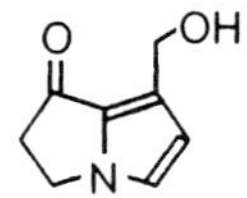

350 loroquine

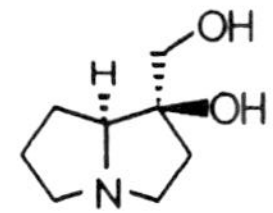

351 curassanecine

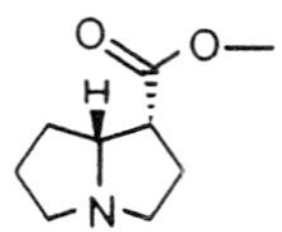

352 chysine A

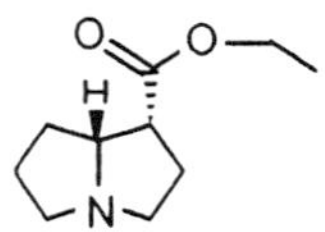

353 chysine B

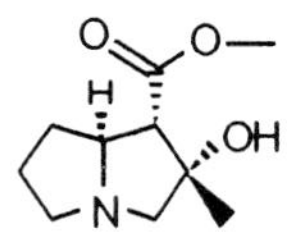

354 tussilagine

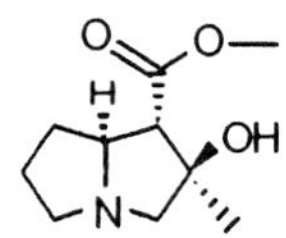

355 isotussilagine

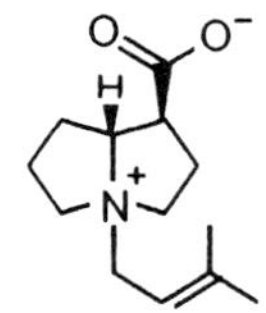

356 anodendrine

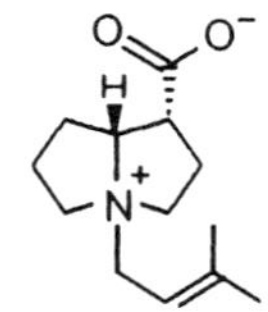

357 alloanodendrine

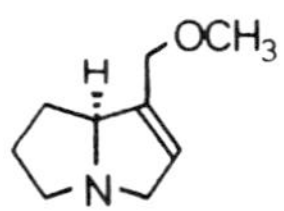

358 O[9]-methyl-(-)-supinidine

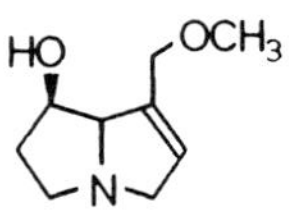

359 O[9]-methylretronecine

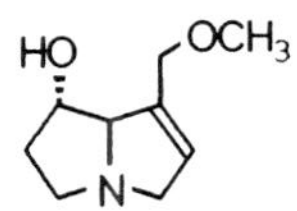

360 O[9]-methylheliotridine

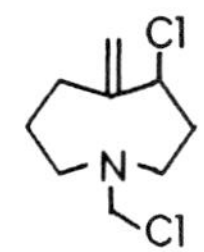

361 oxypterine

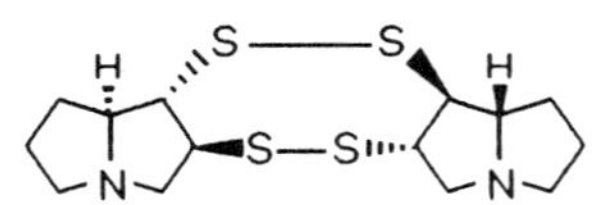

362 cassipurine

Appendix II Loline Type (L)

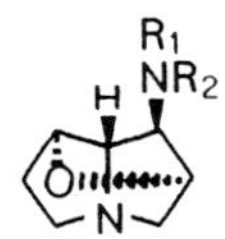

R_1	R_2	
H	H	363 norloline
H	CH_3	364 loline / festucine
CH_3	CH_3	365 N-methylloline
H	CHO	366 N-formylnorloline
H	Ac	367 N-acetylnorloline
CH_3	CHO	368 N-formylloline
CH_3	Ac	369 lolinine / N-acetylloline

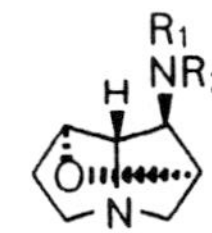

R_1	R_2	
H	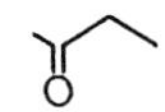	370 decorticasine
H	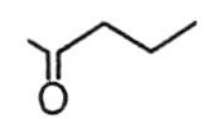	371 N-butyrylnorloline
H	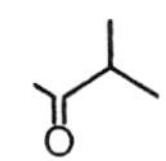	372 N-isobutyrylnorloline
H	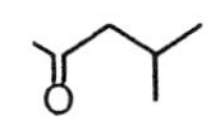	373 N-isovalerylnorloline

Chapter Five

Alkaloids from Cell Cultures of *Aspidosperma Quebracho-Blanco*

P. Obitz and J. Stöckigt
Institute of Pharmacy, Johannes Gutenberg-Universität Mainz, Staudinger Weg 5, 55099 Mainz, Germany

L. A. Mendonza
I.N.T.A., Centro de Investigaciones de Recursos Naturales, Castelar, Argentina

N. Aimi and S.-i. Sakai
Faculty of Pharmaceutical Sciences, Chiba University, 1-33 Yayoi-cho, Inage-ku, Chiba 260, Japan

CONTENTS

1. INTRODUCTION

Within the large plant family *Apocynaceae* an impressive number of plant genera has in the past attained important significance in therapeutical applications [1]. Some of the outstanding examples are *Catharanthus roseus* G. Don, which provides the highly valued dimeric indole alkaloids vinblastine and vincristine used for the treatment of cancer, *Rauwolfia serpentina* (L.) Benth., and other *Rauwolfia* species which are also of industrial importance for the isolation of the monoterpenoid indole alkaloids ajmalicine, yohimbine, ajmaline or reserpine which are all used in the therapy of various diseases. Most of the alkaloid-bearing *Apocynaceae* plants are, however, extremely important in traditional medicine. They are collected from wild forests rather than grown in plantations, making the drug supply rather uncertain. Therefore many attempts have been made to establish cell culture systems of these rare plants which would allow a continuous supply of plant material. One of the recent examples in this field involves the plant *Aspidosperma quebracho-blanco* Schlecht. from which we initiated a cell suspension culture and analyzed it for its pattern of monoterpenoid indole alkaloids.

2. SIGNIFICANCE OF *ASPIDOSPERMA QUEBRACHO-BLANCO* PLANTS

The medicinal plant *Aspidosperma quebracho-blanco* Schlecht. belongs to the *Apocynaceae* family, the subfamily *Plumerioideae* and to the tribe *Alstonieae* (syn. *Plumerieae).* Systematically the genus *Aspidosperma* is currently divided into about 80 species and, of these species, *A. quebracho-blanco* is one of the most important, occurring in South America, especially in Argentina, Brazil and Bolivia. *A. quebracho-blanco* is a tree growing up to more than 20 m high with yellowish-green leaves, 7 - 12 cm in length, and orange-brown bark. The tree is used as a rich source of high quality wood. Extracts of parts of this tree are also important in traditional medicine. The bark extract is employed industrially as a taste modifier for alcoholic and non-alcoholic drinks, for milk products and in food technology in general in concentrations of about 3×10^{-3} % [2]. For traditional medicine the stem bark is used in South America for liver disorders, feverish colds and as an analgetic. It is especially recommended for the treatment of malaria as a febrifuge [3], but is also used against asthma and bronchitis. Extracts of the *Quebracho* bark are frequently used as additives for phytopreparations employed as expectorants.

The tree is not cultivated in large plantations. Although seeds germinate easily the growth of the plant is rather slow. Its distribution is common in South-East Bolivia, Argentina, Southern Brazil and Chile at altitudes of 1500 to 1800 m.

The medicinal components of the tree-bark, consisting of monoterpenoid indole alkaloids, are isolated from the bark by conventional extraction techniques. Although the pharmacological activities of *Quebracho* tinctures and crude extracts have not been investigated in detail, the biological activities of single alkaloids were determined in the past, e. g. for the main alkaloidal components aspidospermine or quebrachidine from *Quebracho* leaves. The demand for this material on the market is of only little significance. A technological development of alkaloid production from this plant was never an important task, although the Eli Lilly company (Indianapolis) isolated significant amounts of some of the alkaloids for pharmacological investigations [4]. Therefore, the significance of this plant is still its application as a source for phytopreparations in traditional medicine. More than 25 monoterpenoid indole alkaloids have been isolated from the plant *Aspidosperma quebracho-blanco* Schlecht [5]. They belong to various structural classes: yohimban-type **1**, sarpagan-type **2**, aspidospermidine-type **3**, aspidospermatine-type **4**, eburnamenine-eburnamonine-type **5**, quebrachamine-type **6**, rhazinilam-type 7. Structures of some of them are shown below.

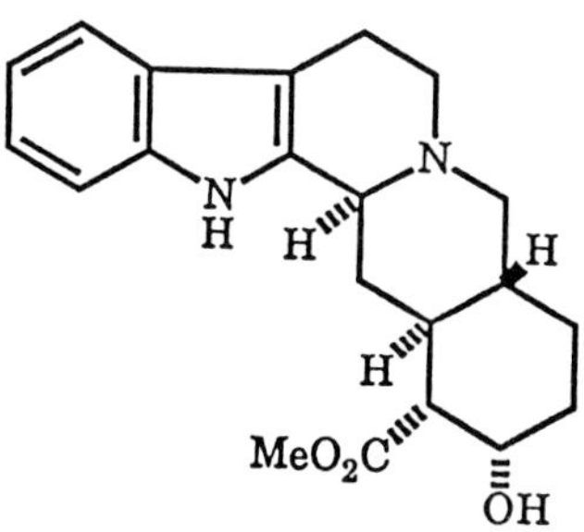

1
(Yohimbine)

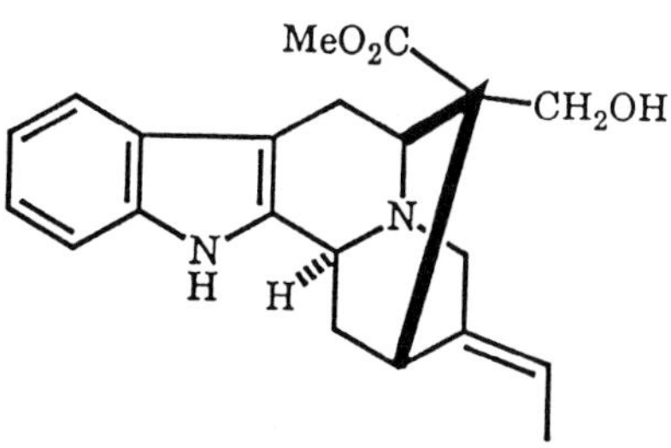

2
(Akuammidine)

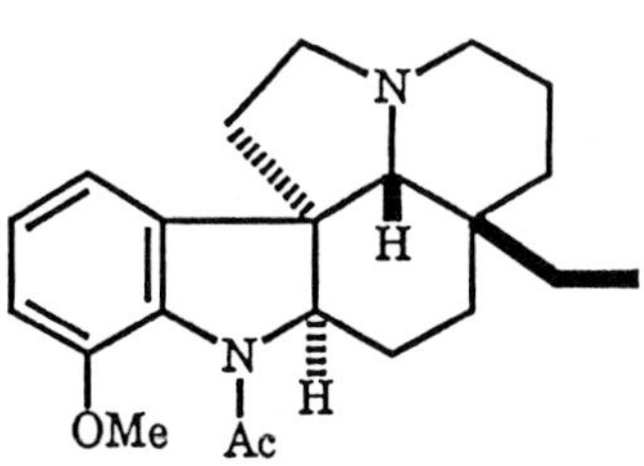

3
(Aspidospermine)

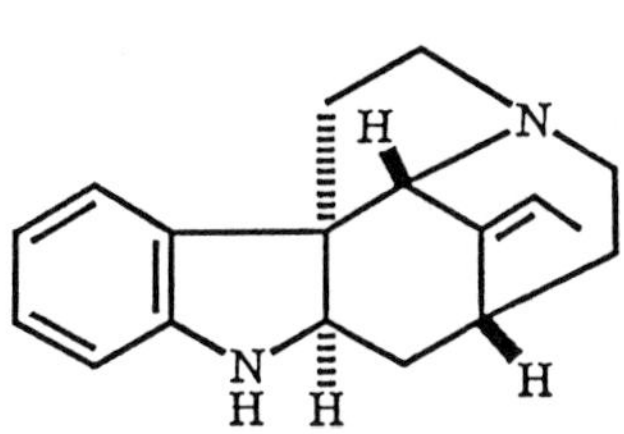

4
(Aspidospermatidine)

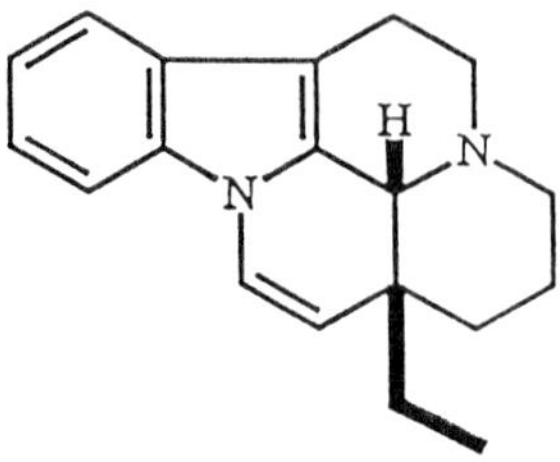

5

(Eburnamenine)

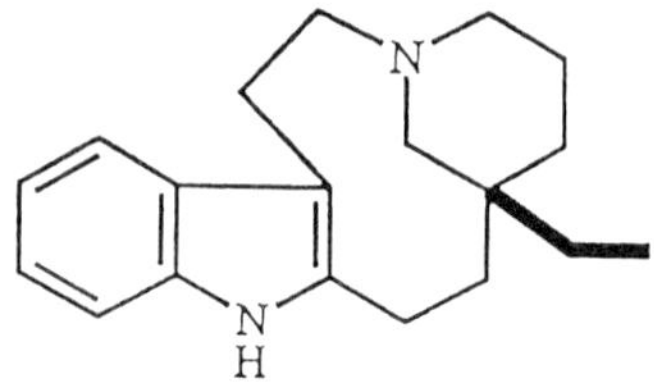

6

(Quebrachamine)

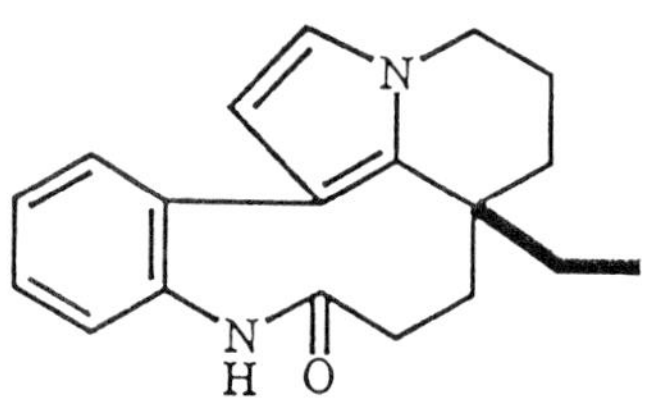

7

(Rhazinilam)

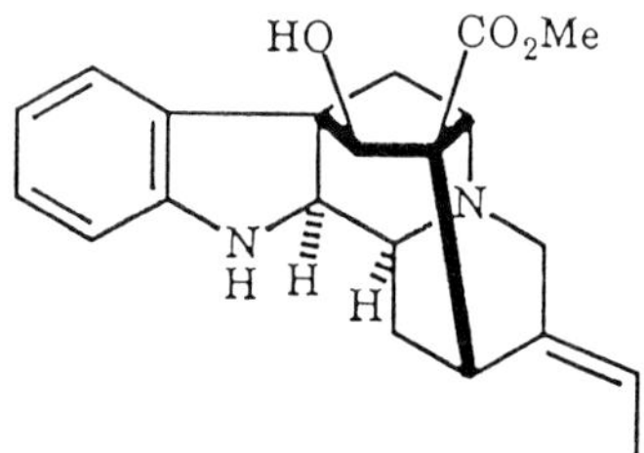

8

(Quebrachidine)

3. ESTABLISHMENT OF TISSUE AND CELL CULTURES

Although a large number of *Apocynaceae* plants have been successfully transformed into tissue and cell suspension cultures during recent years and have also been analysed for their alkaloid content, to our knowledge so far no culture system has been established from *A. quebracho-blanco*. In addition, no special biotechnological investigations on this *Aspidosperma* species have been published.

A. quebracho-blanco contains biogenetically interesting alkaloids, e. g. the ajmalan-type alkaloid quebrachidine **(8)** with a 2 α-configuration instead of a 2 ß-configuration. The latter is found in the *Rauwolfia* alkaloids of the ajmaline group.

We were interested in the establishment of a cell culture system, especially because the *Aspidosperma* plant growth is extremely slow. We also had in mind to perform protoplast fusion of cultivated cells of the genera *Aspidosperma* and *Rauwolfia*. Since the biosynthesis of some of the alkaloids of both genera seems to be closely related, it would be interesting to "join" both biochemical pathways as this could occur in the appropriate hybrid cells. Moreover, our aim was also to identify and compare the alkaloid pattern of both the intact plant and the cultivated cells.

The procedure to obtain the *A. quebracho-blanco* cell suspensions was straight-forward, starting with eight-day old seedlings. The hypocotyl of the seedlings was cut into small pieces (5 - 8 mm), surface sterilized with hypochlorite solution and placed on agar plates supplemented with Linsmaier a2nd Skook (LS)- or 4 x-media [6 - 8]. After 6 - 7 weeks at 25° C callus tissue started to grow and was separated from the differentiated tissue after 3 months and transferred to new solidified media. After three transfers, depending on the growth of the callus tissue, calli were then transferred each 5 weeks to fresh agar media for more than one year, resulting in a stable growth of the yellowish tissue. From these calli a cell suspension culture was established which generated on average 130 g fresh cell material (10.4 g dry weight) per 1 litre nutrition medium within 14 days. This cell material was harvested by suction filtration, frozen with liquid nitrogen and stored at - 25° C before extraction of the alkaloids.

4. ISOLATION AND STRUCTURE OF ALKALOIDS

We studied the alkaloidal constituents of the cultured cells of *A. quebracho-blanco* and isolated four nitrogen-containing constituents. Three were monoterpenoid indole alkaloids and one was a dioxopiperazine. All were novel compounds and one member, aspidochibine, was an alkaloid belonging to a new structural type of monoterpenoid indole alkaloid [9].

4.1 Extraction and Separation

Lyophilized cultured cells of *Aspidosperma quebracho-blanco* were exhaustively extracted with hot methanol. From 98.2 g of cell culture 25.4 g of extract was obtained. The extract was suspended in a mixture of 1 N HCl and ethyl acetate. After the insoluble material had been removed, the ethyl acetate layer and aqueous layer were separated. The aqueous layer was then basified with ammonia to pH 11 and the resulting solution was extracted with chloroform. Removal of the solvent from the chloroform layer gave 129 mg of residue (Fraction A). Removal of solvent from the ethyl acetate layer gave a residue of 1.23 g (Fraction B).

Fraction A was submitted to silica gel flash column chromatography. The fraction eluted with chloroform was then submitted to medium pressure liquid chromatography on normal phase adsorbent (SiO_2). Purification by use of preparative liquid chromatography and preparative TLC gave three compounds having the respective tentative designations, AQC-2 (**9**) [9] (1.4 mg), AQC-3 (**10**) [9] (1.3 mg), and AQC-4 (**11**) [10] (2.2 mg).

Fraction B was fractionated by use of flash chromatography and preparative scale liquid chromatography on silica gel. Repeated fractionation combined with preparative TLC purification as necessary gave a dioxopiperazine (**12**) [10], diethyl malate and di *n*-propyl malate.

4.2 Characterization of the Cellular Metabolites

4.2.1 Aspidochibine

AQC-2, now named aspidochibine (**9**), has the molecular formula $C_{19}H_{22}N_2O_2$ and its UV spectrum was typical of an indole nucleus indicating the absence of any conjugated group or additional chromophore. The ^{13}C-NMR spectrum disclosed the presence of two carbonyl carbons, an amide carbonyl (C-3) at δ 180.0 and a lactone carbonyl (C-5) at δ 175.1. The ^{1}H-NMR spectra indicated the presence of the following partial structures.

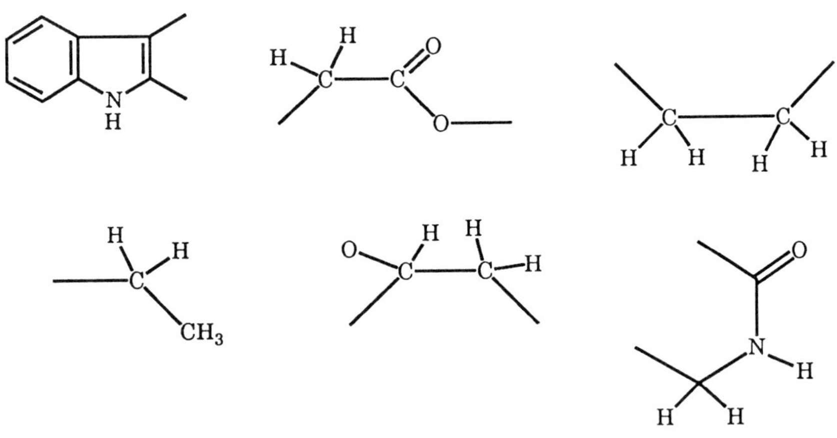

Detailed studies by use of ^{1}H-and ^{13}C-NMR measurements (1D and 2D) enabled us to deduce the structure shown below for aspidochibine (**9**). The conformation of this molecule was then studied by means of ^{1}H-NMR. The amide NH is coupled with 21-Hα with the coupling constant of 11.0 Hz, whereas a quite small coupling constant (J=2.2 Hz) was observed for coupling to 21-Hß. This indicated the dihedral angle between N-H and 21-Hß is close to 90°. A NOESY spectrum showed an NOE between 6-Hα and 9-H, between N*a*-H and

17-Hα, and between 15-Hß and 16-Hß. From these observations and other information the molecular structure of aspidochibine (**9**) was deduced as shown below.

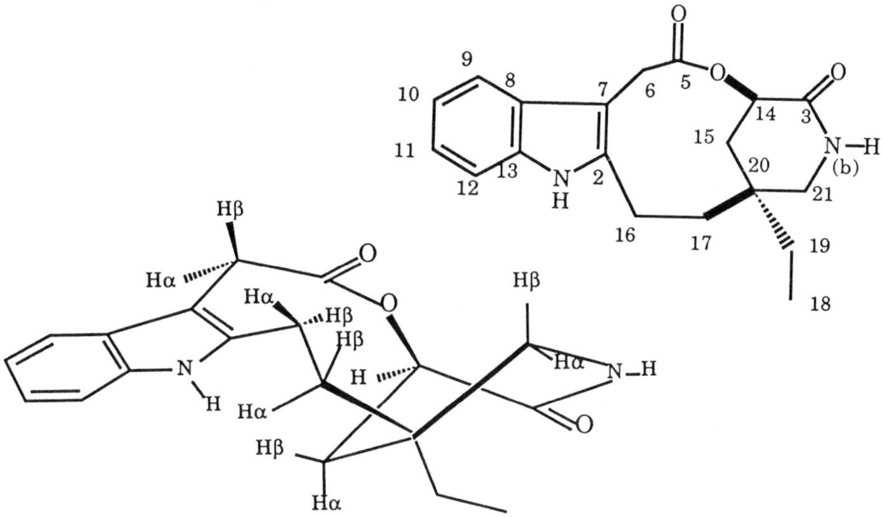

9

The hitherto unknown skeletal frame work of aspidochibine (**9**) could be formulated by reference to the quebrachamine-type molecule as shown below. The absolute configuration is assumed to be as depicted and is based on the known absolute configuration of quebrachamine and aspidospermine which coexist in the same plant.

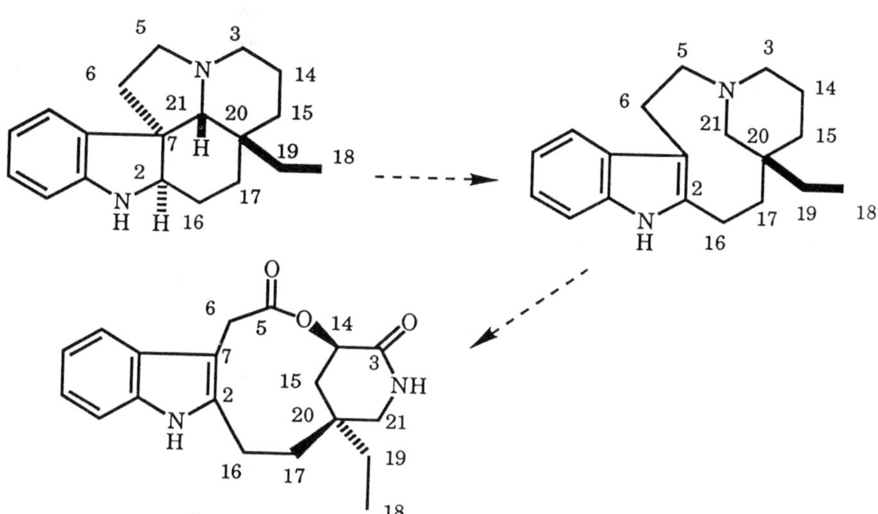

9

4.2.2 3-Oxo-14,15-dehydrorhazinilam

AQC-3 (**10**) has the molecular formula $C_{19}H_{18}N_2O_2$, indicating a high degree of unsaturation of this molecule. The UV spectrum, however, showed only strong end absorption leading to a shoulder at 224 nm (log ε 4.32) and a broad maximum with modest intensity (log ε 3.83) at 264 nm. Close similarity of this UV spectrum to that of rhazinilam [11] was soon noticed. In the molecule of rhazinilam (7) the pyrrole ring, benzene nucleus, and (Ar)-NH-CO-group are held perpendicular to each other in turn. As a result otherwise expected conjugations are not possible, as was evidenced by the lack of characteristic absorption maxima in the UV spectra of rhazinilam (7) and also of AQC-3 (**10**).

The ^{1}H-NMR spectrum showed the presence of an ethyl group on a quaternary carbon (δ 0.71 (3H, 18-H_3), δ 1.29 (19-Hα), and δ 1.81 (19-Hß)) and -C(16)H2-C(17)H2- moiety (δ 1.89 (17-Hα), δ 2.05-2.11 (2H, 17-Hß, 16-Hα), δ 2.49 (16-Hß)).These are all of the protons on sp^3 carbons. Protons on an α,ß-unsaturated amide moiety were observed at δ 6.23 (d, J=9.9 Hz, 14-H) and δ 6.56 (d, J=9.9 Hz, 15-H). Two adjacent protons, 5-H and 6-H, on a pyrrole part were observed at δ 6.03 (d, J=3.3 Hz, 6-H) and δ 7.53 (d, J=3.3 Hz, 5-H). In the spectrum of rhazinilam the chemical shift of 5-H is at δ 6.59. The observed low field shift of 5-H in AQC-3 denoted the presence of a carbonyl group at 3-C. This was supported by the ^{13}C-NMR where the 3-C amide carbonyl carbon was observed at δ 159.6. The above and other observations showed the structure of AQC-3 to be 3-oxo-14,15-dehydrorhazinilam (**10**). Rhazinilam (7) itself was found in *A. quebracho-blanco*, but in this study on the constituents of the cultured cells we have not found it.

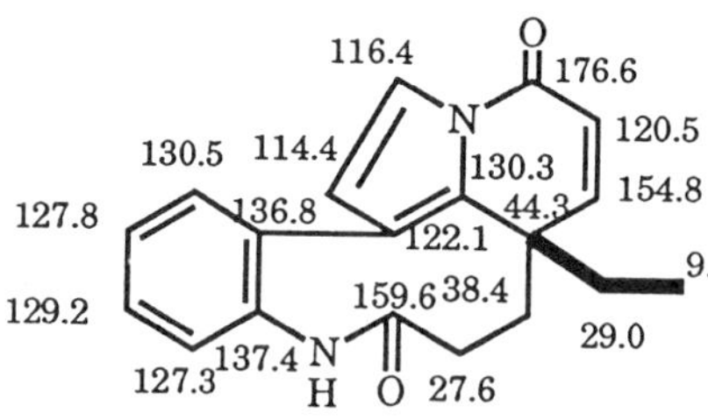

10

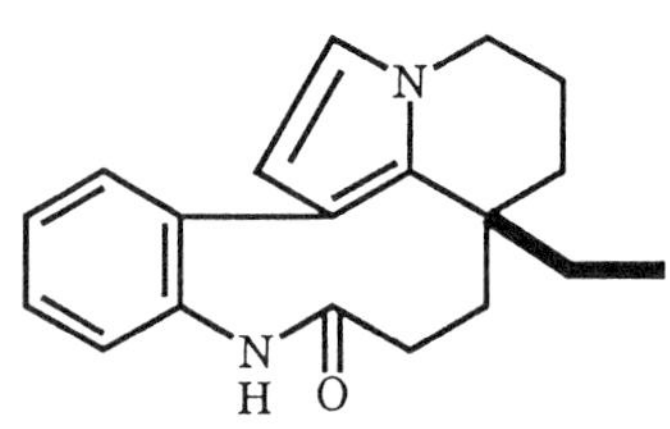

7

4.2.3 11-Hydroxytubotaiwine

The third indolic metabolite AQC-4 (**11**) was found to have the molecular formula, $C_{20}H_{24}N_2O_3$, the high resolution MS showing a pseudo molecular ion having the correct composition at M+1. The UV spectrum clearly revealed that AQC-4 belongs to a group of monoterpenoid indole alkaloids having the ß-anilinoacrylic ester group. Inspection of its EI-MS indicated that this compound was a new member of the tubotaiwine group having an additional oxygen atom to the molecule of tubotaiwine.

To clarify the position of the substituent, comparison of ^{13}C-NMR spectra was made between tubotaiwine and AQC-4 (**11**). The chemical shifts due to the carbons in the non-aromatic part of the molecule are quite similar, thus clearly demonstrating that the additional oxygen is on the benzene ring in a phenolic hydroxyl group. This fact was also supported by the coupling pattern of the aromatic protons in AQC-4 (**11**). To deduce the position of the substituent 10 and 11-substitution calculations of the chemical shifts of aromatic carbons based on the empirical method were made. We found that the observed values agreed well with those expected for 11-substitution. Thus AQC-4 was concluded to be 11-hydroxytubotaiwine (**11**). This is the first time that this compound had been found in nature.

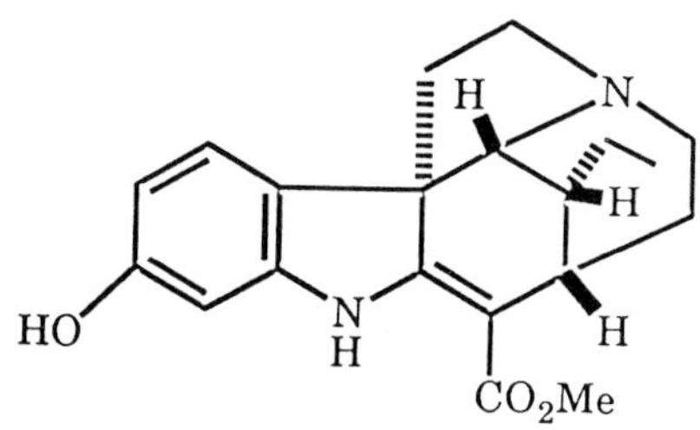

11

4.2.4 Metabolites Closely Related to Primary Metabolism

Dioxopiperazine (**12**). During our investigation on the alkaloids we found several other classes of compounds: AQC-5 has the molecular formula $C_9H_{12}N_2O_2$ as was evidenced by the HR-MS. The ^{1}H-NMR spectrum of this compound showed an amide proton exchangeable to D_2O at δ 7.45 and the protons due to an ethylidene group (δ 1.75 (3H, d, J=7.4 Hz) and δ 6.13 (1H, q, J=7.4 Hz)). A series of seven protons on the proline part was observed to denote the whole structure as shown. These and other spectral data suggested the structure shown below. The ^{13}C-chemical shifts shown below also strongly supported this assumption, although due to scarcity of the sample the signal intensities of the two amide carbons and a quaternary olefinic carbon were under the limit of detection. This molecule is considered to have been formed through the condensation between two amino acids, most likely l-proline and threonine, followed by elimination of one molecule of water. As far as we know this molecule has not been found or synthesized yet. Investigation of a potential synthesis is underway. It is not clear whether this molecule was formed as a real metabolite of the plant cells or as an artificial compound resulting from chemical reaction between additives in the culture media, during the process of heat sterilization or during the process of extraction. An attempt was made to detect the presence of this compound in the control experimental material without plant cells, but we were unable to detect it under these conditions.

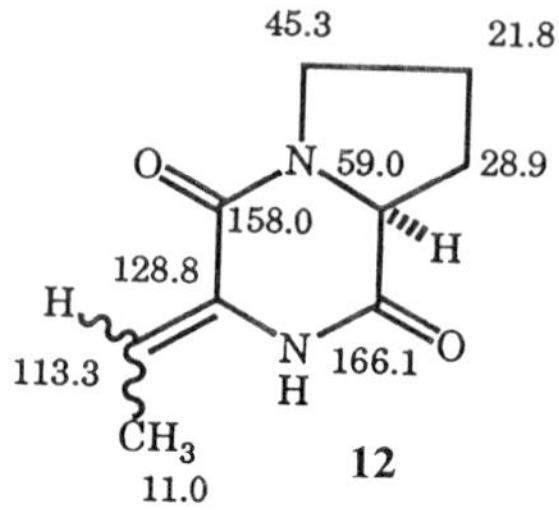

12

Diethyl Malate and Di n-Propyl Malate. From the neutral fraction (Fraction B) of the cell culture extract two known compounds diethyl malate and di *n*-propyl malate were found. It is not clear whether they were formed in the cultured cells or artificially produced in the culture medium while cultivation proceeded or during the usual procedures such as sterilization, harvesting and isolation.

5. CONCLUSIONS AND FUTURE ASPECTS

In summary, the cell suspension culture, which has been generated from seedlings of *A. quebracho-blanco* proved to be an interesting source of novel alkaloids. As well as the newly detected dioxopiperazine, three indole alkoloids have been identified, which have not previously been found in nature. The here described cell culture system demonstrates that undifferentiated plant material such as callus or cell cultures might be an excellent tool in natural product chemistry for detecting novel secondary compounds, including those with completely novel skeletons. It seems that the unorganised stage of plant cells provides synthesis conditions for a relatively large range of new constituents. In addition to some of our recent examples in the field of monoterpenoid indole alkaloids, like the voafrines [12] and raumaclines [13, 14], a summary of new plant constituents from cell and tissue cultures [15, 16] clearly demonstrates their ability to generate a wide variety of new natural products.

In the currently discussed case of a cell suspension of *Aspidosperma quebracho-blanco* the amount of alkaloids formed under the *in vitro* conditions was extremely low, not exceeding one milligram of alkaloid from one litre of cell suspension. We therefore believe that this culture system at this stage does not have any commercial interest, but it might eventually be a good system for investigating the biosynthesis of the newly described alkaloids using tracer techniques and enzymatic studies.

Acknowledgements.- Financial support from the "Deutsche Forschungsgemeinschaft" (Bonn Bad-Godesberg) and from the "Fonds der Chemischen Industrie" (Frankfurt/Main) is gratefully acknowledged. We thank Prof. W. E. Court for linguistic help. Our thanks are also due to the Ministry of Education, Science and Culture, Japan (International Scientific Research Program, joint Research; N. 04044035) for generous support.

References

1 Neuss N (1980) in: Indole and Biogenetically Related Alkaloids (Phillipson J D, and Zenk M H, eds): Academic Press, London - New York - Toronto - Sydney - San Francisco: 293.

2 Leung A Y (1980) in: Encyclopedia of Common Natural Ingredients, Wiley-Interscience Publication, John Wiley & Sons, New York - Chichester - Brisbane - Toronto.

3 Dominguez I A (1932) Ref farm (Buenos Aires) **73**: 82 (CA **26**: 4102).

4 Neuss N, personal communication.

5 Lyon R L, Fong H H S, Farnsworth N R, and Svoboda G H (1973): Biological and Phytochemical Evaluation of Plants XI: Isolation of Aspidospermine, Quebrachidine, Rhazinilam, (-)-Pyridfolidine, and Akuammidine from *Aspidosperma quebracho-blanco* (Apocynaceae). J Pharm Sci **62**: 218 and references cited therein.

6 Linsmaier E M and Skoog F (1965): Physiol Plant **18**: 110.

7 Gamborg O L, Müller R A, and Ojima K (1968): Exp Cell Res **50**: 152.

8 Ulbrich B and Zenk M H (1979): Phytochem **18**: 929.

9 Aimi N, Uchida N, Ohya N, Hosokawa H, Takayama H, Sakai S-i, Mendoza L, Polz L, and Stöckigt J (1991): Novel Indole Alkaloids from Cell Suspension Cultures of *Aspidosperma quebracho-blanco* Schlecht. Tetrahedron Lett **32**: 4949.

10 Manuscript under preparation.

11 a) Banerij A, Majumder P L, and Chatterjee A (1970): Occurence of Geissoschizine and other minor biogenetically Related Alkaloids in Rhazya stricta. Phytochem **9**: 1491.
b) Abraham D J, Rossenstein R D, Lyon R L, and Fong H H S (1972): The Structure Elucidation of Rhazinilam, a New Class of Alkaloid from the Apocynaceae, by X-ray Analysis. Tetrahedron Lett **10**: 909
c) De Silva K T, Ratcliff A H, Smith G F, and Smith G N (1972): Rhazinilam, A Neutral Alkaloid Artefact from Rhazya stricta Decaisne. Tetrahedron Lett **10**: 913

12 Stöckigt J, Pawelka K H, Tanahashi T, Danieli B, and Hull W E (1983): Voafrine A and B, the First Dimeric Indole Alkaloids Isolated from Plant Cell Cultures. Helv Chim Acta **66**: 2525.

13 Polz L, Stöckigt J, Takayama H, Uchida N, Aimi N, and Sakai S-i (1990): Biotransformation of Ajmaline in Plant Cell Cultures of *Rauwolfia serpentina* Benth. Tetrahedron Lett **46**: 6693.

14 Endreß S, Takayama H, Suda S, Kitajima M, Aimi N, Sakai S-i, and Stöckigt J (1993): Alkaloids from *Rauwolfia Serpentina* Cell Cultures Treated with Ajmaline. Phytochem **32**: 725.

15 Ruyter C M, and Stöckigt J (1989): Novel Natural Products from Plant Cell and Tissue Culture - an Update. DIT Fachz Lab **4**: 283.

16 Kreis W, Reinhard E (1989): The Production of Secondary Metabolites by Plant Cells Cultivated in Bioreactors. Planta Med **55**: 409.

Chapter Six

Fumonisins

Richard G. Powell and Ronald D. Plattner
Bioactive Constituents Research,
National Center for Agricultural Utilization Research,
Agricultural Research Service,
U.S. Department of Agriculture,
1815 N. University Street,
Peoria, Illinois 61604
U.S.A.

CONTENTS

1. INTRODUCTION

Fumonisins were first described by South African researchers in 1988 after a decade of work related to the high incidence of esophageal cancer in the Transkei region of southern Africa [1-3]. The discovery of fumonisins, their implication in animal toxicoses [4, 5] and their reported carcinogenicity [6] has stimulated widespread interest in this new group of mycotoxins. The fumonisins are alkaloids produced by *Fusarium moniliforme* and closely related species. The importance of fumonisins is magnified because the fungi that produce them are found throughout the world and *F. moniliforme* is one of the most prevalent fungi associated with human and animal dietary staples such as corn (*Zea mays*) [7]. Contamination of foodstuffs with this fungus has been suspected of being involved in human and animal diseases since its description in the early 1900's.

Species of *Fusarium* produce a number of secondary metabolites that are toxic to plants and animals. The trichothecenes, a family of sesquiterpene compounds that are potent inhibitors of peptidyl transferase and that are acutely toxic to metabolically active cells, are probably the best known of these mycotoxins. *F. moniliforme* and closely related species in section *Liseola* do not produce trichothecenes but do produce several toxic or biologically active secondary metabolites in addition to fumonisins. These additional mycotoxins include fusaric acid, the fusarins, moniliformin, beauvericin, gibberellins and their diterpene precursors, and an array of naphthoquinone and azaanthraquinone pigments that are acutely phytotoxic but of undetermined mammalian toxicity. In laboratory cultures of many toxigenic strains of *F. moniliforme*, or of closely related species that were isolated because of their association with animal or human toxicoses, several highly toxic components can be produced simultaneously at high levels. When cultures of toxigenic isolates of this fungus were fed to various species of animals a variety of lesions were induced including cirrhosis, nodular hyperplasia of the liver in rats, nephrosis and hepatosis in sheep and acute heart failure in baboons [8, 9]. *F. moniliforme* cultures are known to cause equine leucoencephalomalacia (ELEM), a fatal disease of horses in which liquefactive lesions develop in the brain, and porcine pulmonary edema (PPE) in swine. Fumonisin B_1 has produced disease symptoms similar to ELEM and PPE under experimental conditions [4, 5]. It has also been shown to be a potent liver toxin [6, 10] and is suspected of being a carcinogen in rats. The role of other toxic components of *Fusarium* culture materials fed in animal toxicology studies has not yet been rigorously determined.

The 1989 corn crop in the United States was associated with unusually high numbers of cases of ELEM and corn associated PPE. These widespread outbreaks occurred just as analytical methods were becoming available to determine the presence of fumonisins in corn and corn screenings. Analyses of corn and corn screenings associated with disease outbreaks showed that higher levels of fumonisins were usually present in these materials than in control samples [11, 12], but also revealed that fumonisins were often present at detectable levels in apparently sound corn. The discovery of a new group of alkaloid mycotoxins in corn, and concern over the potential for their occurrence in food and feed products based on corn, has led to a veritable explosion of research on fumonisins since 1988. This review summarizes the literature on fumonisins, including some structurally related alkaloids from other sources, through June 1994.

2. FUMONISIN STRUCTURES AND OCCURRENCE

2.1 Structures of the Fumonisins

Structure elucidation of four unusual alkaloids, fumonisins, B_1 (1), B_2 (2), A_1 (3), A_2 (4), was first reported by Bezuidenhout, et al. [2]. Their isolations were guided by a bioassay based on short-term cancer initiation-promotion in rat liver [1], and fumonisin B_1 (1) was found to be the predominate fumonisin. Cultures of *F. moniliforme* Sheldon (MRC 826) on corn were used to isolate the fumonisins and **1** was obtained in excess of 2 g per kg of culture material (>2000 ppm). Laurent et al. [13], independently characterized a compound called macrofusin from a strain of *F. moniliforme* isolated from corn associated with equine leucoencephalomalacia in New Caledonia. Macrofusin and fumonisin B_1 were found to be identical [13]. Initial reports of the isolation and characterization of fumonisin B_3 (5) were by Cawood et al. [14] and by Plattner et al. [15]. Cawood et al. also reported the occurrence of fumonisin B_4 (6) [14].

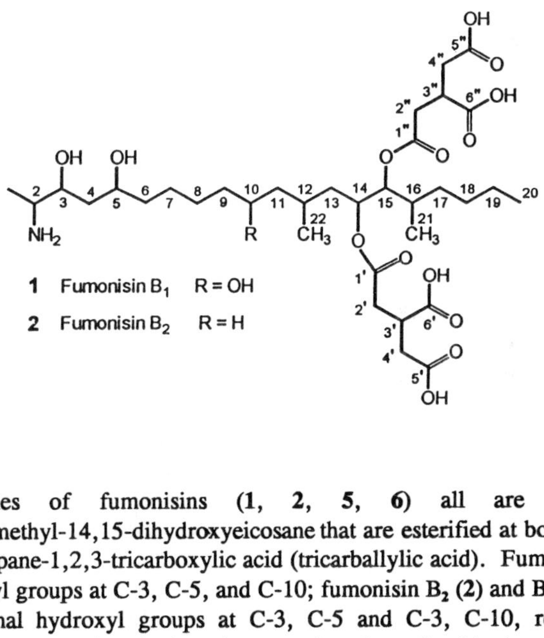

1 Fumonisin B_1 R = OH

2 Fumonisin B_2 R = H

The B series of fumonisins (**1**, **2**, **5**, **6**) all are derivatives of 2-amino-12,16-dimethyl-14,15-dihydroxyeicosane that are esterified at both the 14 and 15 positions with propane-1,2,3-tricarboxylic acid (tricarballylic acid). Fumonisin B_1 (1) has additional hydroxyl groups at C-3, C-5, and C-10; fumonisin B_2 (2) and B_3 (5) are isomers that have additional hydroxyl groups at C-3, C-5 and C-3, C-10, respectively; and fumonisin B_4 (6) has one less hydroxyl group than **2** or **3** with the only unesterified hydroxyl group being at C-3. Fumonisins A_1 (3) and A_2 (4) are the *N*-acetyl derivatives of **1** and **2**, respectively. Fumonisin C_1 (7) has also been described [16], and it differs from **1** only in that **7** lacks the C-1 methyl group characteristic of the other fumonisins.

The two *N*-acetylated fumonisins (**3** and **4**) have been isolated at low levels under chromatographic conditions on silica where acetic acid was used in the elution solvent [14] and it is doubtful that these derivatives occur naturally. Partially methylated fumonisins are observed after treatment of culture material with acidic methanol [14] and partial and total hydrolysis products of the fumonisins, in which either one or both of the tricarballylic acid groups are absent, have also been observed (R. D. Plattner, unpublished and [17]). Fumonisin *N*-acyl derivatives such as **3** and **4** are found only in minor amounts. These compounds, hydrolysis products, and methyl esters are possible isolation artifacts.

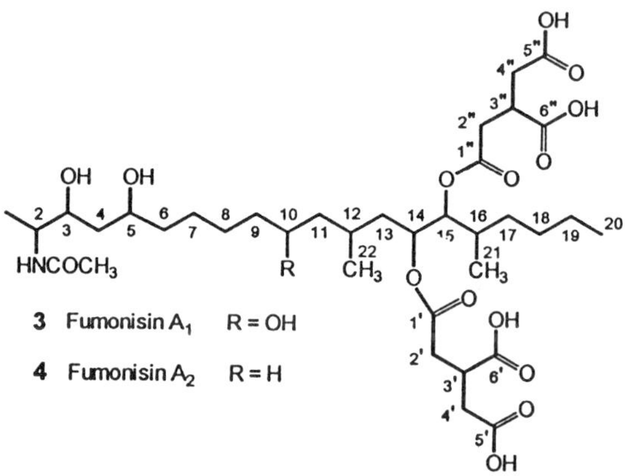

3 Fumonisin A_1 R = OH

4 Fumonisin A_2 R = H

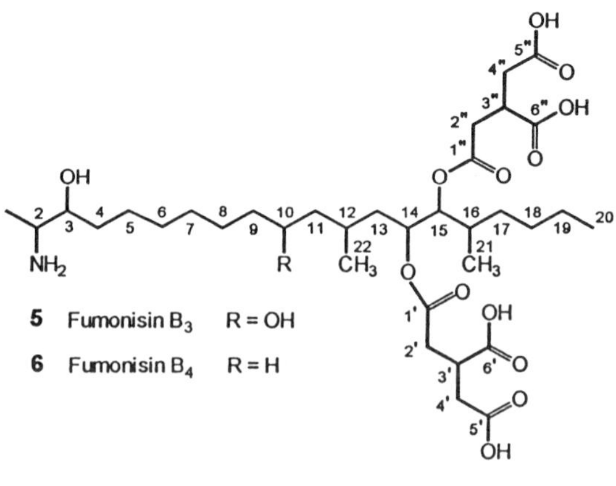

5 Fumonisin B_3 R = OH

6 Fumonisin B_4 R = H

2.2 Fumonisins in *Fusarium moniliforme* Contaminated Food and Feed

Fumonisins **1** and **2** were reported at levels of 150 and 20 mg/kg, respectively, in naturally contaminated corn samples associated with Illinois field cases of ELEM [18]. Analyses of 14 feed samples associated with similar ELEM outbreaks in North and South Carolina gave concentrations of **1** ranging from 1.3-27.0 mg/kg and of **2** ranging from 0.1-12.6 mg/kg [19]. Other reports include **1** in *F. moniliforme* contaminated white corn (125 mg/kg), contaminated corn screenings (236 mg/kg) and laboratory corn cultures (1,557 mg/kg) [20], in feeds from Brazil (**1**, 0.2-38.5 mg/kg; **2**, 0.1-12.0 mg/kg) [21], and up to 9 mg/kg in forage grass in New Zealand [22]. Concentrations of **1** from 175 whole

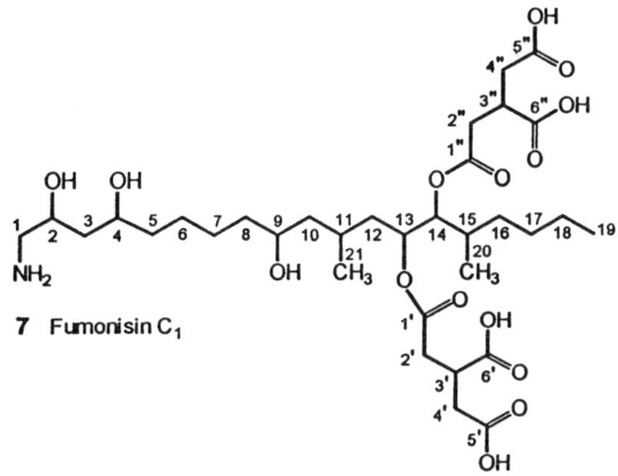

7 Fumonisin C_1

corn samples harvested in the state of Iowa during 1988, 1989, 1990, and 1991 [23] ranged from 0-38 mg/kg with average concentrations between 2.5 and 3.5 mg/kg.

Corn-based human foodstuffs (cornmeal, corn grits, cornflakes, and alkali treated corn) have been examined from commercial sources in Canada, Egypt, Peru, South Africa, and the United States [24]. Fumonisins were detected in the majority of samples tested and their concentrations varied from 0 to about 3 mg/kg (total). Analysis of corn-based products from the Swiss market (corn grits, cornflakes, cornmeal, sweet corn, and poultry feed) gave 0-0.7 mg/kg total **1** and 0-0.2 mg/kg total **2** [25]. Corn from an area of South Africa having a high rate of esophageal cancer, and where corn is a human dietary staple, contained up to 118 mg/kg of **1** and 23 mg/kg of **2** [26]. Fumonisin B_1 can generally be detected in apparently healthy corn, with no symptoms of fungal infection, at levels between 0.05-2.00 mg/kg. Likewise, *F. moniliforme* can usually be isolated from healthy corn. The toxicological significance of this low level of fumonisin contamination in foods and feeds is not known; however, it is clear that fumonisins are present in the diets of animals and humans throughout the world.

3. ANALYSIS, ELUCIDATION OF STRUCTURES AND REACTIONS

3.1 Chromatography

Fumonisins were first isolated from corn cultures by chromatographic separation using silica columns and their separations were guided by a short term bioassay that measured gamma glutamyltranspeptidase positive foci or nodules in rat liver cells [1]. The active fraction contained **1**, which was much more water soluble than lipid soluble, and **1** was not extracted from corn cultures unless water was present in the extraction solvent. Undoubtedly the water soluble nature of fumonisins delayed their discovery because nearly all other toxic compounds isolated from *Fusarium* species are lipid soluble. The water solubility of **1** varies widely with pH (Plattner, unpublished observation) as a result of protonation and deprotonation of the free primary amine and multiple free carboxyl groups. At physiological pH, fumonisin B_1 is expected to occur as an ionic species. Most methods for isolation and analysis for fumonisins, or their derivatives, rely on chromatographic separations on silica or on C_{18} reverse phase media.

3.1.1 Thin Layer Chromatography

Fumonisins can easily be detected at microgram levels by thin layer chromatography (TLC) on C_{18} reverse phase or silica plates. C_{18} plates can be developed in methanol/water (3/1) or acetonitrile/water solvent systems. The fumonisins appear as light to dark purple spots after spraying with *p*-anisaldehyde in methanol and sulfuric acid followed by heating (1). TLC methods for detection of fumonisins in corn and corn products have been reported [27, 28, 29]. Fumonisins can also be detected by color or fluorescence reactions with the free primary amine using reagents such as ninhydrin or fluorescamine [28]. Detection limits are reported to be in the low ng (50-100 ng) level; however, because of interferences caused by other components of the corn extract matrix, an overall detection limit of 1 μg/kg or better requires a sample clean-up step prior to the TLC analysis.

3.1.2 Analytical HPLC

Several methods based on HPLC for the analysis of fumonisins have been reported [1, 30, 31, 32, 33, 34]. Because fumonisins do not possess a UV absorbing or fluorescent chromophore, detection methods involve derivatization of the free amine to form a UV absorbing or fluorescent product. Two HPLC based methods were reported by Sydenham et al. [30]. The first involved derivatization of an extract with maleic anhydride [35] to form a maleyl derivative of the primary amine of fumonisin B_1 which was then detected by UV absorption of the eluate at 250 nm. The second, a more sensitive method, was to treat the extract with fluorescamine [36] which yields strongly fluorescent derivatives of primary amines. This method has the advantage (or disadvantage depending on perspective) that fumonisins each form two well resolved peaks presumably from formation of acid alcohol and lactone derivatives [30]. This method was modified and used by Ross et al. [36] to measure fumonisin B_1 and B_2 concentrations in corn and corn screenings associated with animal problems. The method was collaboratively studied in four laboratories [20].

Shephard et al. [31] and Thiel et al. [37] reported an HPLC method which uses *o*-pthaldialdehyde (OPA) to form strongly fluorescent derivatives with primary amines. The relative instability of OPA derivatives has been discussed and several other fluorescent derivatives have been reported [32, 34]. The OPA method has also been collaboratively studied [37], has been modified slightly by several groups [12, 33] and is widely used as the method of choice for analysis of fumonisins in corn and corn products. Typical HPLC separations of fumonisins and their hydrolysis products are shown in Figure 1.

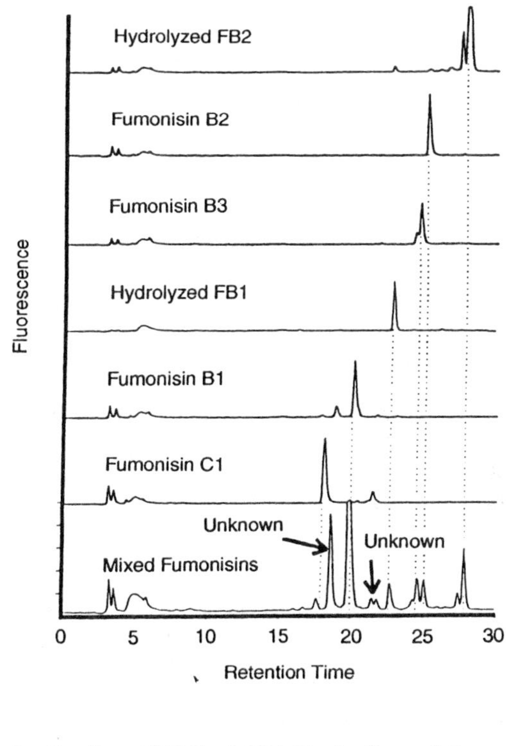

Figure 1. Gradient HPLC of OPA Derivatives of Fumonisins

There are two main concerns with the HPLC methods. First, they all require preparation of fumonisin derivatives that are fluorescent or UV adsorbing; thus, the methods are susceptible to interference from other molecules with functional groups that react to form similarly absorbing or fluorescent derivatives. To achieve low level detection ($\mu g/g$ or lower) these methods require a clean-up step. Secondly, all of the methods reported for preparation of derivatives with the primary amine group in fumonisins are unable to detect fumonisin related compounds that are amides, such as **3** and **4**, or that are secondary amines. There are basically three clean-up strategies. Ross and coworkers [36] use a C_{18} clean-up column while Shephard's method [31] uses strong-anion exchange (SAX) columns. The SAX approach yields much cleaner samples for derivatization, but recoveries following the published procedure can be erratic for an inexperienced analyst and careful work with the inclusion of blank and spiked samples to validate recoveries are an important part of the analytical protocol.

An alternative proposed clean-up method utilizes columns with antibodies to fumonisins [38, 39] to bind the fumonisins. The fumonisins are then eluted, derivatized and analyzed using conventional procedures. Fumonisin antibody columns are available commercially. The columns have sufficient antibody to bind only about one microgram of fumonisin B_1 (and anything that cross reacts with the antibody). Great care must be taken when using these columns to assure that sufficiently small sample aliquots are used so that none of the desired compounds elute prematurely from the column due to saturation of the binding sites.

3.1.3 Preparative HPLC

Reference samples of fumonisin B_1 (**1**) and fumonisin B_2 (**2**) are available commercially while the more minor fumonisins are not. The various fumonisins can be purified from cultures of *F. moniliforme* containing them by extraction with methanol/water, or acetonitrile/water, followed by preparative chromatography on combinations of silica and C_{18} columns. Cawood et al. [14] describe a purification protocol that can be used to obtain pure fumonisins B_1, B_2, and B_3. Alternative approaches have been reported by Vesonder et al. [40] and Nelson et al. [41]. Fumonisins are easily produced at levels in excess of 2 g/kg in corn cultures. Most isolated strains of *Fusarium* are primarily fumonisin B_1 producers (typically about 70% of total fumonisins) with lesser amounts of fumonisin B_2 and/or fumonisin B_3. However, purification of fumonisin B_2, B_3, or B_4 can be greatly facilitated by selecting an isolate that produces predominately fumonisin B_2 and B_4, or fumonisin B_3 and B_4 [36, 42].

3.2 Mass Spectrometry

Mass Spectrometry (MS) has long been an important tool both for structural characterization and for analytical determination of components in complex mixtures. MS techniques have contributed heavily both to the structure determination of fumonisins and to analytical methods developed to measure fumonisins. The traditional ionization method for MS has been electron ionization (EIMS). EIMS has found widespread use in studies of natural products because the technique yields reproducible mass spectra for many organic molecules with predictable fragments that can be related to structure. Because EIMS is highly energetic, extensive fragmentation and rearrangements often occur with resultant loss of structural information. A second disadvantage of EIMS is the requirement that the sample have some volatility.

Chemical ionization mass spectrometry (CIMS), developed in the late 1960s, relies on a higher source pressure in which ionized molecules of a reagent gas collide with and impart charge to sample molecules. This process results in both positively and negatively charged molecular and fragment ions. CIMS, though less commonly used than EIMS, has become an essential tool for the natural product chemist. Like any chemical reaction the process is concentration dependent and is affected by the chemical nature of both the reagent gas ion and the sample molecule. Protonated molecules $(MH)^+$ are usually observed in the positive ion mode and prominent negative ions include $(M)^-$ by electron capture or $(M\text{-}H)^-$ by hydride abstraction. The chemistry of CIMS has been described in a book by Harrison [43]. While these two ionization methods have contributed greatly to the characterization and analysis of many important molecules, the necessity that analyses be conducted in the gaseous state has been a limitation for nonvolatile compounds. This volatility requirement has largely been overcome by the introduction of liquid secondary ion mass spectrometry (liquid SIMS) and fast atom bombardment (FAB) techniques developed in the early 1980's. FAB and liquid SIMS methods produce essentially identical spectra for nonvolatile compounds. These techniques require that the samples be dissolved in a liquid matrix (often glycerol) and then bombarded with a high energy ion (liquid-SIMS) or atom beam (FAB). Compounds that are not ordinarily volatile are desorbed from the matrix as charged ions using these techniques. Mechanisms involved in the desorption process have been studied at length and have not been totally agreed upon, but these techniques have revolutionized MS of nonvolatile organic materials.

3.2.1. Identification and characterization of fumonisins

The molecular weights of *N*-acetyl fumonisins, A_1 (**3**) and A_2 (**4**), were determined from $(M+H)^+$ signals at m/z 820 and m/z 804, respectively, in liquid-SIMS) spectra of the isolated components. Smaller signals from sodiated molecular ions were observed at m/z 842 $(3+Na)^+$, and 826 $(4+Na)^+$. Similarly, fumonisins B_1 (**1**) and B_2 (**2**) had $(M+H)^+$ ion signals at m/z 722 and m/z 706, and $(M+Na)^+$ ion signals at m/z 744 and 728 [2]. Figure 2 shows a typical FAB mass spectrum of **1**. The spectrum was obtained by adding 1 μl of an aqueous solution of fumonisin B_1 (ca. 0.1-1.0 mg/ml) to about 1.0 μl of glycerol on the FAB probe tip and bombarding the sample with 8 kV xenon atoms. The matrix background spectrum of glycerol in the FAB mode is intense and consists of protonated glycerol clusters. Literally there are ion signals at every m/z value. However, signal averaging and background substraction greatly improves the signal to noise ratio in FAB spectra. The largest ions in FAB spectra of fumonisins are protonated molecules $(MH)^+$ and little fragmentation is observed. Small signals at M+22 or M+40 due to sodium or potassium ion attachments are also common. Addition of a small amount of sodium or potassium chloride to the matrix will enhance these signals; however, high concentrations of inorganic salts can greatly suppress FAB ionization. Weak signals at m/z 564, and 546 arise from loss of one tricarballylic acid (TCA) sidechain, while signals at 406, 388, 370, and 334 result from loss of both ester sidechains and subsequent losses of water. The tandem MS (MS/MS) daughter spectrum of protonated fumonisin B_1 (Figure 3) shows signals arising from losses of sidechains and of water [17].

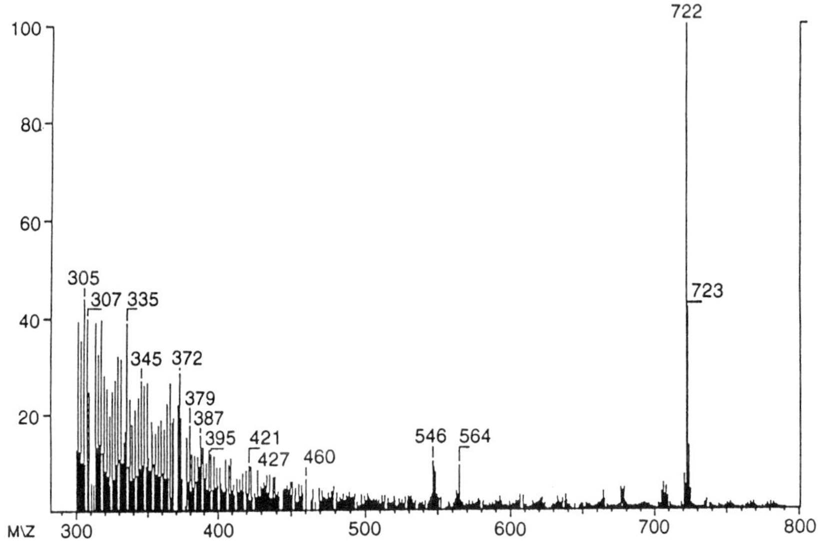

Figure 2. FABMS of Fumonisin B_1 (**1**).

Figure 4 shows the CIMS of partially purified fumonisins B_1 (**1**), B_2 (**2**), and B_3 (**5**). Underivatized fumonisins undergo decomposition upon heating to desorb them from the solids probe and do not give abundant high mass ions indicative of molecular weight. CIMS of fumonisins as their tetramethyl esters do however give abundant protonated molecules (m/z 778 - MH^+ for **1**, Figure 4) when analyzed via the solids probe [15].

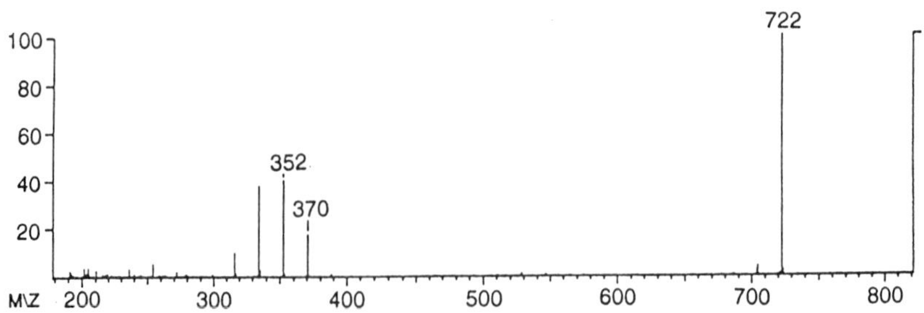

Figure 3. Tandem FABMS (MS/MS) of Protonated Fumonisin B_1 (**1**).

Unhydrolyzed fumonisins are not sufficiently volatile for analysis by GC/MS, even when the free carboxyl groups are esterifed and the hydroxyl groups are derivatized [15]. After hydrolysis of fumonisin B_1 (2N KOH, 1 hr) a signal at m/z 406 $(M+H)^+$ is observed for the amino-pentol backbone, HB_1 (**8**), using either FABMS (Figure 5) or CIMS. After

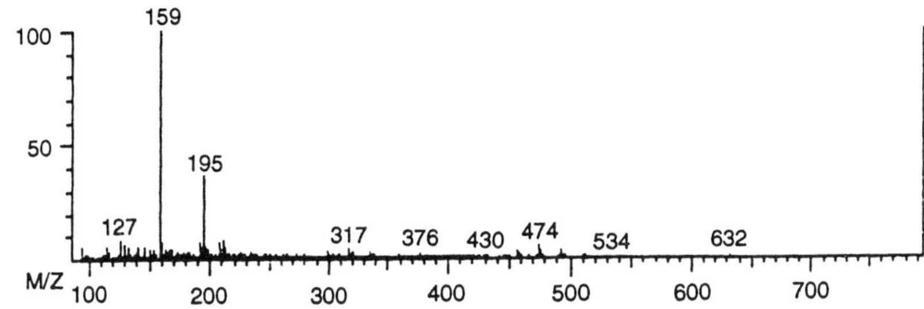

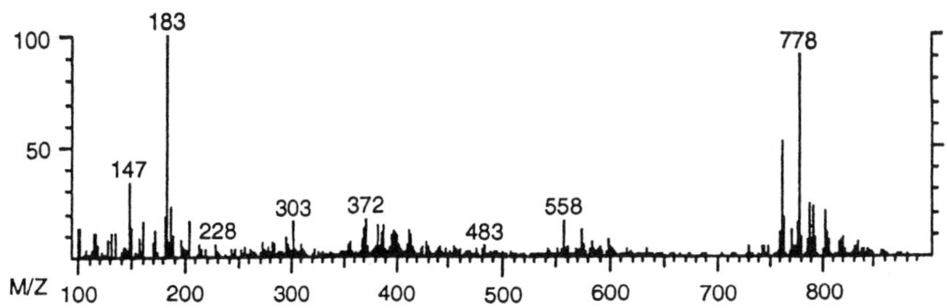

Figure 4. CIMS of Partially Purified Fumonisins $B_1$1 (1), B_2 (2) and B_3 (5) (Upper); CIMS of Tetramethyl Esters of Fumonisins B_1 (1), B_2 (2) and B_3 (5) (Lower).

acidification and removal of solvents, derivatization of the hydrolysis mixture with *N*-methyl-*N*-trimethylsilyl-trifluoroacetamide places trimethylsilyl (TMS) groups on the three carboxylic acids of the tricarballylic acid (TCA) as well as placing TMS groups on the free hydroxyl groups of the C-22 backbone (**8**). Thus, the two hydrolysis products of fumonisins are amenable to GC/MS [15].

The mass spectrum of the first eluting TMS derivative has an abundant fragment at m/z 377 $(M-15)^+$ indicating a molecular weight of 392, and the spectrum and retention times are identical with those of the TMS derivative of authentic TCA. The component eluting next has a mass spectrum consistent with a C-22 amino penta-TMS derivative (Figure 6). The EIMS shows a weak signal at m/z 671 $(M-15)^+$ while the CIMS shows an abundant protonated molecular ion at m/z 766 (data not shown). Abundant fragments at m/z 578 and 187 arise from cleavage between C-14 and C-15 and an abundant fragment at m/z 44 arises from cleavage between C-2 and C-3. A third component had the mass spectrum

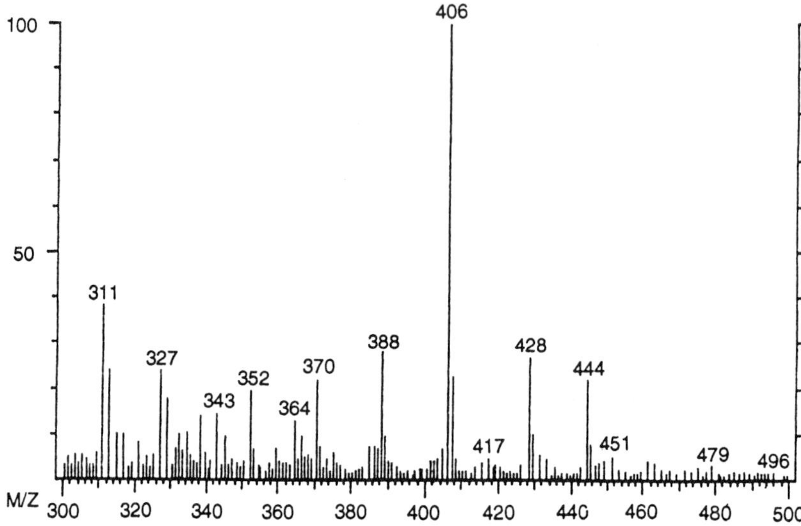

Figure 5. FABMS of Fumonisin B_1 Hydrolysis Product (**8**).

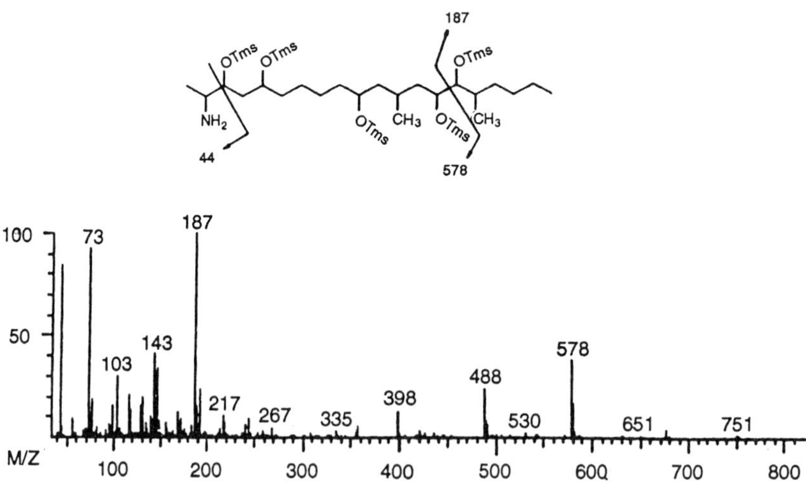

Figure 6. EIMS of the Penta-TMS Derivative of Fumonisin B_1 Hydrolysis Product (**8**).

expected for the completely derivatized amino-pentol containing a TMS group on the amine group at C-2 as well as on the five hydroxyl groups.

In CIMS, the protonated molecule is observed at 72 daltons higher mass (838). The fragments at m/z 44 and 650 are shifted by 72 daltons, to m/z 116 and m/z 650, while the fragment at m/z 187 remains the same. TMS derivatives of the C-22 aminotetraols from hydrolysis of fumonisin B_2 and B_3 (**9** and **10**) coelute from DB-1 capillary columns about a minute before the TMS derivative of **8**. The TMS derivatives of **9** and **10** give nearly identical spectra (Figure 7; spectrum of tetra-TMS derivative of **9**). The CIMS spectra of TMS derivatives of **9** and **10** (hydrolyzed **2** and **5**) show abundant protonated molecular ions at m/z 678 for the tetra-TMS derivatives and at m/z 750 for the penta-TMS derivatives (data not shown). In EIMS of the tetra-TMS derivatives an $(M-15)^+$ ion is observed at m/z 662, and abundant fragments are seen from cleavage between C-14 and C-15 at m/z 490 and 187 in the spectra of TMS derivatives of both **9** and **10**. The cleavage fragment at m/z 490 shifts to m/z 562, in the EIMS of the penta-TMS derivatives, while the fragment from cleavage between C-2 and C-3, with the free amino group, shifts from m/z 44 to m/z 116 when the amine is a TMS derivative. The TMS derivative of hydrolyzed **2** (**9**) can be differentiated from the derivative of hydrolyzed **5** (**10**) by the presence of two minor fragments (ca. 10% of base peak) in the spectrum of the tetra- and

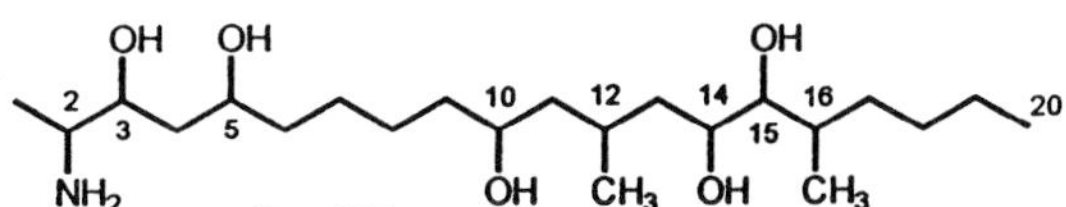

8 HB_1

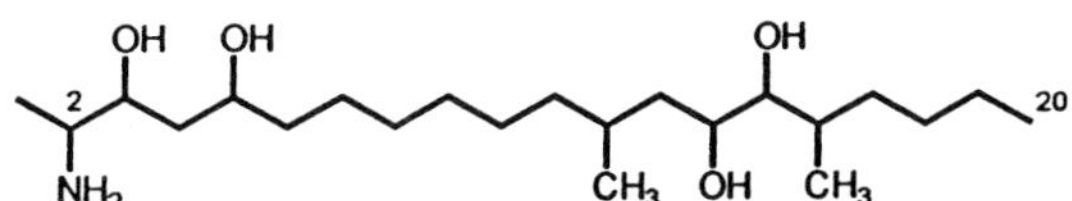

9 HB_2

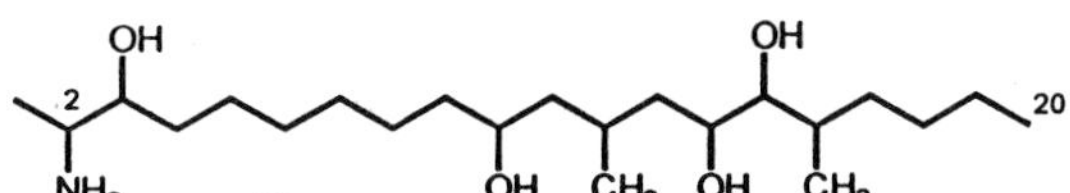

10 HB_3

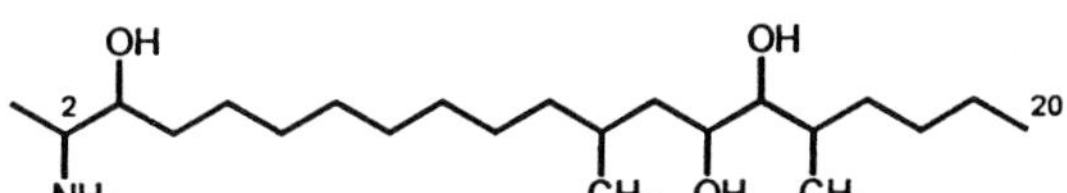

11 HB_4

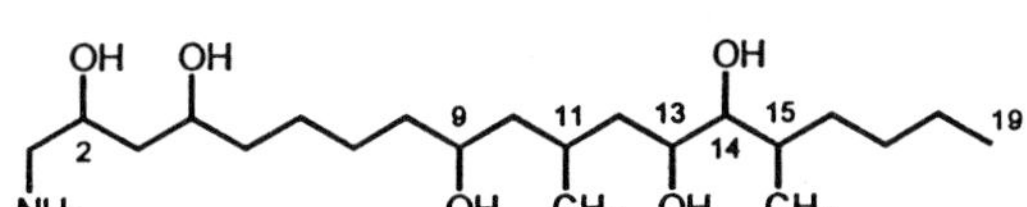

12 HC_1

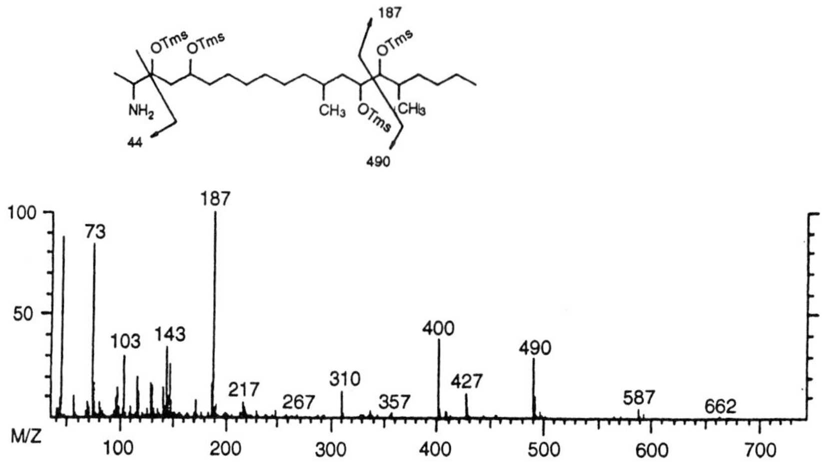

Figure 7. EIMS of the Tetra-TMS Derivative of Fumonisin B_2 Hydrolysis Product (9).

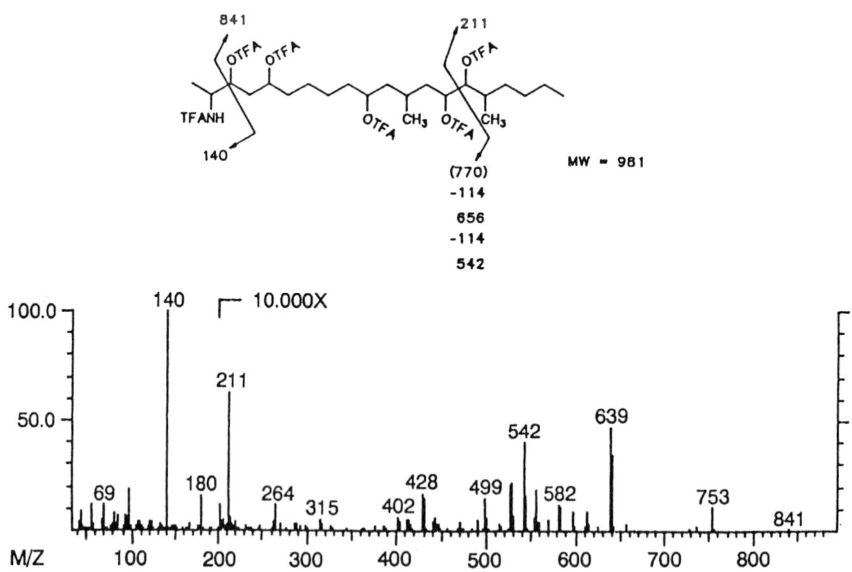

Figure 8. EIMS of the Penta-TFA Derivative of Fumonisin B_1 Hydrolysis Product (8).

penta-TMS derivatives of **2** that presumably arise from cleavage between C-3 and C-4. These characteristic ions are m/z 427 in the tetra-TMS derivative of **2**, which is consistent with cleavage between C-3 and C-4 with loss of C-1 through C-3 and TMSOH, and m/z 218 in the penta-TMS derivative, which is consistent with the same cleavage and charge retention by the C-1 through C-3 fragment. The signal at m/z 427 is not observed in the spectrum of the penta-TMS derivative of hydrolyzed **2** (**9**) and is not present in spectra of either derivative of hydrolyzed **5** (**10**). M/z 218 is also of very low abundance (< 2%) in the spectrum of the penta-TMS derivative of **5**.

Two TMS derivatives are formed from each of the hydrolyzed fumonisin backbones (**8, 9, 10, 11, 12**) and, because the TMS derivatives of **9** and **10** do not resolve on DB-1

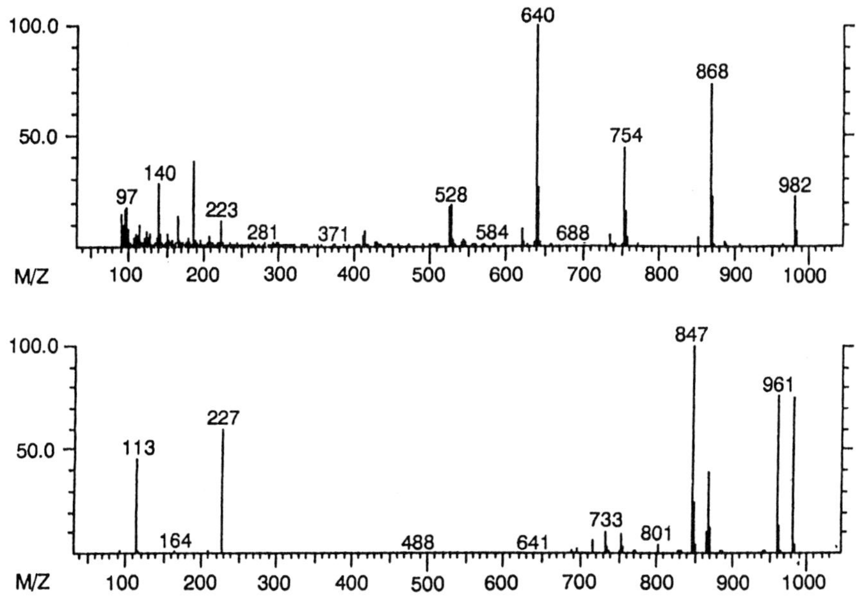

Figure 9. Positive Ion CIMS of the TFA Derivative of **8** (Upper), Negative Ion CIMS of the TFA Derivative of **8** (Lower).

columns, trifluoroacetate (TFA) derivatives were prepared by treating **9** and **10** with *N*-methyl-bis-triflouoroacetamide (MBTFA) or with trifluoroacetic anhydride (12). Both reagents yield the same products; however, MBTFA is the reagent of choice because injection of residues of trifluoroacetic anhydride gives rapid degradation of GC columns. Reaction of the hydrolyzed fumonisin backbones with MBTFA gives a single derivative from each compound that has TFA groups on each hydroxyl and on the primary amine group. Base peaks in the EIMS spectra of these TFA derivatives are at m/z 140 arising from cleavage between C-2 and C-3 (Figure 8, EIMS of the TFA derivative of **8**). The other fragment (m/z 841) in the TFA derivative of **8** is weak but signals from loss of one and two molecules of trifluoroacetic acid are observed (m/z 727 and 614). Cleavages between C-14 and C-15 are not as abundant as those observed in the corresponding TMS derivatives. A fragment at m/z 211 (6%) is from retention of charge by the C-15 terminus. The fragment from C-1 to C-14 (m/z 770) is not evident but loss of trifluoroacetic acid (114) from this ion fragment gives a weak signal at m/z 656. Loss of a second molecule of trifluoroacetic acid is clearly visible at 542 (4%). The most abundant signals observed at higher mass (m/z 753, 640, 639) arise from consecutive losses of trifluoroacetic acid from the molecular ion (981). Although abundances of molecular ions and of high mass fragments are low in EIMS of TFA derivatives, CIMS of these derivatives gives intense signals. The protonated molecule is clearly observed (Figure 9) in positive CIMS, while an abundant molecular anion is seen in the negative CIMS mode

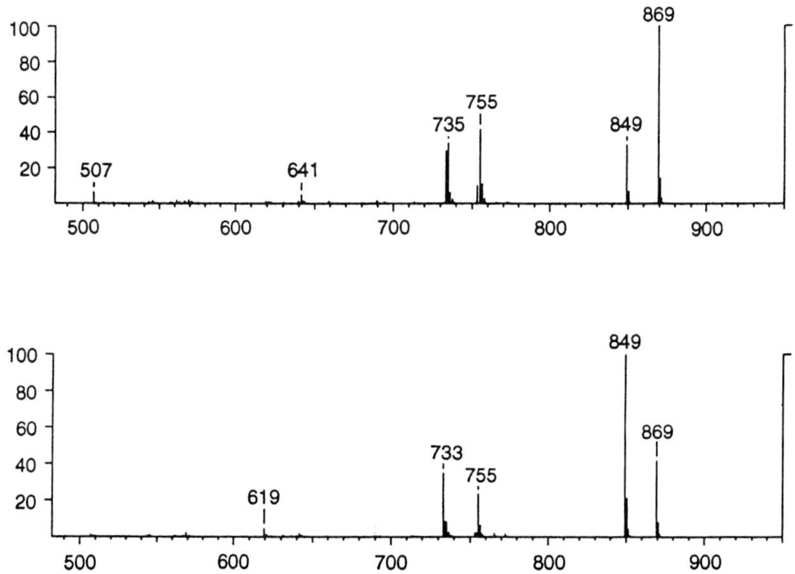

Figure 10. CIMS of the TFA Derivative of Fumonisin B_2 Hydrolysis Product **9** (Upper), CIMS of the TMS Derivative of Fumonisin B_3 Hydrolysis Product **10** (Lower).

(Figure 9). The positive CIMS has prominent signals resulting from consecutive neutral losses of trifluoroacetic acid at m/z 868, 754, and 640. The negative ion spectrum shows losses of HF, trifluoroacetic acid and HF + trifluoroacetic acid (m/z 961, 867, and 847). The cross section of the capture of a thermal electron to form a molecular anion in negative CIMS (43) is large, resulting in high sensitivity in this mode. Excellent signal to noise full scan spectra are obtained when 10-50 ng samples are injected; the detection limit in the full scan mode is well below 1 pg.

TFA derivatives of hydrolyzed fumonisins B_2 (**9**) and B_3 (**10**) have an added advantage over TMS derivatives in that they are separable by GC. On DB-1 columns the fumonisin backbones elute in the following order: **11** (HB_4), **9** (HB_2), **8** (HB_1) and **10**, (HB_3). Elution times for TFA derivatives of the fumonisin B_4 and B_2 backbones differ by a few seconds, while TFA derivatives of the fumonisin B_1 and B_3 backbones elute nearly together and approximately 1 minute later [15]. Separations between critical backbone TFA derivatives are better on DB-5 columns and the elution order is the same.

Figure 10 shows the negative CIMS of the TFA derivatives of the hydrolyzed backbones of fumonisin B_2 and B_3. Both spectra exhibit strong molecular anions and the same fragments (neutral losses of HF and TFAOH), but the TFA derivative of the fumonisin B_3 backbone shows greater loss of HF as the base peak appears at m/z 849. The TFA derivative of the hydrolyzed backbone of fumonisin C_1 (**12**) [16] elutes approximately 2-3 minutes after the TFA derivative of fumonisin B_3 backbone. This derivative from fumonisin C_1 gives an intense molecular anion at m/z 967 in negative CIMS. Fumonisin C_1 is observed in extracts of culture materials and in naturally contaminated samples. It elutes on the leading edge of **1** (HPLC) but is only about 1-2% as abundant as fumonisin B_1.

3.2.2. Analytical methods for fumonisins

Hydrolysis followed by GC/MS of TFA and TMS derivatives has been used for quantitation of fumonisins [15, 17, 44]. Excellent agreement for quantitation of fumonisin B_1 was reported between HPLC with fluorescent derivatives and hydrolysis-GC/MS using the external standard method. However, the GC/MS method reported a higher coefficient of variation which was attributed to poor reproducibility of injection into the GC/MS. Thus, an internal standard MS method was developed [44] to improve precision. The internal standard of choice was a stable isotope labeled fumonisin B_1 having six deuterium atoms incorporated on the two branch chain methyl groups (C-12 and C-16, see section 4 on biosynthesis). Use of this internal standard greatly improved the precision of the analytical method for fumonisin B_1. An advantage realized by adding this internal standard at an early stage of the analysis is that any losses of fumonisins that occur during sample workup are corrected for based on the recovery of the internal standard.

Plattner and Branham [44] also used the labeled internal standard in direct assays of unhydrolyzed and derivatized fumonisins by FAB. While somewhat less precise than hydrolysis-GC/MS and HPLC methods with fluorescent OPA derivatives, this method also gives excellent agreement when applied to extracts from corn and corn products. These two MS based procedures for fumonisin analysis are more labor intensive and require more expensive laboratory equipment than the common HPLC procedures, but they offer important advantages in some instances. Two of the most serious deficiencies of popular HPLC methods are: the possibility of losses or erratic recoveries in the cleanup step on SAX or C_{18} columns that can lead to underestimation of fumonisin concentrations, and the fact that the required fluorescent derivatives form only with primary amines (*N*-acetylated fumonisin derivatives such as the A series thus cannot be detected). The former shortcoming is overcome with both MS based methods. No sample clean-up is necessary in the hydrolysis-GC/MS method; however, losses that occur during sample hydrolysis and work-up are corrected for by addition of the internal standard prior to hydrolysis. Samples do require extensive clean-up using the FABMS method; however the internal standard is added before clean-up allowing corrections for recoveries. The second limitation, inability to measure *N*-acetylated fumonisins, is not completely overcome by MS methods. *N* - acetylated amines are not hydrolyzed in base as are the TFA groups. Thus, in principle, any A type fumonisin should retain the *N*-acetate group and the TFA derivative should elute after the corresponding TFA derivative of the hydrolyzed B type fumonisin. While synthetic fumonisin A_1 does not hydrolyze and no peak elutes for the hexa- TFA backbone derivative, we also have not seen the penta- TFA backbone *N*-acetate peak in GC/MS. Furthermore, fumonisin A_1 can be detected by FABMS as it gives an abundant signal at m/z 764, but the response is not quantitative in the presence of fumonisin B_1 (Plattner, unpublished).

3.3. Nuclear Magnetic Resonance

3.3.1. ^{1}H NMR

Satisfactory and reproducible proton NMR spectra of the fumonisins are relatively difficult to obtain due to their multifunctional character: primary amine, tetra-carboxylic acid and multiple hydroxyl groups. The presence of water, solvent residues, pH influenced ionic character such as residual metallic cations and the solvent selected for obtaining the spectrum all contribute to broaden or shift many of the proton signals. The first definitive NMR spectra were obtained on the tetramethyl ester of fumonisin A_1 (tetramethyl **3**) [2] which was crucial for the initial characterization of fumonisins. Laurent et al. [13, 45] also reported NMR spectra; however, reports of ^{1}H NMR spectra for free fumonisins by most researchers are notably absent in the literature. Full or partial derivatization of fumonisins (such as methylation or acetylation) is highly recommended in order to obtain suitable and reproducible NMR spectra.

3.3.2. ^{13}C NMR

^{13}C NMR of fumonisins are also sensitive to pH changes as described in detail by Plattner et al. [15, 45] for **1**, **2**, and **5**. The most significant shifts evident with changes in pH, as expected, are those of the carboxylic acid sidechains that occur below pH 7.0. These problems are essentially eliminated in ^{13}C spectra of the corresponding backbones (**8**, **9**, **10**) and in derivatives such as the tetramethyl esters.

^{13}C spectra of **1**, **2**, **5**, and the corresponding hydrolysis products **8**, **9**, and **10** are summarized in Table 1. Corresponding data for fumonisin C_1 (**7**) and for fumonisin B_1 fully deuterated at C-21 and C-22 are also available [16, 46].

Table 1. ^{13}C NMR Assignments for Fumonisins B_1 (1), B_2 (2), and B_3 (5) and Their Respective Hydrolysis Products HB_1 (8), HB_2 (9), HB_3 (10)

Carbon Number	Fumonisin (1)	(8)	(2)	(9)	(5)	(10)
1	17.7q	16.8q	16.0q	16.8q	15.9q	16.8q
2	55.5d	53.8d	53.8d	53.8d	53.5d	53.5d
3	71.8d	70.4d	70.4d	70.9d	73.1d	73.1d
4	42.2t	42.9t	41.8t	41.9t	34.6t	34.7t
5	69.9d	68.5d	68.7d	68.6d	26.2t[a]	26.3t[a]
6	39.5t	39.3t	39.3t	39.3t	26.1t[a]	30.7t[a]
7	27.7t	26.8t	27.7t	26.8t	26.8t[a]	30.8t[a]
8	27.6t	26.9t	29.7t[a]	27.9t[a]	29.6t	26.9t
9	39.8t	39.6t	26.7t[a]	31.2t[a]	39.3t	39.6t
10	71.4d	70.7d	30.7t[a]	30.8t[a]	69.9d	70.1d
11	45.1t	44.6t	36.1t	36.9t[b]	44.5t	44.6t
12	27.7d	26.9d	30.2d	30.5d	26.9d	27.0d
13	37.5t	41.6t	36.2t	40.7t[b]	36.4t	41.6t
14	74.7d	70.4d	73.1d	70.4d	73.1d	70.4d
15	80.6d	80.8d	78.9d	80.9d	78.8d	80.8d
16	35.9d	35.9d	34.9d	36.0d	34.9d	35.9d
17	34.3t	31.7t	33.2t	31.9t	33.1t	31.7t
18	30.8t	30.6t	30.8t	30.6t	30.7t	30.7t
19	25.1t	24.2t	23.9t	24.2t	23.8t	24.2t
20	16.3q	14.5q	14.4q	14.5q	14.4q	14.5q
21	17.3q	16.0q	16.0q	16.0q	16.0q	16.0q
22	22.4q	21.4q	20.9q	21.5q	20.6q	21.4q
1'	175.2s[c]	--	173.0s[c]	--	173.1s[c]	--
2'	37.8t[d]	--	36.5t[d]	--	36.6t[d]	--
3'	40.0d[e]	--	38.6d[e]	--	38.6d[e]	--
4'	38.1t[d]	--	36.1t[e]	--	36.1t[e]	--
5'	179.6s[f]	--	276.8s[f]	--	177.0s[f]	--
6'	179.1s[f]	--	175.2s[f]	--	175.2s[f]	--
1"	175.1s[c]	--	172.9s[c]	--	173.0s[c]	--
2"	37.8t[d]	--	36.5t[d]	--	36.6t[d]	--
3"	39.9d[e]	--	38.5d[e]	--	38.6d[e]	--
4"	38.0t[d]	--	36.1t[d]	--	36.1t[d]	--
5"	178.0s[f]	--	176.6s[f]	--	176.6s[f]	--
6"	177.6s[f]	--	175.0s[f]	--	175.0s[f]	--

[a-f]Shift assignments with identical superscripts may be interchanged. Spectra were obtained in D_2O at 75.5 MHz on a Bruker WM-300 instrument and chemical shifts are in ppm from TMS, [15].

3.4. Concerning the Chemistry of Fumonisins

Fumonisins are highly functionalized hydrocarbon derivatives that are readily soluble in polar solvents (i.e. water, methanol, ethanol, etc.) while nearly insoluble in nonpolar solvents (chloroform, hexane). They are easily purified to the 90-95% level by HPLC; however, rigorous purification beyond these levels is more difficult and has not always been successful. The problem in obtaining fumonisins of high purity is that under certain combinations of heat, pH, and solvents they appear to undergo intramolecular reactions or to form dimers, oligomers or polymers [45]. These products seem to involve the amine group as they do not react, or they react only partially, with OPA and other primary amine specific fluorescent reagents. The anomalous and unreactive forms of fumonisins have not yet been fully characterized.

The fumonisins are remarkably stable and they survive the alcohol fermentation and production process [47]. Detoxification studies have included treatments with heat (autoclaving) [48], treatment with ammonia [49, 50], with strong base [51, 52] and with sodium hypochlorite [53]. Such studies describe reductions in fumonisin concentrations but often fail to determine the nature of products formed. These results can be misleading as hydrolyzed fumonisins and acetylated derivatives are active in some bioassays while inactive in others [54, 55, 56].

When stored as dry powders the fumonisins are quite stable; however, when placed in solution they slowly methylate in methanol, hydrolyze and/or transesterify in water, and may revert to other unidentified products depending upon storage conditions (Plattner, unpublished). Freshly prepared solutions of standard fumonisins having known purities are recommended for all biological experiments with these mycotoxins.

4. BIOSYNTHESIS OF THE FUMONISINS

The sphingolipids and their breakdown products are ubiquitous biologically active molecules involved in many aspects of cellular regulation [57, 58]. Structural resemblances between long chain sphingolipid bases such as sphingosine (**13**), sphinganine = dihydrosphingosine (**14**), phytosphingosine (**15**) and fumonisins led to speculation that the two series of compounds may be biosynthesized in a similar manner. Sphingosine biosynthesis occurs through condensation of hexadecanoyl-CoA and serine to produce a 1-hydroxy-2D-amino-3-oxo octadecane which is then reduced to dihydrosphingosine [59]. If fumonisins are similarly biosynthesized then alanine, the analog of serine lacking the hydroxyl group, would be expected to combine with a suitable CoA polyketide ester. This intermediate, or the polyketide precursor, must then be further modified by a methyl transferase addition of branch methyl groups at C-12 and C-16 and by appropriate hydroxylations. Feeding studies in liquid culture with stable isotope (deuterium) labeled L-methionine gave high incorporation of the label (> 90%) into the two branch methyl groups of fumonisin [46]. Branham and Plattner [60] also fed labeled alanine to *F. moniliforme* cultures and demonstrated incorporation of label into fumonisin B_1. This result supports the hypothesis that alanine is the source of C-1 and C-2 in the B series of fumonisins. The apparent inability of *F. moniliforme* cultures to process added hydrolyzed fumonisin backbone or fumonisin B_2 or B_3 into fumonisin B_1, and the discovery of *F. moniliforme* strains that are incapable of making fumonisin B_1 while producing large

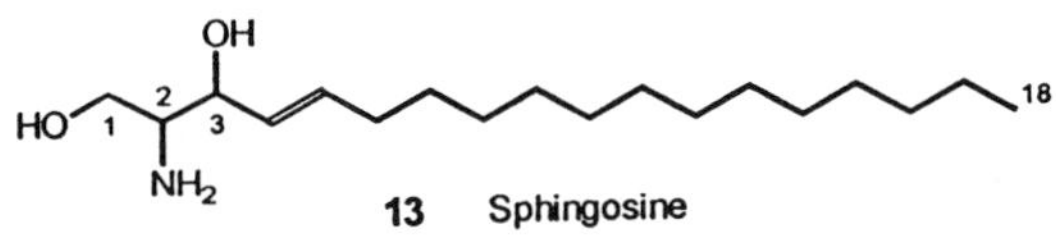

13 Sphingosine

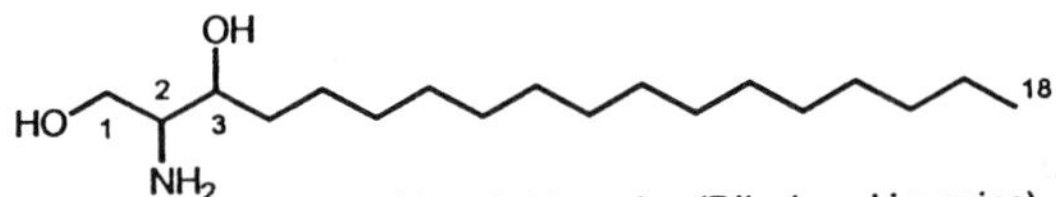

14 Sphinganine (Dihydrosphingosine)

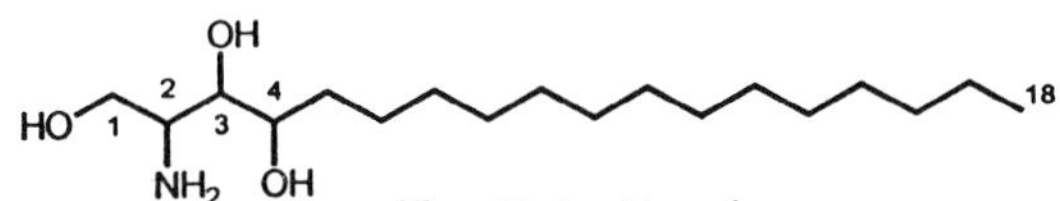

15 Phytosphingosine

amounts of either fumonisin B_2 or B_3 (but not both), suggest that these three main fumonisin analogs may be derived from enzymatically different pathways relatively early in the biosynthesis (Plattner, unpublished). Blackwell et al. [61] report incorporation of ^{13}C and ^{14}C labeled acetate into the fumonisin backbone (C-3 to C-20). They speculate that, from the timing of the acetate incorporations, the backbone is synthesized by a pathway involving four separate pools of primary metabolites. These are presumably acetate (C-3 through C-20), serine (C-1 and C-2), methionine (C-21 and C-22) and glutaric acid (as the source of TCA groups). Lesser incorporations of alanine and other amino acids into C-1 and C-2 were also reported. The order in which these biosynthetic steps occur, whether or not the methyl groups and hydroxyl groups are incorporated during or after synthesis of the fumonisin backbone and at what steps the biosynthesis of fumonisins B_2 and B_3 diverge from that of fumonisin B_1 have not been determined. Similarity of fumonisins to AAL toxins (refer to section 6.1) produced by *Alternaria alternata* cultures, which differ at C-1 and in the length of the backbone chain, suggests that AAL toxins may also be synthesized in a similar manner. The consistent presence of small amounts of fumonisin C_1 ($< 2\%$ of the level of B_1) [16] which, like AAL toxin, lacks C-1 of the normal fumonisin series suggests that the enzyme that combines the CoA ester and amino acid can utilize minor amounts of other substrates (in this case glycine).

5. BIOLOGICAL EFFECTS OF FUMONISINS

Fumonisins were isolated as a result of a series of studies with isolates of *F. moniliforme* that produced compounds toxic to animals (mycotoxins). Members of the genus *Fusarium* also are frequently associated with diseases in plants (phytotoxins). A number of biologically active secondary metabolites including mycotoxins and phytotoxins had been isolated earlier from *F. moniliforme*, but it was the acute liver toxicity associated with feeding culture materials and epidemiological links to esophageal cancer in humans that drove the isolation of the water soluble fumonisins [1, 3, 6]. Deleterious effects of fumonisins on plants and animals have generated substantial interest in these compounds, particularly since their isolation and characterization was accomplished using bioactivity as a guide and these mycotoxins are commonly present at low levels in corn-based foods and feeds.

5.1. Fumonisins and Diseases in Animals

The suspected mode of action of fumonisins in animals is through disruption of the sphingolipid biosynthetic pathway [62]. Animal diseases associated with fumonisins and with *F. moniliforme* have been reviewed [42, 63]. The primary animal diseases that have been demonstrated to be caused by fumonisins are equine leucoencephalomalacia (ELEM) [5] and porcine pulmonary edema (PPE) [64]. Acute liver and kidney toxicity has also been noted in rats, mice and other species in experiments where culture materials containing high levels of fumonisins and purified fumonisins have been fed [6, 10, 65]. The role of other *Fusarium* metabolites in diseases caused by feeding culture materials has not yet been completely elucidated. A long term feeding study designed to obtain data to use for human exposure risk calculations has been planned and should soon be in progress.

Alterations in free sphingosine/sphinganine ratios [63] have been proposed as a sensitive marker for fumonisin exposure in animals. The ratio of free sphinganine (**14**) to sphingosine (**13**) becomes rapidly elevated in tissues exposed to fumonisins at levels well below those that produce observable gross changes. Sphinganine, which nomally is rapidly *N*-acetylated with a fatty acid, accumulates in tissues exposed to fumonisins indicating enzyme inhibition at this step [66]. Concentrations of complex sphingolipids are distinctly lower when animals have been exposed to fumonisins for longer periods of time [63, 67].

5.2. Phytotoxic Effects

Over the years numerous secondary metabolites isolated from *F. moniliforme* and related species of section *Liseola* have been observed to produce pronounced phytotoxic symptoms in plants infected with these fungi. Although intensely phytotoxic metabolites such as fusaric acid and the napthazarine pigments have been isolated from these fungi, it has been difficult to define specific modes of action for the various phytotoxins. Fumonisins are also highly phytotoxic and they produce disease symptoms on tomato leaves very similar to those produced by the structurally similar AAL toxins [68]. Correlations between the ability to produce fumonisins and virulence of the fungus on corn seedlings have also been reported [69]. These virulence correlations are inherited along with the ability to produce fumonisins in a Mendelian manner [69]. Fumonisin B_1, B_2, and B_3 are phytotoxic in a seed germination inhibition bioassay (Plattner unpublished) and in corn seedlings [70]. It is likely that the phytotoxicity of fumonisins in plants is also a result of their ability to alter sphingosine metabolism as others have shown that similar alterations in sphinganine/sphingosine ratios occur in duckweed (*Lemna*) when the plants are exposed to fumonisins. Additional phytotoxic and herbicidal studies relating to fumonisins have appeared in the literature [55, 71, 72, 73].

6. RELATED COMPOUNDS

6.1 *Alternaria alternata* Toxins

Although the fumonisins were identified as a new class of alkaloid mycotoxins in 1988, their structures are remarkably similar to the AAL toxins produced by *Alternaria alternata* (Fr.) Keissler f. sp *lycopersici* [74, 75] whose structures were reported in 1981 [35]. The AAL toxins (**16, 17, 18, 19**) are all monotricarballalyic acid esters of either

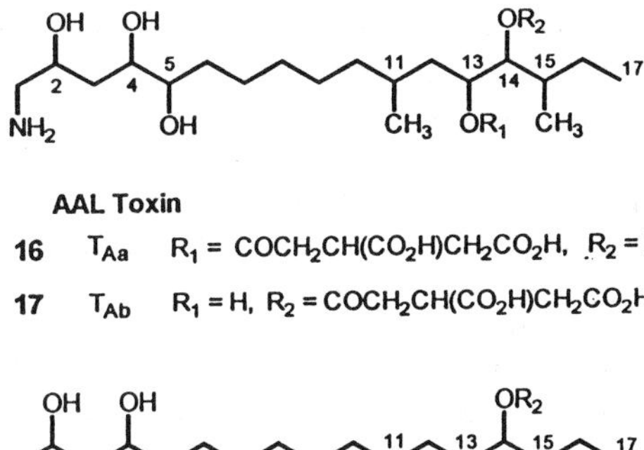

AAL Toxin

16 T_{Aa} $R_1 = COCH_2CH(CO_2H)CH_2CO_2H$, $R_2 = H$

17 T_{Ab} $R_1 = H$, $R_2 = COCH_2CH(CO_2H)CH_2CO_2H$

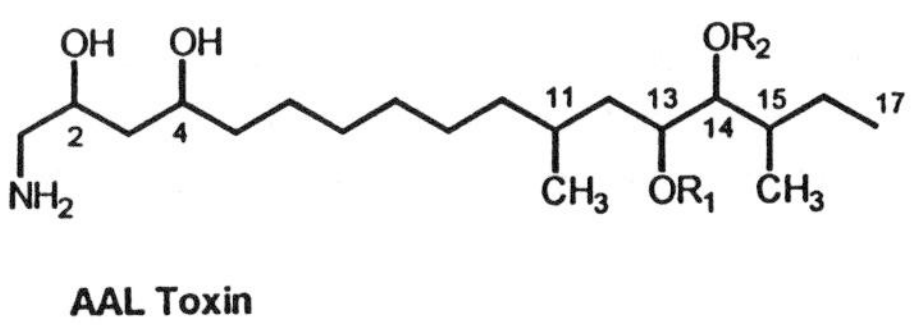

AAL Toxin

18 T_{Ba} $R_1 = COCH_2CH(CO_2H)CH_2CO_2H$, $R_2 = H$

19 T_{Bb} $R_1 = H$, $R_2 = COCH_2CH(CO_2H)CH_2CO_2H$

1-amino-11,15-dimethylheptadeca-2,4,5,13,14-pentol (the T_A series) or 1-amino-11,15-dimethyl-heptadeca-2,4,13,14-tetrol (the T_B series). AAL toxins are two methylene units shorter at the methyl terminus, possess only one TCA functionality and lack a methyl group at the amino terminus, but otherwise they closely resemble fumonisin C_1 (7). Three additional pairs of AAL toxins have been reported [76]. Two of these compounds lack the C-4 hydroxyl group and the corresponding *N*-acyl derivatives of these and of **18** and **19** were also isolated and identified.

AAL toxins are the reported causative agents of stem canker disease in susceptible tomato cultivars [77]. Both the fumonisins and AAL toxins produce equivalent necrotic symptoms in tomato leaflet bioassays [68] and there is evidence that the biosynthetic pathway for production of these toxins is present in several species of both *Alternaria* and *Fusarium*. Peracetylated derivatives of AAL toxins and the fumonisins are inactive in both the tomato bioassay and animal tissue culture systems. Acetylation of the amine group renders AAL toxin inactive. The hydrolysis product of AAL is toxic to the susceptible tomato line whereas that of fumonisin B_1 is not [54]. Comparative toxicities of fumonisin B_1, AAL and yeast sphingolipids have also been reported [78].

6.2 Sphingofungins and Myriocin

A family of antifungal agents, sphingofungins A-D (**20**, **21**, **22**, **23**), were recently isolated from cultures of *Aspergillus fumigatus* (ETC 20857) [79, 80]. Sphingofungins resemble fumonisins and sphingosines with the most significant difference being an additional carboxyl group at C-1 in the sphingofungins. Relative and absolute stereochemistry of the sphingofungins has been determined as 2S, 3R, 4R, 5S. These compounds are claimed to be potent and specific inhibitors of serine palmitoyl transferase, the first enzyme in the *de novo* sphingolipid biosynthesis pathway. Myriocin (**24**), isolated from submerged cultures of *Myriococcum albomyces* [81], is another interesting fumonisin-related compound containing a C-1 carboxyl group.

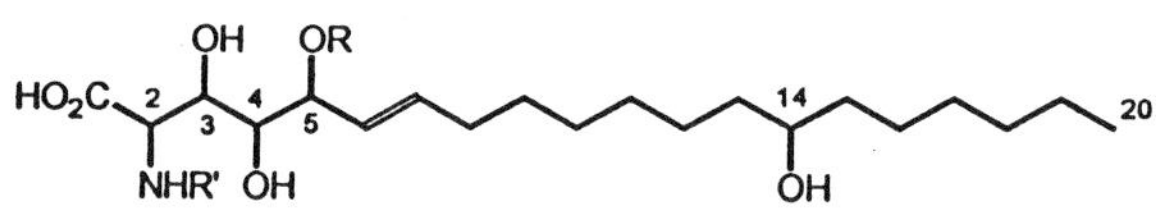

20	Sphingofungin A	R = H; R' = C(NH)NH_2
21	Sphingofungin B	R = H; R' = H
22	Sphingofungin C	R = Ac; R' = H
23	Sphingofungin D	R = H; R' = Ac

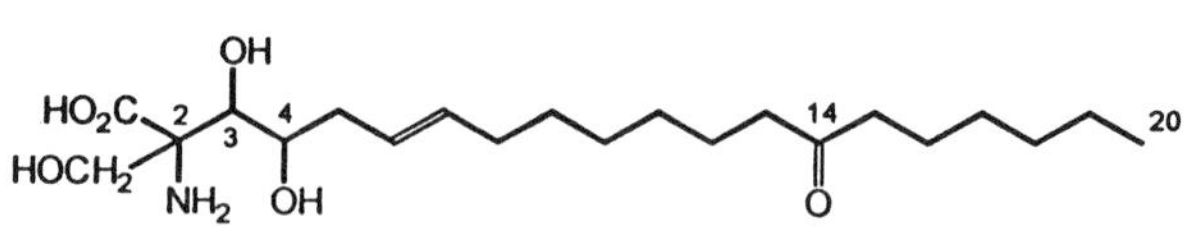

24 Myriocin

6.3. Related Compounds of Marine Origin

Carter and Rinehart [82] isolated aplidiasphingosine (**25**) from a marine tunicate, *Asplidium* species, collected in the Gulf of California. Asplidiasphingosine (2-amino-5,9,13,17-tetramethyl-8,16-octadecadiene-1,3,14-triol or 14-hydroxy-5,9,13,17-

tetramethyl-8,16-sphingadiene) is a terpenoid analog of sphingosine and has inhibitory activity against representative Gram-positive and Gram-negative bacteria, fungi, Herpes virus type I, and cultured tumor cell lines (KB and L1210).

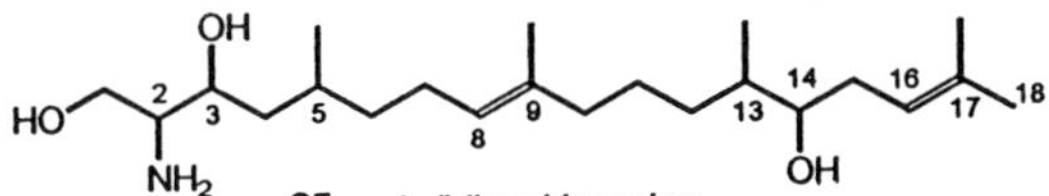

25 Aplidiasphingosine

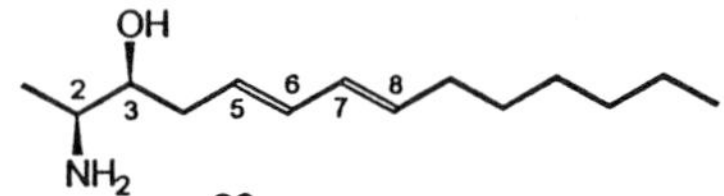

26

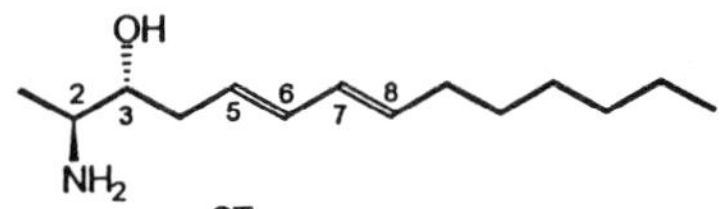

27

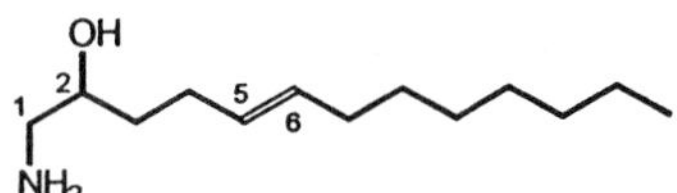

28

Two epimeric amino alcohols, 2(S)-aminotetradeca-5,7-dien-3(S)- and -3(R)-ol (**26** and **27**) were isolated from a Papua-New Guinea sponge, *Xestospongia* sp. (83). Their structures were determined spectrally, relative stereochemistry by derivatization, and absolute stereochemistry by degradation to L-alanine. In contrast to the sphingoid bases, which are derived from fatty acids and serine, these compounds are alanine derivatives more closely related to the fumonisins. Both compounds inhibit the growth of *Candida albicans*.

An ascidian, tentatively *Didemnum* sp., collected on the Great Barrier Reef, Australia, has yielded several linear antifungal amino alcohols [84]. The major component was identified as (2R)-1-amino-5-tridecen-2-ol (**28**).

Novel sphingosine-derived azetidine alkaloids, penaresidin A (**29**) and penaresidin B (**30**) were isolated from a marine sponge of the *Penares* genus [85]. These compounds possess potent actinomycin ATPase-activating activity. Another member of this series, penazetidine A (**31**) has been isolated from *P. sollasi* [86]. Penazetidine A inhibits protein kinase C and is active *in vitro* against several solid tumor cell lines.

6.4. Synthetic Fumonisin Analogs

Relative and absolute stereochemistry of the fumonisins remains unknown; however, attempts to prepare synthetic fumonisin analogs have been reported. Analog studies are

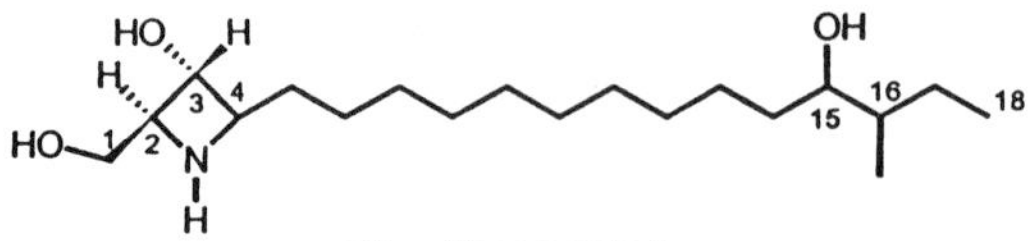

29 Penaresidin A

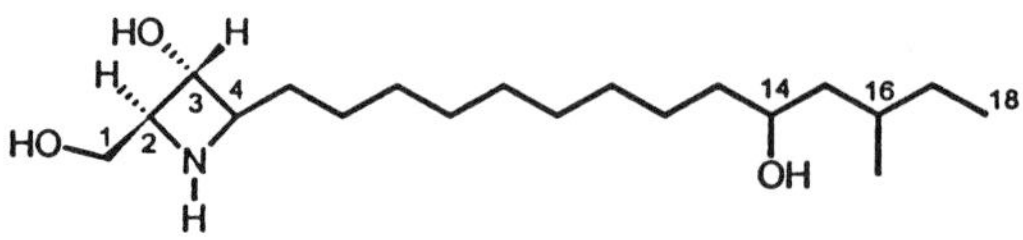

30 Penaresidin B

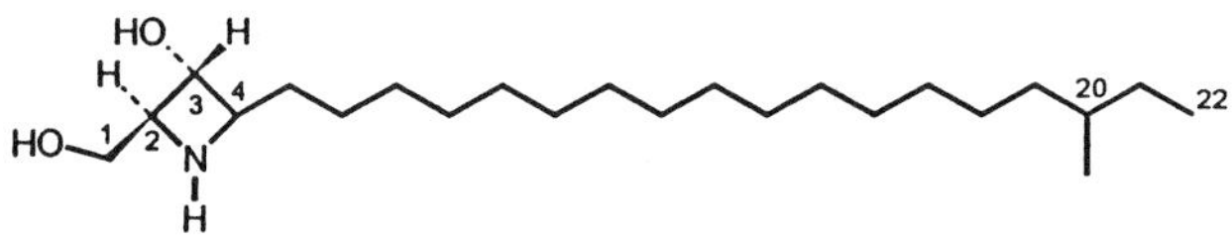

31 Penazetidine A

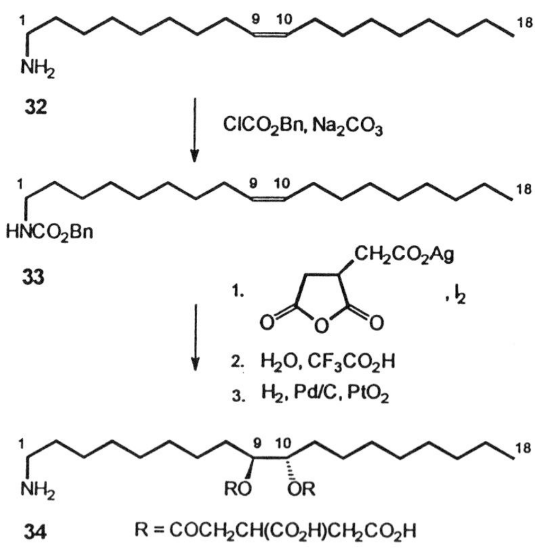

Scheme 1

expected to provide toxicologists with information concerning minimum structural requirements for activity and to enable researchers to prepare inhibitors of the fumonisin biosynthetic pathway or to suggest methods for chemical or biochemical detoxification. Synthetic fumonisin analogs may also provide leads for new pharmaceuticals.

The first reported approach toward synthesis of fumonisin and its analogs began with oleylamine (**32**) [87]. Oleylamine was treated with benzyl chloroformate and sodium carbonate to yield carbamate **33**. The silver salt of tricarballylic anhydride was then allowed to react with **33** to yield a bis anhydride which on hydrolysis followed by removal of the *N*-benzyloxycarbonyl protecting group gave analog **34**.

An additional synthetic approach by the same group [87] involved reaction of ketone **35** with aldehydes **36** and **37** to provide beta-hydroxyketones **38** and **39** which were reduced to the corresponding diols **40** and **41**. Reaction of **40** and **41** with the silver salt of tricarballylic anhydride, as in the previous preparation of **34**, followed by hydrolysis and removal of the protecting group gave fumonisin analogs **42** and **43**. A toxicity study using cultured rhesus monkey cells indicated that **42** and **43** were more toxic than fumonisin B_1 (**1**) and that **34** was comparable in toxicity to fumonisin B_1.

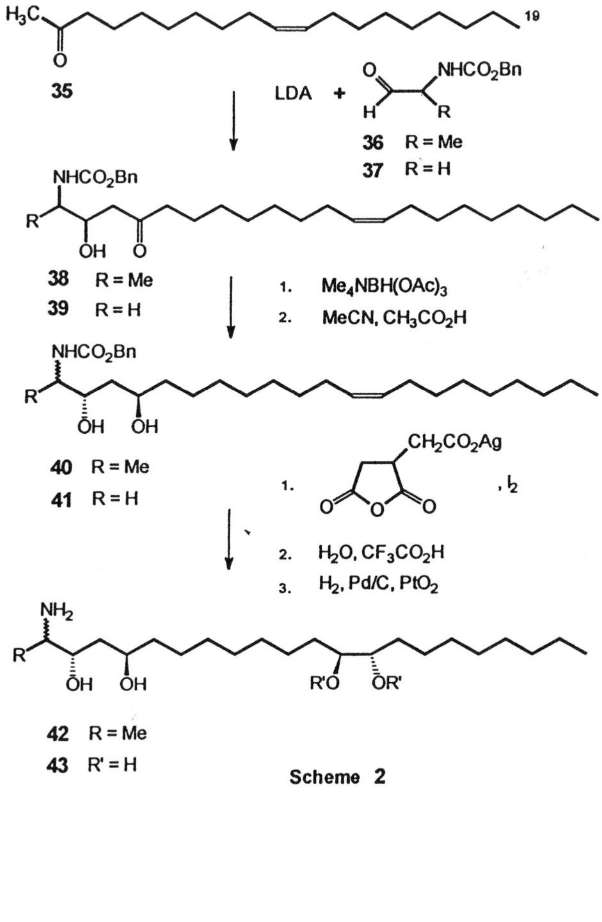

Scheme 2

Additional fumonisin analogs have been synthesized from readily available fatty acids [88]. Typical fatty acids, such as stearic acid, were first converted to the corresponding aldehydes (**44**) and the fatty aldehydes were then treated with nitroethane [89] to yield diastereomeric mixtures of products such as **45**. The nitro group was subsequently reduced to the corresponding amine (**46**) with Pd/C and ammonium formate. The resulting amino-hydroxy diastereomeric mixtures were readily converted to oxazoline derivatives, the *erythro* and *threo* isomers easily separated by TLC and their relative stereochemistry confirmed by NMR. The method should be general for preparation of fumonisin analogs depending on availability of appropriate fatty acid precursors.

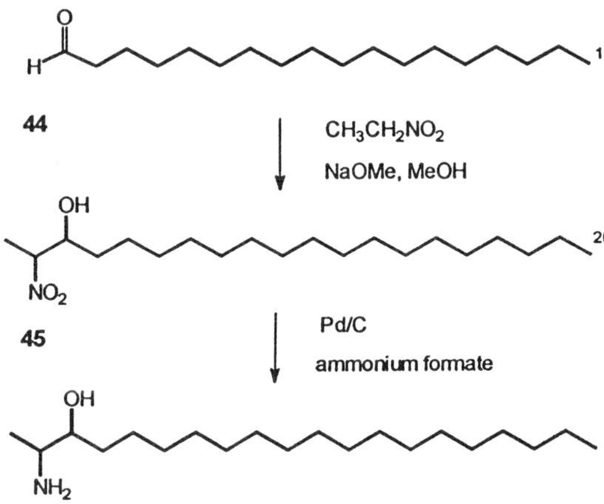

Scheme 3

SUMMARY

Fumonisins are a newly recognized group of alkaloid mycotoxins produced by certain strains of *Fusarium moniliforme*. Under suitable culture conditions they are easily produced at levels exceeding 2000 ppm [1]. When fumonisins are present at high levels in food or feeds they are associated with serious diseases in animals, with the liver and kidney as apparent primary target organs. In rats, ^{14}C labeled fumonisin B_1 is rapidly excreted in the feces [90]. Fumonisins are also associated with cancer and they appear to be complete carcinogens [91]. Most animals seem to tolerate fumonisin levels normally encountered in the environment and serious difficulties are noted only after ingesting highly contaminated feeds; or, upon direct injection. Their mode of action involves disruption of sphingolipid metabolism which, in turn, may lead to disruption of normal cellular regulation. Riley et al. [67, 92] and Merrill et al. [93] have summarized current knowledge of fumonisins as inhibitors of sphingolipid biosynthesis.

Analytical methods for the fumonisins continue to appear in the literature [94-98] as do reports of their production [99, 100], occurrence [101, 102, 103] and chemistry [104, 105]. Structural similarities of the fumonisins to toxins produced by *Alternaria alternata* and to other compounds produced by a wide variety of organisms, including sphingosines and phytosphingosines, suggests that further chemical and biochemical investigations of these alkaloids would be appropriate. One of the most pressing questions, now that reliable methods for analysis of these mycotoxins are available, is to determine what levels of the fumonisins should be considered as safe and acceptable in the food supply.

REFERENCES

1. WCA Gelderblom, K Jaskiewicz, WFO Marasas, PG Thiel, RM Horak, R Vleggaar and NPJ Kriek, Appl Environ Microbiol 54:1806 (1988).
2. SC Bezuidenhout, WCA Gelderblom, CP Gorst-Allman, RM Horak, WFO Marasas, G Spiteller and R Vleggaar, J Chem Soc Chem Commun 743 (1988).
3. WFO Marasas, K Jaskiewicz, FS Venter and DJ van Schalkwyk, S Afr Med J 74:110 (1988).
4. LR Harrison, B Colvin, JT Green, LE Newman and JR Cole, J Vet Diagn Invest 2:217 (1990).
5. WFO Marasas, TS Kellerman, WCA Gelderblom, JAW Coetzer, PG Thiel and JJ van der Lugt, Onderstepoort J Vet Res 55:197 (1988).
6. WCA Gelderblom, NPJ Kriek, WFO Marasas and PG Thiel, Carcinogenesis 12:1247 (1991).
7. WFO Marasas, PE Nelson and TA Toussoun, Toxigenic *Fusarium* species, identity and mycotoxicology, The Pennsylvania State University Press, University Park, Pennsylvania, 1984.
8. NPJ Kriek, TS Kellerman and WFO Marasas, Onderstepoort J Vet Res 48:129 (1981).
9. NPJ Kriek, WFO Marasas and PG Thiel, Food Cosmet Toxicol 19:447 (1981).
10. KA Voss, WJ Chamberlain, CW Bacon and WP Norred, Natural Toxins 1:222 (1993).
11. PF Ross, LG Rice, RD Plattner, GD Osweiler, TM Wilson, DL Owens, HA Nelson and JL Richard, Mycopathologia 114:129 (1990).
12. F Ross, LG Rice, JC Reagor, GD Osweiler, TM Wilson, HA Nelson, DL Owens, RD Plattner, KA Harlin, JL Richard, BM Colvin and MI Banton, J Vet Diagn Invest 3:238 (1991).
13. D Laurent, N Platzer, F Kohler, MP Sauviat and F Pellegrin, Microbiologie, Aliments, Nutrition 7:9 (1989).
14. ME Cawood, WCA Gelderblom, R Vleggaar, Y Behrend, PG Thiel and WFO Marasas, J Agric Food Chem 39:1958 (1991).
15. RD Plattner, D Weisleder, DD Shackelford, R Peterson and RG Powell, Mycopathologia 117:23 (1992).
16. BE Branham and RD Plattner, J Nat Prod 56:1630 (1993).
17. WA Korfmacher, MP Chiarelli, JO Lay, Jr, J Bloom, M Holcomb and KT McManus, Rapid Communications in Mass Spectrom 5:463 (1991).
18. RD Plattner, WP Norred, CW Bacon, KA Voss, R Peterson, DD Shackelford and D Weisleder, Mycologia 82:698 (1990).
19. PG Thiel, GS Shephard, EW Sydenham, WFO Marasas, PE Nelson and TM Wilson, J Agric Food Chem 39:109 (1991).
20. RD Plattner, PF Ross, J Reagor, J Stedelin and LG Rice, J Vet Diagn Invest 3:357 (1991).

21. EW Sydenham, WFO Marasas, GS Shephard, PG Thiel and EY Hirooka, J Agric Food Chem 40:994 (1992).
22. CJ Mirocha, CG Mackintosh, UA Mirza, W Xie, Y Xu and J Chen, Appl Environ Microbiol 58:3196 (1992).
23. PA Murphy, LG Rice and PF Ross, J Agric Food Chem 41: 263 (1993).
24. EW Sydenham, GS Shephard, PG Thiel, WFO Marasas and S, Stockenstrom, J Agric Food Chem 39:2014 (1991).
25. A Pittet, V Parisod and M Schellenberg, J Agric Food Chem 40:1352 (1992).
26. JP Rheeder, WFO Marasas, PG Thiel, EW Sydenham, GS Shephard and DJ van Schalkwyk, Phytopathology 82:353 (1992).
27. T Ackermann, J Appl Toxicol 11:451 (1991).
28. GE Rottinghaus, CE Coatey and HC Minor, J Vet Diagn Invest 4:326 (1992).
29. J Dupuy, P LeBars, J LeBars and H Boudra, J Planar Chrom 6:476 (1993).
30. EW Sydenham, GS Shephard and PG Thiel, J AOAC Int 75:313 (1992).
31. GS Shephard, EW Sydenham, PG Thiel and WCA Gelderblom, J Liquid Chrom 13:2077 (1990).
32. PM Scott and GA Lawrence, J AOAC Int 75:829 (1992).
33. ME Stack and RM Eppley, J AOAC Int 75:834 (1992).
34. M Holcomb, JB Sutherland, MP Chiarelli, WA Korfmacher, HC Thompson, Jr, JO Lay, Jr, LJ Hankins and CE Cerniglia, J Agric Food Chem 41:357 (1993).
35. DJ Siler and DG Gilchrist, J Chromatog 238:167 (1982).
36. PF Ross, LG Rice, GD Osweiler, PE Nelson, JL Richard and TM Wilson, Mycopathologia 117:109 (1992).
37. PG Thiel, EW Sydenham, GS Shephard and DJ van Schalkwyk, J AOAC Int 76:361 (1993).
38. JI Azcona-Olivera, MM Abouzied, RD Plattner, WP Norred and JJ Pestka, Appl Environ Microbiol 58:169 (1992).
39. JI Azcona-Olivera, MM Abouzied, RD Plattner and JJ Pestka, J Agric Food Chem 40:531 (1992).
40. R Vesonder, R Peterson, R Plattner and D Weisleder, Mycotoxin Research 6:85 (1990).
41. PE Nelson, RD Plattner, DD Shakelford and AE Desjardins, Appl Environ Microbiol 57:2410 (1991).
42. PE Nelson, AE Desjardins and RD Plattner, Ann Rev Phytopathology 31:233 (1993).
43. AG Harrison, Chemical Ionization Mass Spectrometry, CRC Press, Boca Raton, Florida, 1983, 156 pp.
44. RD Plattner and BE Branham, J AOAC Int 77:525 (1994).
45. D Laurent, M Lanson, N Goasdoue, F Kohler, F Pellegrin and N Platzer, Analysus 18:172 (1990).
46. RD Plattner and DD Shackelford, Mycopathologia 117:17 (1992).
47. RJ Bothast, GA Bennett, JE VanCauwenberge and JL Richard, Appl Environ Microbiol 58:233 (1992).
48. J Depuy, P Le Bars, H Boudra and J Le Bars, Appl Environ Microbiol 59:2864 (1993).
49. WD Norred, KA Voss, CW Bacon and RT Riley, Food Chem Toxicol 29:815 (1991).
50. DL Park, SM Rua, Jr, CJ Mirocha, EAM Abd-Ala and CY Weng, Mycopathologia 117:105 (1992).
51. EC Hopmans and PA Murphy, J Agric Food Chem 41:1655 (1993).
52. S Hendrich, KA Miller, TM Wilson and PA Murphy, J Agric Food Chem 41:1649 (1993).
53. RF Vesonder, personal communication.
54. CJ Mirocha, DG Gilchrist, WT Shier, HK Abbas, Y Wen and RF Vesonder, Mycopathologia 117:47 (1992).
55. T Tanaka, HK Abbas and SO Duke, Phytochemistry 33:779 (1993).

56. WCA Gelderblom, ME Cawood, SD Snyman, R Vleggaar and WFO Marasas, Food Chem Toxicol 31:407 (1993).
57. YA Hannun and RM Bell, Science 243:500 (1989).
58. K-A Karlsson, Lipids 5:878 (1970).
59. GA Thompson, Regulation of Membrane Lipid Metabolism, CRC Press, Boca Raton, Florida (1980).
60. BE Branham and RD Plattner, Mycopathologia 124:99 (1993).
61. BA Blackwell, JD Miller and ME Savard, J AOAC Int 77:506 (1994).
62. E Wang, WP Norred, CW Bacon, RT Riley and AH Merrill, J Biol Chem 266:14486 (1991).
63. RT Riley, Ann Rev Nutrition 13:167 (1993).
64. BM Colvin and LR Harrison, Mycopathologia 117:79 (1992).
65. KA Voss, WP Norred, RD Plattner and CW Bacon, Fd Chem Toxicol 27:89 (1989).
66. E Wang, PF Ross, TM Wilson, RT Riley and AH Merrill, J Nutrit 122:1706 (1992).
67. RT Riley, N-H An, JL Showker, H-S Yoo, WP Norred, WJ Chamberlain, E Wang, AH Merrill, Jr, G Motelin, VR Beasley and WM. Haschek, Toxicol Appl Pharmacol 118:105 (1993).
68. DG Gilchrist, B Ward, V Moussato and CJ Mirocha, Mycopathologia 117:57 (1992).
69. AE Desjardins, RD Plattner, DD Shackelford, JF Leslie and PE Nelson, Appl Environ Microbiol 58:2799 (1992).
70. RB Wolf, GF Spencer, and WF Kwolek, Weed Science 32:612 (1984).
71. HK Abbas, CD Boyette, RE Hoagland and RF Vesonder, Weed Science 39:673 (1991).
72. HK Abbas and CD Boyette, Weed Technology 6:548 (1992).
73. RF Vesonder, RE Peterson, D Labeda and HK Abbas, Arch Environ Contam and Toxicol 58:2799 (1992).
74. AT Bottini and DG Gilchrist, Tetrahedron Letters 22:2719 (1981).
75. AT Bottini, JR Bowen and DG Gilchrist, Tetrahedron Letters 22:2723 (1981).
76. ED Caldas, AD Jones, B Ward, CK Winter and DG Gilchrist, J Agric Food Chem 42:327 (1994).
77. DG Gilchrist and RG Grogan, Phytopathology 66:165 (1982).
78. RF Vesonder, RE Peterson, D. Labeda and HK Abbas, Arch Environ Contam Toxicol 23:464 (1992).
79. F VanMiddlesworth, RA Giacobbe, M Lopez, G Garrity, JA Bland, K Bartizal, RA Fromtling, J Polishhook, M Zweerink, AM Edison, W Rozdilsky, KE Wilson and RL Monaghan, J Antibiotics 45:861 (1992).
80. F VanMiddlesworth, C Dufrensne, FE Wincott, RT Mosley and KE Wilson, Tetrahedron Letters 33:297 (1992).
81. JF Bagli, D Kluepfel and M St-Jacques, J Org Chem 38:1253 (1973).
82. GT Carter and KL Rinehart, J Amer Chem Soc 100:7441 (1978).
83. NK Gulavita and PJ Scheuer, J Org Chem 54:366 (1989).
84. PA Searle and TF Molinski, Abstract P:61, 34th Annual Meeting, American Society of Pharmacognosy, July (1993).
85. J Kobayashi, J Cheng, M Ishibashi, MR Waichii, S Yamamura and Y Ohizumi, J Chem Soc Perkin Trans I 1135 (1991).
86. KA Alvi, W Palmer and P Crews, Abstract O-41, 33rd Annual Meeting, American Society of Pharmacognosy, July (1992).
87. GA Kraus, JM Applegate and D Reynolds, J Agric Food Chem 40:3331 (1992).
88. GK Poch, RG Powell and RD Plattner, Abstract P:90, 34th Annual Meeting, American Society of Pharmacognosy, July (1993).

89. G Rosini, The Henry (Nitroaldol) Reaction, in Comprehensive Organic Synthesis, BM Trost and I Fleming Eds, vol 2, pp 321-340, Pergamon Press, New York (1991).
90. WP Norred, RD Plattner and WJ Chamberlain, Natural Toxins 1:341 (1993).
91. RT Riley, KA Voss, HS Yoo, WCA Gelderblom and AH Merrill, Jr, J Food Protection 57:000 (1994, in press).
92. RT Riley, E Wang and AH Merrill, Jr, J AOAC Int 77:533 (1994).
93. AH Merrill, Jr, E Wang, DG Gilchrist and RT Riley, Adv Lipid Res 26:215 (1993).
94. RA Thakur and JS Smith, Rapid Communications in Mass Spectrometry 8:82 (1994).
95. JJ Peska, JI Azcona-Olivera, RD Plattner, F Minervini, MB Doko and A Visconti, J Food Protection 57:169 (1994).
96. GA Bennett and JL Richard, J AOAC Int 77:501 (1994).
97. MA Dombrink-Kurtzman, GA Bennett and JL Richard, J AOAC Int 77:512 (1994).
98. JC Young and P Lafontaine, Rapid Communications in Mass Spectrometry 7:352 (1993).
99. J Le Bars, P Le Bars, J Dupey, H Boudra and R Cassini, J AOAC Int 77:517 (1994).
100. PE Nelson, JH Juba, PF Ross and LG Rice, J AOAC Int 77:522 (1994).
101. A Visconti and MB Doko, J AOAC Int 77:546 (1994).
102. RA Shelby, Plant Disease 78:582 (1994).
103. V Sanchis, M Abadias, L Oncins, N Sala, I Vinas and R Canela, Appl Environ Microbiol 60:2147 (1994).
104. PM Scott and GA Lawrence, J AOAC Int 77:541 (1994).
105. JW ApSimon, BA Blackwell, OE Edwards, A Fruchier, JD Miller, M Savard and JC Young, Pure Appl Chem 66:000 (1994, in press).

Subject Index

Organism Index